Genetically Engineered

Mice Handbook

Research Methods for Mutant Mice Series

Series Editor
John P. Sundberg

Systematic Approach to Evaluation of Mouse Mutations
John P. Sundberg and Dawnalyn Boggess

Systematic Evaluation of the Mouse Eye:
Anatomy, Pathology, and Biomethods
Richard S. Smith, Simon W. M. John,
Patsy M. Nishina, and John P. Sundberg

Genetically Engineered Mice Handbook
John P. Sundberg and Tsutomu Ichiki

Genetically Engineered

Mice Handbook

Taylor & Francis
Taylor & Francis Group
Boca Raton   London   New York

A CRC title, part of the Taylor & Francis imprint, a member of the
Taylor & Francis Group, the academic division of T&F Informa plc.

Published in 2006 by
CRC Press
Taylor & Francis Group
6000 Broken Sound Parkway NW, Suite 300
Boca Raton, FL 33487-2742

International Standard Book Number-10: 0-8493-2220-0 (Hardcover)
International Standard Book Number-13: 978-0-8493-2220-4 (Hardcover)
Library of Congress Card Number 2005041899

Library of Congress Cataloging-in-Publication Data

Genetically engineered mice handbook / edited by John P. Sundberg and Tsutomu Ichiki.
     p. cm. -- (Research methods for mutant mice series)
  Includes bibliographical references and index.
  ISBN 0-8493-2220-0 (alk. paper)
  1. Transgenic mice--Handbooks, manuals, etc. I. Sundberg, John P. II. Ichiki, Tsutomu. III. Series.

QH470.M52G463 2005
616'.027333--dc22   2005041899

Taylor & Francis Group
is the Academic Division of T&F Informa plc.

Visit the Taylor & Francis Web site at
http://www.taylorandfrancis.com

and the CRC Press Web site at
http://www.crcpress.com

We wish to thank our wives, Beth A. Sundberg and Noriko Ichiki,
for their support and encouragement in conception to completion of this book.

Preface

The laboratory mouse has unquestionably become the animal model of choice in biomedical research. Three decades ago this was not obvious unless one counted the numbers of models by species. Mice were always first, rats second, and domestic companion animals third. Transgenic technology followed by targeted mutagenesis and many variations thereof, collectively known as genetic engineering, rapidly brought the mouse to the head of the line as the premier model organism. The problem that arose was a severe lack of expertise in what became known as phenotyping. This is a modern term for detailed clinical and pathologic evaluation of the patient, in this case, the mouse. To address this need, several of us organized a meeting at the National Institutes of Health (NIH) in Bethesda, Maryland, titled *Pathology of Genetically Engineered Mice*. This large open meeting was followed by a series of similar meetings held in Stockholm, Sweden and then Kumamoto, Japan on alternate years. The topic spread from being focused on pathology of mice with spontaneous or genetically engineered mutations, primarily, if not exclusively, single gene mutations, to a broader topic initially on large scale phenotyping to an overview approach of the entire field. This book, *Genetically Engineered Mice Handbook*, is an extension of the meeting in Japan and attempts to bring the technology together into a centralized resource as an aid to those entering the field.

The first meeting at NIH resulted in a book titled *Pathology of Genetically Engineered Mice*. Much of the information revolves around traditional technology that is still accurate and useful today. An obvious change is the predominance in this current book of information focusing on publicly accessible databases. The database approach merges the value of a hard-copy book with the plasticity of web-based information resources, which, if maintained, will be regularly updated and expanded. Knowing where to start and how to use these is the key. Complaints on the pathology book focus on lack of coverage of some organ systems and its being static and therefore not having information on new models. The current book provides information on how to access databases that have this information in a constantly expanding and updated format. Mouse pathology Websites are now publicly accessible. While in the past finding and working with an expert on a particular organ system in mice was ideal, now groups of experts work together to generate these Websites, discuss and argue the data, and provide a resource easily accessible. While this does not eliminate the need for a trained pathologist to interpret the histopathology of a disease, these Websites provide reference materials for those lacking expertise in particular anatomic structures.

The lack of adequate numbers of veterinarians trained in mouse clinical medicine and pathology resulted in development of many imaging devices as well as test apparatuses. Downsizing imaging tools used in human medicine has now become commonplace. Many centers, particularly the mutagenesis centers, developed testing systems for various organ systems, which, over time, have been developed to high levels of accuracy. Technicians with relatively little training can now perform many of these tasks.

Husbandry and microbiological monitoring are critical aspects of dealing with any animal model system. These too have evolved to such a high level of sophistication that specific microflora are needed to optimize experimental results or else the presence of some agents can severely and negatively affect the outcome of experiments. Examples of these approaches are provided.

The technology to generate mutant mice remains the same but also changes very quickly. Spontaneous mutations remain the gold standard as we learn that manipulation of the genome often results in unexpected findings that are artifacts and not true representations of real life. The chemical mutagenesis approach, a method of choice in decades past, became a major focus during the last 5 years. Large scale ethylnitrosourea (ENU) chemical mutagenesis programs on several continents were very active and generated many novel mutant mice or allelic mutations of spontaneous mutant mice already characterized. The value of this approach from a cost benefit perspective was questioned in recent years resulting in the downsizing and closing of such projects in the United States and Germany. Newer approaches are to generate repositories of embryonic stem cells (ES cells) that carry null mutations with reporter constructs.

Genetic engineering methods have evolved into production line science. High throughput phenotype screening programs have evolved to meet this demand to some degree. Ultimately, scientists trained in clinical medicine, particularly in pathology, are key to defining the real value of these mutant mice as models for specific human diseases. Dramatic breakthroughs continue to emerge from studies with these mouse models on the basic genetics, and to some degree the mechanisms of disease. Now these models need to be used aggressively as preclinical models to have a more direct impact on human patients. All of this is happening and we can expect more exciting discoveries in the coming years.

Editors

John P. Sundberg, D.V.M., Ph.D. is a senior staff scientist at The Jackson Laboratory in Bar Harbor, Maine. Dr. Sundberg graduated summa cum laude from the University of Vermont in 1973, and received his doctor of veterinary medicine degree in 1978 from Purdue University. After working in private practice he received his Ph.D. in 1981 in virology from the University of Connecticut where he also did a residency in anatomic pathology, passing the specialty board examination in 1982. He worked as an assistant professor and staff pathologist at the University of Illinois College of Veterinary Medicine from 1981–1986 after which he joined The Jackson Laboratory as a staff scientist and diagnostic pathologist. He served as head of pathology there from 1989–2000. In 1998 he became a senior staff scientist.

He currently is a principal investigator studying genetically-based skin diseases in laboratory mice as models for genetic diseases in humans. He is also involved in setting up large scale drug efficacy trials at The Jackson Laboratory's West Sacramento, California facility for the mouse dermatology models he developed. His basic genetic research focuses on autoimmune (alopecia areata), proliferative (psoriasis-like), and hair follicle developmental studies. He has published nearly 300 scientific papers and more than 100 book chapters. This is his sixth book on the biology, pathology, and management of laboratory mice.

Tsutomu Ichiki, Ph.D. is a general manager of the Safety Assessment Lab at Dainippon Ink and Chemicals, Inc. in Sakura, Chiba, Japan. Dr. Ichiki graduated in 1979 from Kyusyu University in Fukuoka with a bachelor of science degree in animal husbandry and obtained his masters degree in 1981 from Kyusyu University.

Dr. Ichiki completed his residency in pathology, and earned his Ph.D., from Fukuoka University School of Medicine. Dr. Ichiki was a manager of the Pathology Division at Panapharm Laboratories Co., Ltd. from 1985 to 2000. In 2000 he joined the staff scientist of Transgenic Inc. In 2001 he served a sabbatical as visiting investigator of The Jackson Laboratory until 2003.

Dr. Ichiki is a diplomate of Japanese Society of Toxicologic Pathology. He is a member of the Society of Toxicologic Pathology in United States, and a member of Planning and Managing Panel of the Long-range Research Initiative in the Japan Chemical Industry Association.

Dr. Ichiki's research interests focus on laboratory animals — especially on mice and rats — for the health effects of various chemicals.

Contributors

Kimi Araki, Ph.D., Institute of Molecular Embryology and Genetics, Kumamoto University School of Medicine, Kumamoto, Japan.

Jonathan B. L. Bard, Ph.D., Department of Biomedical Sciences, Edinburgh University, Edinburgh, United Kingdom.

Carol J. Bult, Ph.D., The Jackson Laboratory, Bar Harbor, Maine, United States.

Hsueh-Ping Chu, Ph.D., Institute of Molecular Biology, Academia Sinica, Nankang, Taipei, Taiwan and Graduate Institute of Biochemistry, National Yang-Ming University, Taipei, Taiwan.

Bon-chu Chung, Ph.D., Institute of Molecular Biology, Academia Sinica, Nankang, Taipei, Taiwan.

Pascal Dollé, M.D., Ph.D., Institut de Genetique et de Biologie Moleculaire et Cellulaire, CNRS/INSERM/Universite Louis Pasteur, CU de Strasbourg, France.

David Einhorn, B.S.E., L.L.B., The Jackson Laboratory, Bar Harbor, Maine, United States.

Janan T. Eppig, Ph.D., The Jackson Laboratory, Bar Harbor, Maine, United States.

James R. Fahey, Ph.D., D.V.M., American College of Veterinary Microbiologists, The Jackson Laboratory, Bar Harbor, Maine, United States.

Martin D. Fray, Ph.D., Mammalian Genetics Unit, Medical Research Council, Harwell, United Kingdom.

Jabier Gallego Llamas, M.Sc., Institut de Genetique et de Biologie Moleculaire et Cellulaire, CNRS/INSERM/Universite Louis Pasteur, CU de Strasbourg, France.

Peter H. Glenister, Mammalian Genetics Unit, Medical Research Council, Harwell, United Kingdom.

Carroll-Ann W. Goldsmith, Sc.D., The Jackson Laboratory, Bar Harbor, Maine, United States.

Noomen Ben El Hadj, Ph.D., Institute of Molecular Biology, Academia Sinica, Nankang, Taipei, Taiwan.

Meng-Chun Hu, Ph.D., Graduate Institute of Physiology, National Taiwan University College of Medicine, Taipei, Taiwan.

Tsutomu Ichiki, Ph.D., Japanese Society of Toxicologic Pathology, Safety Assessment Laboratory, Dainippon Ink and Chemicals, Inc., Chiba, Japan.

Kimiko Inoue, Ph.D., Bioresource Center, RIKEN, Tsukuba, Ibaraki, Japan.

Kikuji Itoh, Ph.D., Laboratory of Veterinary Public Health, Graduate School of Agricultural and Life Sciences, The University of Tokyo, Tokyo, Japan.

Takehito Kaneko, Ph.D., Division of Reproductive Engineering, Center for Animal Resources and Development (CARD), Kumamoto University, Kumamoto, Japan.

Shuichi Kani, Ph.D., Department of Genome Sciences, Kobe University Graduate School of Medicine, Kobe, Japan.

Hideki Katoh, Ph.D., Institute for Experimental Animals, Hamamatsu University School of Medicine, Hamamatsu, Japan and ICLAS Monitoring Center, Central Institute for Experimental Animals, Kanagawa, Japan.

Isabelle Le Roux, Ph.D., Institut de Génétique et de Biologie Moléculaire et Cellulaire, CNRS/INSERM/Universite Louis Pasteur, CU de Strasbourg, France.

Hiromi Miki, Bioresource Center, RIKEN, Ibaraki, Japan.

Yasuhiro Minami, M.D., Ph.D., Department of Genome Sciences, Kobe University Graduate School of Medicine, Kobe, Japan.

Tsutomu Mizoshita, M.D., Ph.D., Division of Oncological Pathology, Aichi Cancer Center Research Institute, Nagoya, Japan.

Naomi Nakagata, Ph.D., Center for Animal Resources and Development, Kumamoto University, Kumamoto, Japan.

Seiko Narushima, Ph.D., Food Functionality Research Institute, Meiji Dairies Corporation, Kanagawa, Japan.

Karen Niederreither, Ph.D., Departments of Medicine and Molecular and Cellular Biology, Center for Cardiovascular Development, Baylor College of Medicine, Houston, Texas, United States.

H. Nishida, Center for Animal Resources and Development (CARD) and Graduate School of Molecular and Genomic Pharmacy, Kumamoto University, Kumamoto, Japan.

Y. Ogino, Ph.D., Center for Animal Resources and Development (CARD) and Graduate School of Molecular and Genomic Pharmacy, Kumamoto University, Kumamoto, Japan.

Narumi Ogonuki, Bioresource Center, RIKEN, Ibaraki, Japan.

Atsuo Ogura, Ph.D., D.V.M., Bioresource Center, RIKEN, Ibaraki, Japan.

T. Ohba, M.D., Ph.D., Center for Animal Resources and Development (CARD) and Graduate School of Molecular and Genomic Pharmacy, Kumamoto University, Kumamoto, Japan.

S. Ohta, Ph.D., Center for Animal Resources and Development (CARD) and Graduate School of Medical Sciences, Kumamoto University, Kumamoto, Japan.

Isao Oishi, Ph.D., Department of Genome Sciences, Kobe University Graduate School of Medicine, Kobe, Japan.

Vanessa Ribes, M.Sc., Institut de Génétique et de Biologie Moléculaire et Cellulaire, CNRS/INSERM/Universite Louis Pasteur, CU de Strasbourg, France.

Martin Ringwald, Ph.D., The Jackson Laboratory, Bar Harbor, Maine, United States.

Stephen F. Rockwood, B.S., The Jackson Laboratory, Bar Harbor, Maine, United States.

Björn Rozell, D.D.S., Ph.D., D.V.M., Clinical Research Center, Department of Laboratory Medicine, Karolinska Institutet, Karolinska University Hospital, Stockholm, Sweden.

Y. Satoh, Ph.D., Center for Animal Resources and Development (CARD) and Graduate School of Medical Sciences, Kumamoto University, Kumamoto, Japan.

Paul Schofield, M.A., D.Phil., University of Cambridge, Department of Anatomy, Cambridge, United Kingdom.

John P. Sundberg, D.V.M., Ph.D., American College of Veterinary Pathologists, The Jackson Laboratory, Bar Harbor, Maine, United States.

K. Suzuki, Ph.D., Center for Animal Resources and Development (CARD) and Graduate School of Molecular and Genomic Pharmacy, Kumamoto University, Kumamoto, Japan.

Makoto Mark Taketo, M.D., Ph.D., Department of Pharmacology, Graduate School of Medicine, Kyoto University, Kyoto, Japan.

Masae Tatematsu, M.D., Ph.D., Division of Oncological Pathology, Aichi Cancer Center Research Institute, Nagoya, Japan.

Tetsuya Tsukamoto, M.D., Ph.D., Division of Oncological Pathology, Aichi Cancer Center Research Institute, Nagoya, Japan.

Julien Vermot, Ph.D., Institut de Génétique et de Biologie Moléculaire et Cellulaire, CNRS/INSERM/Université Louis Pasteur, CU de Strasbourg, France.

Y. Wada, Ph.D., Department of Urology Graduate School of Medical and Kumamoto University School of Medicine, Kumamoto, Japan.

Teruhiko Wakayama, Ph.D., Laboratory for Genomic Reprogramming, Center for Developmental Biology, RIKEN Kobe, Kobe, Japan.

Leo Chi-Kuang Wang, Ph.D., Institute of Molecular Biology, Academia Sinica, Nankang, Taipei, Taiwan.

Jerrold M. Ward, D.V.M., Ph.D., American College of Veterinary Pathologists, National Institutes of Health, Maryland, United States.

Y. Xu, Ph.D., Center for Animal Resources and Development (CARD), and Graduate School of Molecular and Genomic Pharmacy, Kumamoto University, Kumamoto, Japan.

G. Yamada, Ph.D., Center for Animal Resources and Development (CARD) and Graduate School of Molecular and Genomic Pharmacy, Kumamoto University, Kumamoto, Japan.

Yukiko Yamazaki, Ph.D., Center for Genetic Resource Information, National Institute of Genetics, Mishima, Japan.

Akinori Yoda, Ph.D., Department of Genome Sciences, Kobe University Graduate School of Medicine, Kobe, Japan.

Y. Zhang, Ph.D., College of Bioengineering, Fujian University, Fuzhou, Fujian, China.

Table of Contents

1 Genetically Engineered Mice: Past, Present, and Future

John P. Sundberg, Tsutomu Ichiki,
Jerrold M. Ward, and Björn Rozell

TABLE OF CONTENTS

Mutant mice have changed from curiosities in the nineteenth century, being primarily of concern to mouse fanciers,[1] to become the premier animal model for human diseases, mammalian genetics, and biomedical research in the twenty-first century. Archives of spontaneously occurring mutations, a valued resource for over fifty years, are being supplemented and supplanted by genetically engineered mice (GEMs) created by transgenesis, targeted mutagenesis, inducible mutagenesis, gene trap, and chemical mutagenesis methods at an overwhelming rate. Evaluation of these mice to determine clinical and physiological deviations from normal for a particular inbred strain (i.e., phenotyping) has become a major undertaking in the evaluation of these new strains. Current methods to create, archive, distribute, and phenotype mice were the focus of a meeting, "Pathology of Genetically Engineered Mice: Large Scale Phenotyping of Mutant Mice," held in October 2003 in Kumamoto, Japan. Leaders in the field presented the current strategies, which are summarized in this book. An overview of the field is given here.

GENERATION OF MUTANT MICE

Transgenesis, the process of introducing novel genes or overexpressing known genes into the mouse genome, has been the basis for the genetics revolution since the first successes in the early 1980s.[2] Another important development, the blastocyst-derived embryonic stem cells,[3,4] later led to the technique of targeted mutagenesis, whereby known genes could be inactivated using homologous recombination of genes containing molecular tags (*LacZ* or neomycin cassettes) to interfere with transcription of specific exons producing so-called gene "knock-outs" or "hypomorphs" for those genes not totally inactivated.[5,6] Many of these targeted mutations resulted in phenotypes similar to human diseases, but equal numbers had no phenotype or were embryonic or neonatal lethals. Stressing the system under investigation sometimes revealed differences from the background strains.[7] Inasmuch as embryonic or perinatal lethality are fairly common consequences of targeted mutagenesis, alternate gene constructs were developed to study these genes in adult animals.

Conditional mutagenesis or the so-called "gene switch" approaches were subsequently developed using *Cre*-recombinase together with *LoxP* or *Frt* sites. Furthermore, inducible systems, based on tetracycline (tet-on/off), or progesterone antagonists, led to even greater refinements whereby genes could be activated or inactivated by adding exogenous systems.[8,9] In this manner, effects of gene inactivation could be studied either throughout the organism in selected organs, or even in part of an organ. The neomycin (*Neo*) gene often leads to a suppression of the mutant allele, resulting in a phenotype not different from normal. Thus, making a construct where the *Neo* cassette is flanked by *LoxP* sequences enables local and restricted excision of the *Neo* gene in skin, thereby activating the targeted allele, thus enabling the creation of mouse models of a selected number of human genetic diseases.[10–13]

As a complement to the technique of targeted mutagenesis, gene trap methods provide another tool to approach the problem of identifying novel genes within the genome. Genetic constructs with marker sequences are used to immediately map mutant gene loci once an interesting phenotype is produced.[14,15] Different modalities of the gene trap technique are available.[16] Because many more genes are coded for in the human and mouse genome than are currently known (www.celera.com), other approaches were needed to augment the mutations derived by targeted mutagenesis or that occurred spontaneously. Spontaneous mutations occur at a frequency 1/100000 offspring, however, this is not enough to achieve a saturation of mutations throughout the genome. To increase the frequency of mutations, chemical mutagenesis, specifically using ethylnitrosourea (ENU), randomly produces single base pair changes, and increases the mutation rate 100-fold. These old protocols are now used in large-scale mutagenesis projects, potentially providing a wealth of novel mutant mice that can subsequently have the mutant locus mapped and defined.[17–24] A new version of this technique uses ethylmethanesulfonate on embryonic stem cells[25] with similar results. Back-crossing the chimeric mice to C57BL/6 immediately provides a rapid tool to map the mutant gene locus.

Contrary to the techniques employing homologous recombination (gene-driven approach), the discovery of the chemically induced mutations are phenotype driven (*vide infra*). This requires a setup whereby mutations affecting different organ systems can be specifically detected by screening assays. Furthermore, although dominant mutations are easily detected in a variety of screens, recessive mutations require special breeding protocols, thus slowing down the process. However, in some screens, such as the one at MRC Harwell, sperm freezing was adopted as the method of choice to preserve identified mutants. An advantage of this procedure is that sperm can be used in an *in vitro* fertilization (IVF) procedure to obtain a large colony suitable for mapping the mutation, thereby circumventing slow and costly breeding.[26–33]

Both dominant and recessive mutations must also be mapped. Because some of the recessive mutations discovered already resemble phenotypic copies of mutant mice previously described, these can be tested by breeding (allelism testing) to see if the new alleles are noncomplementary and therefore allelic with the well-characterized mutant mouse.[34,35] When novel phenotypes are found, the mutated locus must first be mapped using standard genetic techniques (*vide infra*), before positionally cloning the affected gene.[36]

GENE MAPPING

Finding abnormal mice, at least those with dominant mutations, is often relatively simple. Figuring out what the defects are and whether the lesions resemble those found in human or domestic animal diseases is described below. Mapping the mutant gene locus is relatively easy for single gene mutations. A variety of methods and approaches is available. In mice, unlike humans, it is possible to set up breedings between inbred strains carrying the mutated gene and different strains carrying two copies of the wild-type gene. Recessive mutations will be expressed in the second, F2 generation, from such an intercross. The mouse genome can then be rapidly screened using simple sequence length polymorphism (SSLP) and single nucleotide polymorphism (SNP) technologies with molecular markers at regular intervals (15–20 cM).[36,37] A shift in genes from 50 percent of both inbred strains to the strain that originally carried the mutated gene indicates linkage. The gene can be localized by refining this approach with additional markers (saturation mapping) within the suspected interval. The gene can be identified by sequencing through known genes in that interval, or positional cloning of the gene, then creating targeted mutations of the gene or rescuing it by transgenesis with the known gene or BAC clones containing the interval.[36,38] The availability of anchored full-length mouse cDNA clones and other genetic tools is of great help in determining candidate genes for positional cloning.

MAINTENANCE AND ARCHIVING

Production of all these mutant mice and the normal inbred strains that they occur on naturally or are created on has become a daunting task. Live mouse archives were pioneered by Drs. Elizabeth Russell and George Snell at The Jackson Laboratory in the 1940s.[39] Dr. Russell began moving the mutated genes onto a standard inbred background, the C57BL/6J mouse, to create congenic lines. Intercrossing various mutations all maintained on a single inbred background enabled scientists to investigate the interaction of the two genes in question without worrying about the effects of background modifier genes.[36] This approach and creation of repositories to house these mice at The Jackson Laboratory, National Institutes of Health, Oak Ridge National Laboratory, Harwell, The Institut Pasteur, and other institutions around the world were adequate until the early 1990s when transgenic technology became relatively common.

Commercial repositories charge high prices, often have follow-through agreements, and long waiting periods; therefore this was not acceptable to the academic or commercial scientific community. The first public repository was the Induced Mutant Resource at The Jackson Laboratory in 1992.[40,41] Several commercial vendors soon followed as well as other public repositories around the world including the joint European effort, European Mouse Mutant Archive (EMMA). A nodal system in Europe and North America has now been developed to share technology and the burden of housing all these mice, and also to reduce redundancy on those rarely used.

Alternatives to live colonies were necessary whereby mice could still be obtained years later. Building on technology developed for domestic species, cryopreservation of mouse embryos has become the *de facto* standard for archiving potentially important lines. More current work focuses on cryopreservation of sperm and their use for *in vitro* fertilization. An added benefit is the use of this technology for rapidly expanding colonies, rapidly setting up large breedings for gene mapping, and transporting lines in a pathogen-free environment between institutions.[26–33,42–46]

PHENOTYPING APPROACHES

Classic approaches to phenotyping utilize basic medical techniques, and these are discussed in detail in several chapters in this book. Physical examination of the normal and mutated mice may reveal obvious changes. Collection of blood, serum, plasma, urine, and other body fluids can be analyzed using equipment modified from human clinical pathology laboratories.[47] Blood values for commonly used strains are published.[48,49] The Mouse Phenome Project (www.jax.org/phenome) was created to actively obtain a wide variety of data on individual mouse strains from researchers around the world and put them online.[50] This is a major effort to make these types of data on major inbred strains of mice freely available to the biomedical research community.

Careful, systematic, and detailed necropsies of mutant and control mice at defined ages with histopathology remain a standard for defining lesions. Comparative pathology between human and domestic animals provides insight into whether the mutant mice are potential models for specific diseases or unique tools to dissect the function of a gene or organ. Numerous textbooks were published recently providing overviews on these methods and the types of lesions commonly found in aging and mutant mice.[51–54] Many Web sites are online and are constantly being updated with similar information. How to access and use these are described in detail in other chapters in this book. Approaches to setting up such phenotyping programs are also now available.[55]

Very specialized equipment was developed for each organ system. Much effort is being put into tools to better define neurological abnormalities in mice.[52] The SHIRPA protocol was developed to provide a panel of clinical evaluations and analytical tests to determine if mice had neurological abnormalities.[56] These are being adapted by many of the ENU programs worldwide as well as by startup companies providing phenotyping services. Specialty books on various organ systems are also being published.[57–59]

Because anatomic and clinical pathologists with mouse expertise are relatively few and far between, there has been downscaling of traditional diagnostic equipment with intense development of very sophisticated equipment to provide rapid, high-throughput, low-labor information. X-ray machines, originally designed for scanning electronic boards, work extremely well with mammography film to provide flat-plate images of mice (www.faxitron.com).[60] CAT scans, Pet scans, MRIs, and other methods are now available to provide real-time, three-dimensional images of mouse organs or to test physiologic parameters (www.scanco.ch, www.imtekinc.com, www.optiscan.com, www.colinst.com). New technologies are constantly under development.[61–65]

SUMMARY

The laboratory mouse is a domestic animal. Veterinarians, physicians, and biomedical researchers can now use both traditional technology and microanalytical tools to evaluate mutant or experimentally manipulated mice to answer fundamental medical and biological questions. The area is reviewed here to provide key references and Web sites from which specific information and resources can be found.

REFERENCES

1. Gaskoin, J. S., On a peculiar variety of *Mus musculus*, *Proc Zool Soc London* 24, 38–40, 1856.
2. Gordon, J. W., Scangos, G. A., Plotkin, D. J., Barbosa, J. A., and Ruddle, F. H., Genetic transformation of mouse embryos by microinjection of purified DNA, *Proc Natl Acad Sci USA* 77 (12), 7380–7384, 1980.
3. Martin, G. R., Isolation of a pluripotent cell line from early mouse embryos cultured in medium conditioned by teratocarcinoma stem cells, *Proc Natl Acad Sci USA* 78 (12), 7634–7638, 1981.
4. Evans, M. J. and Kaufman, M. H., Establishment of pluripotential cells from mouse embryos, *Nature* 292 (5819), 154–156, 1981.
5. Thomas, K. R. and Capecchi, M. R., Site-directed mutagenesis by gene targeting in mouse embryo-derived stem cells, *Cell* 51 (3), 503–512, 1987.
6. Doetschman, T., Maeda, N., and Smithies, O., Gene transfer with subsequent removal of the selection gene from the host genome, *Proc Natl Acad Sci USA* 88 (23), 10558–10562, 1991.

7. Liaw, L., Birk, D. E., Ballas, C. B., Whitsitt, J. S., Davidson, J. M., and Hogan, B. L., Altered wound healing in mice lacking a functional osteopontin gene, *J Clin Invest* 101 (7), 1468–1478, 1998.

8. Dale, E. C. and Ow, D. W., Gene transfer with subsequent removal of the selection gene from the host genome, *Proc Natl Acad Sci USA* 88 (23), 10558–10562, 1991.

9. Schwenk, F., Baron, U., and Rajewsky, K., A *Cre*-transgenic mouse strain for the ubiquitous deletion of *LoxP*-flanked gene segments including deletion of germ cells, *Nucleic Acids Res* 23 (24), 5080–5081, 1995.

10. Wang, X.-J., Liefer, K. M., Tsai, S., O'Malley, B. W., and Roop, D. R., Development of gene-switch transgenic mice which inducibly express transforming growth factor b1 (TGFb1) in the epidermis, *Proc Natl Acad Sci USA* 96, 8483–8488, 1999.

11. Berton, T. R., Wang, X.-J., Zhou, Z., Kellendonk, C., Schutz, G., Tsai, S., and Roop, D. R., Characterization of an inducible, epidermal-specific knockout system: Differential expression of *LacZ* in different *Cre* reporter mouse strains, *Genesis* 26, 160–161, 2000.

12. Arin, M. J., Longley, M. A., Wang, X.-J., and Roop, D. R., Focal activation of a mutant allele defines the role of stem cells in mosaic skin disorders, *J Cell Biol* 152, 645–649, 2001.

13. Cao, T., Longley, M. A., Wang, X.-J., and Roop, D. R., An inducible mouse model for epidermolysis bullosa simplex: Implications for gene therapy, *J Cell Biol* 152, 651–656, 2001.

14. Salminen, M., Meyer, B. I., and Gruss, P., Efficient poly A trap approach allows the capture of genes specifically active in differentiated embryonic stem cells and in mouse embryos, *Dev Dyn* 212 (2), 326–333, 1998.

15. Cecconi, F. and Meyer, B. I., Gene trap: A way to identify novel genes and unravel their biological function, *FEBS Lett* 480 (1), 63–71, 2000.

16. Mitchell, K. J., Pinson, K. I., Kelly, O. G., Brennan, J., Zupicich, J., Scherz, P., Leighton, P. A., Goodrich, L. V., Lu, X., Avery, B. J., Tate, P., Dill, K., Pangilinan, E., Wakenight, P., Tessier-Lavigne, M., and Skarnes, W. C., Functional analysis of secreted and transmembrane proteins critical to mouse development, *Nat Genet* 28 (3), 241–249, 2001.

17. Soewarto, D., Fella, C., Teubner, A., Rathkolb, B., Pargent, W., Heffner, S., Marchall, S., Wolf, E., Balling, R., and Hrabe de Angelis, M., The large-scale Munich ENU-mouse mutagenesis screen, *Mamm Genome* 11 (7), 507–510, 2000.

18. Nadeau, J. H. and Frankel, W. N., The roads from phenotypic variation to gene discovery: Mutagenesis versus QTLs, *Nat Genet* 25 (4), 381–384, 2000.

19. Nadeau, J. H., Muta-genetics or muta-genomics: The feasibility of large-scale mutagenesis and phenotyping programs, *Mamm Genome* 2000 (7), 603–607, 2000.

20. Nolan, P. M., Peters, J., Strivens, M., Roges, D., Hagan, J., Spurr, N., Gray, I. C., Vizor, I., Brooker, D., Whitehill, E., Washbourne, R., Hough, T., Greenaway, S., Hewitt, M., Liu, X., McCormack, S., Pickford, N., Selley, R., Wells, C., Tymowska-Lalanne, Z., Roby, P., Glenister, P., Thornton, C., Thaung, C., Stevenson, J. A., Arkell, R., Mburu, P., Hardisty, R., Kiernan, A., Erven, A., Steel, K. P., Voegeling, S., Guenet, J. L., Nickols, C., Sadri, S., Nasse, M., Isaacs, A., Davies, K., Browne, M., Fisher, E. M., Martin, J., Rastan, S., Brown, S. D., and Hunter, J., A systematic, genome-wide, phenotype-driven mutagenesis programme for gene function studies in the mouse, *Nat Genet* 25 (4), 440–443, 2000.

21. Nolan, P. M., Peters, J., Vizor, I., Strivens, M., Washbourne, R., Hough, T., Wells, C., Glenister, P., Thornton, C., Martin, J., Fisher, E., Rogers, D., Hagan, J., Reavill, C., Gray, I., Wood, J., Spurr, N., Browne, M., Hunter, J., and Brown, S. D., Implementation of a large-scale ENU mutagenesis program: Towards increasing the mouse mutant resource, *Mamm Genome* 11 (7), 500–506, 2000.

22. Coghill, E. L., Hugill, A., Parkinson, N., Davison, C., Glenister, P., Clements, S., Hunter, J., Cox, R. D., and Brown, S. D., A gene-driven approach to the identification of ENU mutants in the mouse, *Nat Genet* 30 (3), 255–256, 2002.

23. Quwailid, M. M., Hugill, A., Dear, N., Vizor, L., Wells, S., Horner, E., Fuller, S., Weedon, J., McMath, H., Woodman, P., Edwards, D., Campbell, D., Rodger, S., Carey, J., Roberts, A., Glenister, P., Lalanne, Z., Parkinson, N., Coghill, E. L., McKeone, R., Cox, S., Willan, J., Greenfield, A., Keays, D., Brady, S., Spurr, N., Gray, I., Hunter, J., Brown, S. D., and Cox, R. D., A gene-driven ENU-based approach to generating an allelic series in any gene, *Mamm Genome* 15 (8), 585–591, 2004.

24. Hrabe de Angelis, M., Flaswinkel, H., Fuchs, H., Rathkolb, B., Soewarto, D., Marschall, S., Heffner, S., Pargent, W., Wuensch, K., Jung, M., Reis, A., Richter, T., Alessandrini, F., Jakob, T., Fuchs, E., Kolb, H., Kremmer, E., Schaeble, K., Rollinski, B., Roscher, A., Peters, C., Meitinger, T., Strom, T., Stekler, T., Holsboer, F., Klopstock, T., Gekeler, F., Schindewolf, C., Jung, T., Avraham, K., Behrendt, H., Ring, J., Zimmer, A., Schughart, K., Pfeffer, K., Wolf, E., and Balling, R., Genome-wide, large scale production of mutant mice by ENU mutagenesis, *Nat Genet* 25 (4), 444–447, 2000.

25. Munroe, R. J., Bergstrom, R. A., Zheng, Q. Y., Libby, B., Smith, R., John, S. W. M., Schimenti, K. J., and Browning, V. L., Mouse mutants from chemically mutagenized embryonic stem cells, *Nat Genet* 24, 318–321, 2000.

26. Thornton, C. E., Brown, S. D., and Glenister, P. H., Large numbers of mice established by *in vitro* fertilization with cryopreserved spermatozoa: Implications and applications for genetic resource banks, mutagenesis screens, and mouse backcrosses, *Mamm Genome* 10 (10), 987–992, 1999.

27. Glenister, P. H. and Thornton, C. E., Cryoconservation—Archiving for the future, *Mamm Genome* 11 (7), 565–571, 2000.

28. Sztein, J. M., Farley, J. S., Young, A. F., and Mobraaten, L. E., Motility of cyropreserved mouse spermatozoa affected by temperature of collection and rate of thawing, *Cryobiology* 35 (1), 46–52, 1997.

29. Crister, J. K. and Mobraaten, L. E., Cryopreservation of murine spermatozoa, *ILAR J* 41 (4), 197–206, 2000.

30. Sztein, J. M., Noble, K., Farley, J. S., and Mobraaten, L. E., Comparison of permeating and nonpermeating cyroprotectants for mouse sperm cryopreservation, *Cryobiology* 42 (1), 503–512, 2001.

31. Boggess, D., Silva, K. A., Landel, C., Mobraaten, L., and Sundberg, J. P., Approaches to handling, breeding, strain preservation, genotyping, and drug administration for mouse models of cancer. In *Mouse Models of Human Cancer*, Holland, E. C., ed. John Wiley & Sons, Hoboken, NJ, 2004, pp. 3–14.

32. Marschall, S. and HrabedeAngelis, M., Cryopreservation of mouse spermatozoa: Double your mouse space, *Trends Genet* 15 (4), 128–131, 1999.

33. Marschall, S. and Hrabe de Angelis, M., *In vitro* fertilization/cryopreservation, *Methods Mol Biol* 209, 35–50, 2003.

34. Sundberg, J. P., Boggess, D., Sundberg, B. A., Eilersten, K., Parimoo, S., Filippi, M., and Stenn, K., Asebia-2J (*Scdl^{ab–2J}*): A new allele and a model for scarring alopecia, *Am J Pathol* 156 (6), 2067–2075, 2000.

35. Sundberg, J., Boggess, D., Bascom, C., Limberg, B. J., Shultz, L. D., Sundberg, B. A., King, L. E., and Montagutelli, X., Lanceolate hair-J (*lah^J*): A mouse model for human hair disorders, *Exp Dermatol* 9, 206–218, 2000.

36. Montagutelli, X., Determining the genetic basis of a new trait. In *Systematic Approach to Evaluation of Mouse Mutations*, Sundberg, J. P. and Boggess, D., eds. CRC Press, Boca Raton, FL, 2000, pp. 15–33.

37. Petkov, P. M., Ding, Y., Cassell, M. A., Zhang, W., Wagner, W., Sargent, E. E., Asquith, S., Crew, V., Johnson, K. A., Robinson, P., Scott, V. E., and Wiles, M. V., An efficient SNP system for mouse genome scanning and elucidating strain relationships, *Genome Res* 14, 1806–1811, 2004.

38. Montagutelli, X., Effect of the genetic background on the phenotype of mouse mutations, *J Am Soc Nephrol* 16, S101–105, 2000.

39. Holstein, J., *The First Fifty Years at the Jackson Laboratory*. The Jackson Laboratory, Bar Harbor, ME, 1979.

40. Sharp, J. J., Linder, C. C., and Mobraaten, L. E., Genetically engineered mice. Husbandry and resources, *Methods Mol Biol* 158, 381–396, 2001.

41. Davisson, M. T. and Sharp, J. J., Repositories of mouse mutations and inbred, congenic, and recombinant inbred strains. In *Systematic Characterization of Mouse Mutations*, Sundberg, J. P. and Boggess, D., eds. CRC Press, Boca Raton, FL, 2000, pp. 177–190.

42. Glenister, P. H., Whittingham, D. G., and Wood, M. J., Genome cryopreservation: a valuable contribution to mammalian genetic research, *Genet Res* 56 (2–3), 253–258, 1990.

43. Sztein, J., Sweet, H., Farley, J., and Mobraaten, L., Cryopreservation and orthotopic transplantation of mouse ovaries: New approach in gamete banking, *Biol Reprod* 58 (4), 1071–1074, 1998.

44. Sztein, J. M., McGregor, T. E., Bedigian, H. J., and Mobraaten, L. E., Transgenic mouse strain rescue by frozen embryos, *Lab Anim Sci* 49 (1), 99–100, 1999.

45. Sztein, J. M., Farley, J. S., and Mobraaten, L. E., *In vitro* fertilization with cryopreserved inbred mouse sperm, *Biol Reprod* 63 (6), 1774–1780, 2000.

46. Sztein, J. M., O'Brien, M. J., Farley, J. S., Mobraaten, L. E., and Eppig, J. J., Rescue of oocytes from antral follicles of cryopreserved mouse ovaries: Competence to undergo maturation, embryogenesis, and development to term, *Hum Reprod* 15 (3), 567–571, 2000.

47. Car, B. and Eng, V., Special considerations in the evaluation of the hematology and homeostasis of mutant mice, *Vet Pathol* 38, 20–30, 2001.

48. Loeb, W. F., Das, S. R., Harbour, L. S., Turturro, A., Bucci, T. J., and Clifford, C. B., Clinical biochemistry. In *Pathobiology of the Aging Mouse*, Mohr, U., Dungsworth, D. L., Capen, C. C., Carlton, W. W., Sundberg, J. P., and Ward, J. M., eds. ILSI Press, Washington DC, 1996, pp. 3–19.

49. Hough, T. A., Nolan, P. M., Tsipouri, V., Toye, A. A., Gray, I. C., Goldsworthy, M., Moir, L., Cox, R. D., Clements, S., Glenister, P. H., Wood, J., Selley, R. L., Strivens, M. A., Vizor, L., McCormack, S. L., Peters, J., Fisher, E. M., Spurr, N., Rastan, S., Martin, J. E., Brown, S. D., and Hunter, A. J., Novel phenotypes identified by plasma biochemical screening in the mouse, *Mamm Genome* 13 (10), 595–602, 2002.

50. Paigen, K. and Eppig, J. T., A mouse phenome project, *Mamm Genome* 11 (9), 715–717, 2000.
51. Frith, C. H. and Ward, J. M., *Color Atlas of Neoplastic and Non-neoplastic Lesions in Aging Mice.* Elsevier, Amsterdam, 1988.
52. Ward, J., Mahler, J., Maronpot, R., and Sundberg, J. P., *Pathology of Genetically Engineered Mice.* Iowa State University Press, Ames, 2000.
53. Maronpot, R. R., Boorman, G. A., and Gaul, B. W., *Pathology of the Mouse. Reference and Atlas.* Cache River Press, Vienna, IL, 1999.
54. Percy, D. H. and Barthold, S. W., *Pathology of Laboratory Rodents & Rabbits.* Iowa State University Press, Ames, 2001.
55. Sundberg, J. P., Haschek, W. M., Hackman, R. C., and HogenEsch, H., Developing a comprehensive mouse pathology program, *Comp Med* 54, 615–619, 2004.
56. Rogers, D. C., Peters, J., Martin, J. E., Ball, S., Nicholson, S. J., Witherden, A. S., Hafezparast, M., Latcham, J., Robinson, T. L., Quilter, C. A., and Fisher, E. M., SHIRPA, a protocol for behavioral assessment: Validation for longitudinal study of neurological dysfunction in mice, *Neurosci Lett* 306 (1–2), 89–92, 2001.
57. Sundberg, J. P., *Handbook of Mouse Mutations with Skin and Hair Abnormalities. Animal Models and Biomedical Tools.* CRC Press, Boca Raton, FL, 1994.
58. Smith, R. S., John, S. W. M., Nashina, P. M., and Sundberg, J. P., *Systematic Evaluation of the Mouse Eye. Anatomy, Pathology, and Biomethods.* CRC Press, Boca Raton, FL, 2002.
59. Chan, L. S., *Animal Models of Human Inflammatory Skin Diseases.* CRC Press, Boca Raton, FL, 2004.
60. Mahler, M., Rozell, B., Mahler, J. F., Merlino, G., Devor-Henneman, D., Ward, J. M., and Sundberg, J. P., Pathology of the gastrointestinal tract of genetically engineered and spontaneous mutant mice. In *Pathology of Genetically Engineered Mice,* Ward, J. M., Mahler, J. F., Maronpot, R. R., and Sundberg, J. P., eds. Iowa State University Press, Ames, 2000, pp. 269–297.
61. Klaunberg, B. A. and Lizak, M. J., Considerations for setting up a small-animal imaging facility, *Lab Anim* 33 (3), 28–34, 2004.
62. Hoffman, J. M. and Croft, B. Y., Future directions in small animal imaging, *Lab Anim* 30 (3), 32–35, 2001.
63. Paulus, M. J., Gleason, M. S. S., Easterly, E., and Foltz, C. J., A review of high-resolution X-ray computed tomography and other imaging modalities for small animal research, *Lab Anim* 30 (3), 36–45, 2001.
64. Easterly, M. E., Foltz, C. J., and Paulus, M. J., Body condition scoring: Comparing newly trained scorers and microcomputed tomography imaging, *Lab Anim* 30 (3), 46–49, 2001.
65. Weissleder, R., Scaling down imaging: Molecular mapping of cancer in mice, *Nat Rev Cancer* 2, 11–18, 2002.

2 Sharing Research Tools: The Laboratory Mouse

David Einhorn

TABLE OF CONTENTS

This chapter examines the historic change in perception of the laboratory mouse from that of a research tool, freely exchanged among scientists, to its being considered intellectual property. This transformation was triggered by the confluence of three milestones: (1) the U.S. Supreme Court decision allowing patenting of genetically altered life; (2) passage of the Bayh–Dole Act; and (3) development of technologies to genetically engineer mice. Three examples are discussed that illuminate the tension between the scientific sharing ethic and the demands of commercialization: (1) the P53 knock-out mouse, (2) the "Harvard Mouse" Patent, and (3) the Cre-lox technology. We consider how the National Institutes of Health (NIH) dealt with this tension and review its policy statement, "Sharing Biomedical Research Resource: Principles and Guidelines for Recipients of NIH Research Grants and Contracts." "Reach-through" and other problematic provisions in licenses and material transfer agreements are addressed. As a model, The Jackson Laboratory's approach to facilitating the availability and distribution of mouse strains is considered. From the overall perspective on how the patent system may inhibit research in academia, we also consider a recent judicial decision regarding the "experimental use" exemption.

The remarkable advances in molecular biology over the past quarter century have been a challenge to the traditional ethic in academia of encouraging the free flow of ideas and information. Universities and nonprofit research institutions (hereafter "Academia") have always viewed themselves as fundamentally different from for-profit companies (hereafter "For-Profits"), inasmuch as the *raison d'être* of Academia is research and education and not the pursuit of lucre on behalf of shareholders. However, the tremendous advances in biology, and the commercial opportunities created in the new industry of biotechnology, have presented a challenge to the tradition of openness and freedom of exchange in Academia, because efforts to protect and exploit intellectual property developed in Academia through patenting or licensing inevitably involve some secrecy and restriction of availability.[1]

The Jackson Laboratory (hereafter "Jackson") is a nonprofit research institution situated on an idyllic island in the state of Maine. As well as being a preeminent mammalian genetics research and educational institution, Jackson has been for many years and continues to be the most important source in the world for strains of laboratory research mice. For 75 years, Jackson has been identifying and characterizing spontaneous mutant mice in its colonies, importing spontaneous mutant mice from scientists worldwide, creating congenic strains carrying these mutations, developing inbred and other mouse lines, and in the past decade serving as an international repository of strains of genetically engineered mice (also called induced mutant mice) developed by scientists from virtually every major research institution in the world. The laboratory mouse, and more recently the genetically altered mouse, is generally understood to be the single most important research tool for studying human disease. It is therefore not surprising that genetically altered mice have been the trigger points of public debate over research access and commercialization, with Jackson a principal player in the drama. To understand this debate we need to take a short historical journey.

1980

The year 1980 was significant in this story. First, it was the year the U.S. Supreme Court decided Diamond v. Chakrabarty.[2] Before that landmark decision, there was a widely accepted legal assumption that living organisms were unpatentable, because they were considered products of Mother Nature and not of man. In Chakrabarty, the Supreme Court (in a sharply divided 5–4 decision) ruled that a living genetically engineered microorganism was patentable subject matter, and that the litmus test of patentability was not between living and nonliving matter, but between products of Nature and "anything under the sun made by man."

That same year of 1980, the U.S. Congress passed the Bayh–Dole Act.[3] Prior to Bayh–Dole, the federal government owned, but rarely patented or commercialized, inventions that arose from federally funded research in Academia. Out of realization that the federal government was not suited to further develop technologies, Bayh–Dole allowed Academia to retain ownership of inventions developed with federal funds, the expectation being that the turnover of ownership would spur the commercialization of inventions that would in turn benefit the public with new drugs and therapies.

The third event at the start of the 1980s was the development of recombinant DNA technology, which involves the altering of the genetic information of cells by combining the DNA of different organisms. In the case of the mouse, this allows for the creation of transgenic mice. The term "transgenic mouse" technically means a mouse that has a foreign gene introduced randomly into all of its cells, but it has commonly been used to describe any genetically altered mouse. This technology has been succeeded by the ability to target specific genes to "knock-in" or "knock-out" activity of these genes in mice as well as many other technologies, such as the so-called "gene switch" systems where genes can be turned on or off in specific anatomic sites. These new technologies have made the laboratory mouse an even more powerful research tool for studying human diseases.

COLD SPRING HARBOR

The first major manifestation of the tension between sharing and commercialization erupted into the public arena at a meeting of 300 scientists at the Cold Spring Harbor Meeting on Mouse Molecular Genetics, where Nobel laureate Dr. Harold Varmus [then a researcher at the University of California, San Francisco, and later Director of the National Institutes of Health (hereafter NIH)] moderated a heated discussion over the P53 knock-out mouse. This mouse, with the tumor suppressor gene P53 inactivated, had been developed with federal funds by Baylor University and licensed to a private biotechnology company. The company then attempted to license the mice on what was considered by many scientists to be terms and conditions that were too costly and restrictive: title to the mice was reserved, annual payments were required, and there were restrictions on internal breeding. These terms had never before been encountered by researchers when trying to obtain a mouse. What concerned Dr. Varmus and others was that scientists receiving funding grants from the federal government should be able to readily obtain federally funded research tools, such as the P53 mouse, and not be thwarted by unreasonable restrictions.

The following year a workshop was held at the National Academy of Sciences entitled "Sharing Laboratory Resources: Genetically Altered Mice" and a report was issued finding that severe proprietary restrictions on the distribution and breeding of genetically altered mice would impede the progress of biomedical research. In response to these events, Jackson—first with the help of private healthcare foundations and then with funding from the NIH—established the Induced Mutant Resource (IMR) to import and provide a repository for the cryopreservation, maintenance, and distribution of transgenic, targeted, and chemically induced mutant mice, with the assurance by Jackson that these new models would be made available to the whole research community with as few restrictions as possible.

THE "HARVARD MOUSE"

The next controversy involved the "Harvard Mouse". Harvard University filed and was able to obtain the first patent on a mammal, a mouse with a human oncogene that made the mouse more likely to develop cancer. Claim I of the patent is very broad: "A transgenic non-human mammal all of whose germ cells and somatic cells contain a recombinant activated oncogene sequence introduced into said mammal, or an ancestor of said mammal, at an embryonic stage."[4] The funding of the research effort that produced the mouse was provided by the Dupont Company, and Harvard gave Dupont an exclusive license to the patent. The issue of whether an advanced mammal should be subject to a patent is a philosophic and moral question which for many years has been the subject of heated debate in Europe, and last year the Supreme Court of Canada ruled that advanced life forms such as mammals were not patentable subject matter under Canadian law. In the United States, the "Harvard Mouse" patent did not engender the same degree of philosophical debate, perhaps because Chakrabarty had been seen as putting that issue to rest.[5] In contrast to Europe and Canada, in the United States the "Harvard Mouse" has instead been controversial, like the P53 mouse, because the licensing terms requested by Dupont were considered onerous and were resisted by Academia. Dupont's initial terms included limited internal breeding; no use of the mice in corporate-sponsored research; reservation of title; an annual report of research results; and a royalty on mice, products, or processes resulting from the use of the mice (initially a royalty not to exceed 7.5% and, in a subsequent license iteration, an unspecified amount of royalty). In addition, Dupont has taken the position that the breadth of the patent includes not just transgenic mice expressing any oncogene, but also mice with targeted mutations, such as the P53 knock-out mouse, and Dupont licenses all of these mice under its trademark, Oncomouse®.

CRE-LOX

While the resistance by Academia to the license terms for Oncomouse® was still unresolved, another technology, Cre-lox, was developed and patented by Dupont.[6] Cre-lox proved to be a very powerful and ubiquitously used technology because it allowed a researcher to knock out genes on a tissue-specific basis. This technology involves flanking the gene of interest with Lox sites, and when Cre is introduced with a tissue-specific promoter, it deletes the flanked DNA. Cre-lox allows researchers to knock out genes in selective tissues when otherwise knocking out all expression would be embryonically lethal. Later developments allowed a researcher to turn gene expression on and off at will. Dupont's terms for a license to use and share mice made with Cre-lox included an advance view of any publications resulting from using the technology and unspecified royalties on any products resulting from the use of mice containing Cre or Lox.

NIH AND THE MEMORANDA OF UNDERSTANDING (MOU)

A sharing crisis emerged as researchers were inhibited by the Cre-lox patent and licensing terms from exchanging their new Cre-lox mouse models with researchers in other institutions or from having the mice distributed to the research community through Jackson. Because of the licensing terms under the Harvard Patent, there were also inhibitory concerns over the use and sharing of mice genetically altered to develop cancer. In the context of negotiations with Jackson over Cre-lox, Dupont belatedly informed Jackson that it considered a number of the cancer mouse strains that Jackson had been distributing to be subject to the Harvard Patent. As the national mouse repository, the sole source of most genetically altered mice strains, and as a leading voice for the unfettered availability of the mouse as a research tool, Jackson (perhaps inevitably) now found itself a principal player in this drama.

It was apparent to Jackson that this issue concerned not only Jackson but was an important public policy concern that affected the entire biomedical research community, so Jackson sought the assistance of the NIH and Dr. Varmus (who had since become head of the NIH). Jackson worked with the NIH in addressing this impasse, and Dr. Varmus made a personal appeal to Dupont on behalf of the research community. Even though these technologies were reportedly not developed with any federal funds, they had become the poster-child examples of the difficulties involved in sharing and obtaining research tools, and Dr. Varmus used his bully pulpit to convince Dupont that it should reconsider its licensing policy. The result was a Memorandum of Understanding (MOU) in 1998 between the NIH and Dupont covering the use of Cre-lox,[7] followed the next year by another MOU dealing with research conducted under the Harvard Patent.[8]

These MOUs allow use of the patent rights for noncommercial biomedical research without any rights to fees or royalties, rights to review publications, and, most significantly, any rights to developments made using the respective patent rights. Transfers of any materials covered by the patent rights may be made by one nonprofit to another nonprofit institution using a Material Transfer Agreement (hereafter MTA) subject essentially to the same terms set forth in the MOUs between the NIH and Dupont. If a nonprofit institution wishes to transfer such materials to a for-profit company, it may do so using a license or an MTA, but it must also notify the for-profit of the existence of Dupont's patent rights, notify Dupont of the identity of the for-profit and the material to be transferred, and further notify the for-profit that use of the material will require a license and a fee from Dupont. What is unprecedented about the MOUs is that NIH negotiated terms for using Cre-lox technology and the "Harvard Mouse," not only for intramural federal government scientists, but also for extramural nonprofit recipient institutions of federal grants. The MOUs provide that Dupont shall under separate license agreements make the patent rights available to nonprofit grantees "... in accordance with the terms and conditions ..." of the MOUs. The MOUs were and remain a unique example of a resolution by government, Academia, and a company addressing the tension between availability of, and proprietary rights to, research tools.[9]

NIH POLICY

No doubt prompted by the issues surfaced by the P53 Mouse, "Harvard Mouse", and Cre-lox controversies, Dr. Varmus requested that a Working Group of the Advisory Committee to the Director look into concerns over the dissemination and use of unique research tools, the competing interests of the owners and users of these tools, and how the NIH might provide a policy response. The Report of the Working Group found that patent and licensing restrictions could stifle the broad dissemination of research tools and thereby frustrate basic research and the development of products for the benefit of public health. The Report found, on the other hand, that reasonable restrictions might be necessary to protect intellectual property rights and preserve incentives for further research and commercial development. The challenge, of course, was to balance these competing interests. After submission for public comment of a draft policy, the NIH adopted a final document entitled "Sharing Biomedical Research Resources: Principles and Guidelines for Recipients of NIH Research Grants and Contracts" (hereafter "NIH Sharing Policy"),[10] setting forth four basic principles.

1. *Ensure Academic Freedom and Publication:* The NIH Sharing Policy pronounces as a matter of first principle that academic freedom to conduct research and publish are "... at the heart of the scientific enterprise," and that NIH grantees must ensure that interactions with the private sector do not involve veto rights over publication or excessive publication delay beyond that necessary to permit the filing of patent applications or to protect from disclosure of confidential information.

2. *Ensure Appropriate Implementation of Bayh–Dole:* Recipients of federal funding must ensure that there is compliance with the statutory goals of Bayh–Dole to promote utilization, commercialization, and public availability of inventions. The NIH Sharing Policy finds that patenting and licensing are not the only or necessarily appropriate means of achieving these goals. If there is no further private investment needed to develop and utilize a research tool, then the goals of Bayh–Dole can be met by a combination of dissemination approaches, such as publication, deposit in a repository, or widespread nonexclusive licensing. "Restrictive licensing of such an invention, such as to a for-profit sponsor for exclusive internal use, is antithetical to the goals of the Bayh–Dole Act," because by definition, limiting the use to one

company precludes the potential fruitful use of the research tool by other companies. Even when an exclusive license might be appropriate, the NIH Sharing Policy concludes that it should be used only if consistent "… with the goal of ensuring widespread and appropriate distribution of the final product tool."

With respect to genetically altered mice, it is generally understood that they can be fully used without any further research, development, or investment. However, grantees have in the past on occasion given internal exclusive-use licenses to companies, typically when the mice were developed through corporate-sponsored or collaborative research agreements. After publication of research using these mice, other researchers in Academia would request the mice, and some grantees then found themselves in the embarrassing circumstance of having to renegotiate such licenses if they were to comply with NIH regulations. Indeed, well before the NIH Sharing Policy, NIH regulations consistently over the years required that research tools be made readily available in accord with the "public availability" goal of Bayh–Dole,[11] and more recently, Bayh–Dole was amended to promote its statutory goals "without unduly encumbering future research and discovery in the spirit of the NIH Research Tools Policy."

3. *Minimize Administrative Impediments to Academic Research:* Prolonged negotiations over the terms of MTAs or licenses for research tools delay the process of getting research tools into the hands of a researcher and put to use. Research tools should therefore be freely exchanged in Academia using either no formal agreement, a cover letter, a Simple Letter Agreement for the Transfer of Materials (hereafter SLA), or the Uniform Biological Materials Transfer Agreement (hereafter UBMTA).[12] Because academic institutions are both developers and providers as well as users of research tools, it makes no sense for academics to spend unnecessary time negotiating among themselves over research tools that in any event have all been funded by the same source, the federal government. Academia should be mindful of the truism that "What goes around, comes around."

4. *Ensure Dissemination of Research Resources Developed with NIH Funds:* The NIH Sharing Policy calls on grantees to manage their interactions with For-Profits so as not to restrict the grantees' ability to disseminate federally funded research tools. Examples are potentially problematic situations involving the co-mingling of federal funding with commercially sponsored funding or acceptance of research tools from providers, wherein the funding sponsor or tool provider imposes restrictions on dissemination that conflict with the grantee's obligations under the NIH Sharing Policy. In the instance where the research tool is developed solely by the grantee, the NIH Sharing Policy encourages grantees to simplify the transfer of a research tool not just to Academia

but also to For-Profits, because For-Profits as well use research tools in basic research. Accordingly, the NIH Sharing Policy encourages grantees to avoid seeking license terms that include royalties and option rights when the intended use of the research tool by the For-Profit is internal research use. Needless to say, it would be disingenuous for Academia to resist such terms when they are imposed by For-Profits and then take the opposite position when the tables are turned.

ISSUES RAISED IN NIH SHARING POLICY

DEFINITION OF RESEARCH TOOLS

The NIH Sharing Policy uses the term "unique research resource" and states that it should be used broadly to include the full range of research tools that scientists use in basic research, including cell lines, reagents, and of course animal models. It suggests that the test of whether something is a research tool is (1) whether its primary usefulness is as a tool for discovery and not an end product; (2) whether it is useful broadly or only for a narrow purpose; and (3) whether it can be readily used or instead needs further development. Despite the effort to define a research tool, experience teaches us that whether something is a research tool is in the "eye of the beholder."[13] The developer of a research tool will consider it the greatest thing since sliced bread, and the potential user will invariably see it as just another pedestrian tool among many necessary to conduct a research project. Despite this truth about human nature, most researchers would objectively agree that genetically altered mice fit the definitional test as research tools. Indeed, the NIH has recently submitted for public comment a draft policy entitled, "NIH Statement On Sharing And Distributing Mouse Resources."[14] The Statement reaffirms the NIH Sharing Policy and leaves no doubt that the NIH considers the mouse a research tool that is of such significance that it deserves a separate policy statement.

REACH-THROUGH RIGHTS

The major issue in the P53 Mouse, "Harvard Mouse," and Cre-lox controversies was the objection by the research community to "reach-through rights." Reach-through rights are claims of option rights to commercialize inventions, or claims to royalties on products. In the context of the P53 Mouse it was a reservation of title to the mouse and, in the "Harvard Mouse" and Cre-lox situations, it was the specific claim to any product that was developed in the course of using the mice. There are arguments that can be made for and against reach-through rights. The positive view of reach-through is that it should be considered desirable to the user to obtain a mouse at only a nominal or no upfront cost, and desirable to the provider because it allows it to obtain rights in inventions and products that it would not otherwise have through its ownership, or patent coverage, of the mouse alone. To the provider it is an attractive trade-off: forgoing a reasonable charge for the use of a mouse in return for the potential of the "big pay-off," a royalty payment on a drug that was identified, selected, or otherwise developed through the use of the mouse. Sometimes the provider insists on outright ownership of any future discoveries made using a mouse, but it is the more common practice in industry that research tools, including mice, are only made available to academics under MTAs that provide for first option license rights to intellectual property developed using the research tool.

The arguments against reach-through rights are that (1) they are an impediment to the dissemination of research tools because of the reluctance of the user of the tool to agree to such rights; (2) first option license rights given to future inventions can only be given away once, thus preempting a researcher from obtaining other research tools needed in the same research project but containing similar reach-through option claims; and (3) multiple royalty claims to products can result in what is known as "royalty stacking" (too high a stack of royalty obligations can make the development of therapies and drugs prohibitively expensive as well as impede further research). Use of a particular research tool may only carry a 1% royalty on a product, but many research tools are utilized in the course of developing a product. Royalty obligations can quickly add up and potentially constitute a prohibitive tax on commercial development of therapies. In addition, stacking of license royalty rights also creates the potential for the "hold-out" problem. If a patent holder knows that it holds the last patent rights necessary to produce or further develop a technology or product, that patent holder is in a blocking position to make unreasonable demands.[15]

The concept of reach-through sounds pejorative and in practice is generally seen as such, because it is a tool-based claim that seeks to reach beyond the boundary of the use of the *tool qua tool*, and into the *terra incognita* of possible future inventions that may have very little to do with the use of the research tool. At its heart, reach-through involves the issue of who owns future discoveries made in the course of using the mouse in a research project: the provider, the user, or some as yet unknown party in the last leg of a long journey of drug development.

Professor Rebecca Eisenberg, who headed the NIH Working Group Report, has used the metaphor of the "tragedy of the anticommons" to describe the danger of reach-through claims on research tools. Over thirty years ago Garrett Hardin had used the descriptive term "tragedy of the commons"[16] to describe a situation where a resource is overused when too many owners have rights to a common resource and no one has the right to limit or exclude others, for example, the Grand Banks when individual fishermen initially had no fishing limits and quickly depleted the fishing grounds. Eisenberg warns that in biomedical research there is the reverse danger, that overprivatization of upstream rights in research tools can inhibit downstream development of products for improving human health.[17] In short, the argument is that product development becomes more problematic when there are increasing numbers of proprietary rights holders on research tools laying claim to a piece of the product revenue pie. As we have seen, the NIH Sharing Policy endorses this concern by specifically discouraging the claims of reach-through rights on research tools developed with federal funding.

PATENTING MICE

We know from the "Harvard Mouse" that it is possible to obtain patents on genetically altered mice, but those patents that have since been granted are generally for a particular mouse with a specific genetic alteration. It is unlikely that a patent will be issued again in the United States with broad claims to a whole class of mice, as was the case with the "Harvard Mouse." Of course there are many factors to be weighed in deciding whether to file a patent application on a mouse (cost, significance of the mouse, etc.). Because the mouse is clearly a research tool, and as such does not need the protection of a patent to spur further commercial investment, it is hard to see in most circumstances what purpose is served in patenting a genetically altered mouse. Moreover, although a mouse patent gives the patent owner the right to prevent anyone else from making, using, or selling a mouse covered by the patent claims, this monopoly protection is not generally necessary to obtain value for the mouse inasmuch as most companies are more than willing to obtain a license to use a mouse rather than go through the expensive and time-consuming effort of duplicating the mouse.[18] In short, Academia has generally realized that patenting a mouse is not a prerequisite for obtaining some financial return for the development of the mouse, and at Jackson we find it increasingly uncommon to see patented mice.

SHARING MICE WITHIN ACADEMIA

Once research has been published involving a federally funded research tool, the research tool must be made "readily available" to other researchers (and as well the major journals have the same expectation) so that others can validate the research results. In the case of a new mouse strain, the researcher who has developed the strain has the option of providing the mouse to other academic researchers directly or through a mouse breeder distributor. If the researcher chooses to supply the mouse directly, universities typically use their own MTA, UBMTA, or the Simple Letter Agreement (SLA) recommended by the NIH to effectuate the transfer of research tools within Academia.

The SLA is aptly described because it is only one page and simple. The SLA refers to a research tool as "Material", the developer of the tool as the "Provider", and the receiver of the tool as the "Recipient". The salient points of the SLA are (1) the Material is defined and described as the property of the Provider; (2) the Material will be used for teaching or not-for-profit research purposes only; and (3) the Material will not be further distributed to others without the Provider's written consent. If the SLA is used to transfer a mouse, it is customary to define Material to include any progeny of the mice, first because a Recipient needs the right to breed the mice in order to conduct various experiments and make full use of the mouse as a research tool, and second because the Provider has a right to expect that proprietary rights in the mouse include any progeny. Because Recipients as well often cross-bred mice to other mouse strains, conduct experiments that further genetically alter the mouse (such as receiving a transgenic mouse and then knocking out another gene to make a double mutant), or derive biological materials from the mouse (such as a cell line), it becomes somewhat of a challenge to embrace in the definition of Material what the Provider has the reasonable expectation should be included in that proprietary definition without reach-through claims to the significant intellectual contribution by the Recipient.[19] Where Recipient's use of the mouse has added significant research value to the mouse, both Provider and Recipient have proprietary rights and need to negotiate in good faith how the mouse can be further shared with the research community.

The other salient provisions of the SLA are relatively clear: that the mice must be used for not-for-profit purposes (commercial purposes generally understood to include the sale of the mice or their use by a For-Profit) and that the mice cannot be distributed to third parties by the Recipient without permission of the Provider. An issue that sometimes arises is whether a mouse used by an academic Recipient in a research project funded in whole or in part with corporate-sponsored research funding is, in the context of the SLA, a not-for-profit or a for-profit use. It is often the case that a research project may be funded by governmental, institutional, private foundation, and corporate monies, which are co-mingled and impossible to tie to the use of a particular research tool such as a mouse. The UBMTA tackles this issue by defining "Commercial Purposes" as: contract research, screening of compound libraries, or producing or manufacturing products for sale. However, the UBMTA provides that "… industrially sponsored academic research…" shall not be considered "… per se …" a use for "Commercial Purposes" unless it meets the definition. This exclusion is wise because there is substantial and ubiquitous funding by industry of research in Academia, and treating such funding as covered by "Commercial Purposes" would severely limit the sharing of these mice among researchers.

JACKSON MOUSE REPOSITORY

Most researchers do not want to be burdened by animal room responsibility and expense, nor do they generally have sufficient laboratory space to be able themselves to provide their genetically altered mice to researchers in other institutions. As well, the researchers' institutions do not have the marketing and distribution infrastructure to provide this service. A central repository for distribution of mouse strains has also been necessary for health reasons, because with many hundreds of new genetically altered strains created yearly, the exchange of these strains between research laboratories with mouse colonies of varying health status would pose a threat to the health of mouse research colonies worldwide. A repository also protects against genetic drift by adhering to strict breeding protocols backed by extensive pedigree recordkeeping and genetic quality control. Finally, to protect against the accidental loss of these mutants, there must be embryo, egg, or sperm cryopreservation. Jackson's Mouse Repository addresses all of these needs.

Over the past decade, literally thousands of genetically altered mouse strains have been screened for their research value by Jackson's Genetic Resources Committee and hundreds then imported from donor institutions all over the world.

We refer to such provider institutions as "donors" because they understand that the vast majority of mouse strains are distributed in small numbers and therefore must be subsidized by public and private grants; and appreciating this economic reality, donors generally do not expect or ask Jackson for any royalty or other payments for distributing their mice.

Jackson has over the years worked very closely with the NIH in promoting an ethos of sharing that would not take a back seat to commercialization. With respect to the repository, Jackson has successfully discouraged the necessity of MTAs or licenses when mice are distributed to scientists in Academia. Researchers in Academia can (for most strains) pick up a phone and order a strain from Jackson without the delay of having to obtain an MTA or license agreement. Instead, Jackson distributes JAX® mice under a notice of "Conditions of Use" that have long been used by Jackson and are set forth in its "Catalogue of JAX®Mice," its publications, and on its invoices to purchasers of mice. These conditions are essentially the same as the three salient conditions provided for in the SLA recommended by NIH, but avoid the inevitable delays occasioned by even the exchange of an SLA.

Genetically altered mice are expensive to develop and are often the product of substantial effort, therefore it is reasonable that the donor or provider of these mice should have the opportunity to have some financial return from For-Profits. Accordingly, in those circumstances where a mouse developed with federal funds has already been exclusively licensed to a For-Profit by the donor or provider institution, Jackson will accept the strain into its repository and respect the exclusive license by restricting any distribution to For-Profits, but will insist on being able to make the mouse available to Academia. However, in most cases the donor or provider wishes to obtain some financial return from For-Profits using nonexclusive provider licensing as recommended by the NIH Sharing Policy; or if the donor or provider is a For-Profit, it may wish to preclude distribution to its competitors by controlling any distribution to other For-Profits. In these circumstances Jackson will facilitate the licensing needs of the donor or provider by restricting distribution to only those For-Profits for which the donor or provider has given permission to receive the mice. Although the donor/provider institution is free to negotiate whatever terms it chooses in licensing For-Profits who wish to obtain the mouse, both the NIH Sharing Policy and Jackson discourage Academia from making demands for reach-through rights on research tools, whether the recipient or user of the mouse be a researcher in Academia or a For-Profit.

THE RESEARCH EXEMPTION

A court ruling last year has caused a great deal of stir in Academia over the freedom to conduct basic research without the concern of being charged with patent infringement. The case involved John Madey, who as a professor at Stanford University had developed laser patent technology, and when he left Stanford for Duke University took along ownership rights to the patents. Madey and Duke came to a parting of the ways; but Duke continued its laser program and Madey sued Duke for patent infringement. Duke's primary defense was the research (or experimental use exemption) defense. This defense finds its earliest expression in an opinion by one of the leading jurists of the nineteenth century, Justice Joseph Story. Justice Story used the now-anachronistic term "philosophical" to describe activity we would now refer to as "scientific," opining that "It could never have been the intention of the legislature to punish a man, who constructed such a machine merely for philosophical experiments, or for the purpose of ascertaining the sufficiency of the machine to produce its described effects."[20] Shortly there-after, Justice Story amplified this doctrine by stating that, for an offense under patent law, there "... must be the making with the intent to use for profit." [21]

Academia has traditionally understood the experimental use exemption to protect nonprofit research from claims of patent infringement, particularly when that research is for noncommercial, that is, academic, purposes, relying on this nineteenth-century case law and the common sense maxim, *de minimis non curat lex*. The Madey litigation challenges for Academia this comfortable assurance. Indeed, it was factually an unusual case because it involved, not a company crying foul over a university's unlicensed use of the company's patent, but instead a professor suing his university for unlicensed use of the professor's patents. Although the trial court judge ruled in favor of Duke, finding that plaintiff Madey had failed to prove that Duke had not used the laser equipment solely for exper-imental and nonprofit purposes, the U.S. Court of Appeals for the Federal Circuit reversed the trial court, ruling: (1) that the burden of proof was on Duke and not Madey to prove that the use was experimental; (2) that the experimental use defense is "very narrow and limited"; (3) that the profit or nonprofit status of the alleged infringer was not determinative; and most significantly, (4) that the dis-tinction between commercial and noncommercial research was also not determi-native. The Appeals Court also opined that scientific research at universities serves the universities' "business objectives," which business it described as their edu-cational mission.[22] By treating "research projects with arguably no commercial application whatsoever" as well as "education and enlightening students and faculty" as the business objectives of a university, the Madey decision arguably eviscerates for all practical purposes the experimental use exemption.[23]

The impact of the Madey case on the culture of basic research in Academia may be substantial.[24] Academia may have to confront the financial and administra-tive burdens of patent searches, infringement legal opinions, licensing negotiations, litigation, as well as the consequential delays in conducting research occasioned by these new burdens. Academia does not have, as do pharmaceutical and bio-technology companies, in-house patent counsel to guide the direction of research so that it does not run afoul of patents. Even if Academia tries to comply with Madey, it is probably unrealistic to expect researchers, who have long been used to using whatever is necessary to conduct research, to begin asking for a legal evaluation of possible infringement every time they decide to use a tool, method, process, or technique in the laboratory. Moreover, in Academia researchers are anxious to publish as soon as possible, whereas drug and biotech companies invariably maintain strong internal secrecy procedures that delay publication until long after a patented research tool has been utilized and when a patentee can do little to prevent or exact payment for any further use of the tool.[25] Ironically, Madey may embolden owners of patents on research tools to become more aggressive and litigious against Academia as compared to For-Profits. Madey also threatens to undermine the NIH Sharing Policy, which implicitly if not explicitly endorses the public policy interest behind the experimental use exemption.[26]

On the other hand, we should not be entirely surprised by the Madey decision. Since Bayh–Dole, Academia has increasingly looked more "commercial."[27] Universities are increasingly patenting their technologies, and are not shy about bringing infringement suits of their own, and forming startup companies in which they retain equity.[28] Indeed, it was probably only a matter of when, and not if, the experimental use exemption was challenged. Nonetheless, it is undoubtedly the case that despite the growing commercialization at universities, the fact remains that most researchers in Academia still conduct research for purely academic and experimental reasons, and the challenge in the future will be to retain for that research, either in Congress or the courts, the obvious benefits to society of a limited experimental use exemption.

REFERENCES

1. Eisenberg, R. S., Proprietary rights and the norms of science in biotechnology research, *Yale Law J* 97 (2), 177–231, 1987.
2. Diamond v. Chakrabarty. 16 Jun 1980, *US Rep US Supreme Court* 447, 303–22, 1980.
3. 35 U.S.C. ##200–211 and rules promulgated in 37 C.F.R. 401.
4. Leder, P. and Stewart, T., United States Patent No. 4,736,866, 1988, Transgenic non-human mammals. April 12, 1988.
5. Katz, A. P., Patentability of living with traditional Jewish law; "Is the Harvard mouse kosher?", *AIPLA Quart J* 1, 117, 1993.
6. Sauer, B. L., United States Patent No. 4,959,317, 1990, Site-specific recombination of DNA in eukaryotic cells. September 25, 1990.
7. NIH, Memorandum of Understanding between DuPont Pharmaceuticals Company and Public Health Service, U. S. Department of Health and Human Services, *http://ott.od.nih.gov/textonly/cre-lox.htm*, 1998.
8. NIH, Memorandum of Understanding between E.I. DuPont de Nemours and Company and Public Health Service, U.S. Department of Health and Human Services, http://ott.od.nih.gov/textonly/oncomous.htm, 1999.
9. NIH, Memorandum of Understanding between E.I. DuPont de Nemours and Company and Public Health Service, U.S. Department of Health and Human Services, 1999. As of the writing of this chapter there are still some major research institutions that have not signed off with DuPont on a license agreement covering the Harvard Patent because of differing interpretations of the terms and conditions of the MOU, including whether the language " ... will not be used for any commercial purpose or for the direct benefit of any for-profit institution ... " encompasses the typical corporate-sponsored research agreements, and whether a for-profit must be first licensed by DuPont before it receives from a nonprofit any mice or other material covered by the patent rights. 1999.

10. Principles and guidelines for recipients of NIH research grants and contracts on obtaining and disseminating biomedical research resources, Department of Health and Human Resources, National Institutes of Health, http://ott.od.nih.gov/NewPages/RTguide_final.html, 1999.

11. NIH, National Institute of Health Grants Policy Statement (03/01), http://grants1.nih.gov/grants/policy/nihgps_2001/index.htm.

12. UBMTA, Implementing Letter: The UBMTA was a cooperative effort of the NIH and Academia and can be found at Association of University Technology Managers. Once an institution is a signatory to the UBMTA, it can document transfers by a short "Implementing Letter".

13. Stewart, J. P., Analogously the U.S. Supreme Court has had a history of difficulty in defining "obscenity" and Justice Potter Stewart famously gave up trying and opined, "I know it when I see it."

14. NIH, Statement on Sharing and Distributing Mouse Resources, http://www.nih.gov/science/models/mouse/sharing/index.html.

15. Clark, J., Tyson, K., and Stanton, B., Patent Pools: A Solution to the Problem of Access in Biotechnology Patents? http://www.uspto.gov/web/offices/pac/dapp/opla/patpoolcover.html, 2001.

16. Hardin, G., The tragedy of the commons. The population problem has no technical solution; it requires a fundamental extension in morality, *Science* 162 (859), 1243–8, 1968.

17. Heller, M. A. and Eisenberg, R. S., Can patents deter innovation? The anticommons in biomedical research, *Science* 280 (5364), 698–701, 1998.

18. Abrams, I. and Kaiser, M., Licensing transgenic mice: A short tutorial, *J Assoc Univ Technol Managers* XII, 2000.

19. NIEH, Uniform Biological Material Transfer Agreement: The UBMTA makes the definitional distinction between "Unmodified Derivatives", which it defines as "Substances created by the Recipient which constitute an unmodified functional subunit or product expressed by the original material", and "modifications" defined as "Substances created by the recipient which contain/incorporate the material." http://www.niehs.nih.gov/techxfer/ubmta.htm.

20. Whitmore v. Cutter, 21 F. Cas at 554, 1813.

21. Sawin v. Guild, 21 F. Cas at 555, 1813.

22. Madey v. Duke University, 307 F.3d, 1351 (Fed. Cir. 2002).

23. Madey v. Duke University, "Major research universities, such as Duke, often sanction and fund research projects with arguably no commercial application whatsoever. However, these projects unmistakably further the institution's legitimate business objectives, including educating and enlightening students and faculty participating in these projects ... . In short, regardless of whether a particular institution or entity is engaged in an endeavor for commercial gain, so long as the act is in furtherance of the alleged infringer's legitimate business and is not solely for amusement, to satisfy idle curiosity, or for strictly philosophical inquiry, the act does not qualify for the very narrow and strictly limited experimental use defense." Mady, citation supra, at 1351, 1362, 2002.

24. Amici Curiae Brief: Association of American Medical Colleges and other organizations and universities supporting petition for certiorari to the United States Supreme Court, at 3 (the petition was denied by the Court).

25. Maebius, S. and Wegner, H., Ruling on research exemption roils universities, *Nat Law J*, December 16, 2002.

26. It is interesting to note that the U.S. Justice Department opposed the granting of certiorari on the grounds, inter alia, that Bayh–Dole has encouraged the commercialization of academic research and that "There is nothing in the current patent laws to suggest that modern universities—many of which have themselves taken advantage of patent protection and entered into licensing arrangements—are somehow outside the class of potential infringers because of an asserted non-commercial status." Brief for the United States as Amicus Curiae, at 12–13.

27. Bok, D. C., *Universities in the Marketplace: The Commercialization of Higher Education*, Princeton University Press, Princeton, NJ, 2003.

28. Meredith, D., Putting a patent on research; http://www.dukemagazine.duke.edu/dukemag/issues/010203/patent1.html, *Duke Mag* 89 (2), 2003.

3 Managing Success: Mutant Mouse Repositories

Stephen F. Rockwood, Martin D. Fray, and Naomi Nakagata

TABLE OF CONTENTS

INTRODUCTION

Technological advancements are frequently accompanied by less-heralded con-sequential problems that eventually require resolution, or less desirably, manage-ment. This is especially so in the field of genetic engineering. New techniques applied to the creation of mutant mice provided models of human disease and insights into the functional aspects of proteins and nucleic acids. As the number of useful mutant mouse models increases, the associated managerial problems increase proportionally. How should so many genetically unique mice, carefully designed, created with public funds, and frequently possessing a fragile constitution, be maintained? One approach is to utilize centralized repositories that serve as distribution centers. This chapter examines some of the practical issues to consider when interacting with mouse mutant repositories, including a survey of the major repositories that accept submissions from the scientific community.

SETTING THE STAGE

Before the development of recombinant DNA technology, the laboratory mouse enjoyed its status as one of the more prevalent model organisms utilized in biological studies. The domestication of mice has a well-documented history, with its earliest experimental use being recorded during the seventeenth century.[1-4] By the first half of the twentieth century, a period that could justifiably be termed the Dawn of Mouse Genetics, efforts by Castle, Little, Snell, and others demonstrated the value of utilizing inbred mice. After twenty generations of sequential brother–sister matings, littermates are completely (or nearly completely) genetically identical. It is this uniform genetic composition, minimizing experimental variability, that makes the use of inbred strains so highly favored in most mouse genetics laboratories.[1-5]

By the latter half of the twentieth century, the laboratory mouse was recognized as an ideal model organism for research applications. A brief gestation period, short life expectancy, rapid development, relatively low space and sustenance requirements for a healthy existence, and literature offering a rich characterization of the organism's biological features all contribute to its versatility. Many scientists utilized laboratory mice in their research prior to 1980, but after several catalytic developments, it became evident that the problems of mouse management were problems of scale.

In 1980, it was demonstrated that it was possible to inject exogenous DNA into the pronuclei of fertilized mouse oocytes and that upon analysis, the exogenous DNA would be found incorporated into the genome of the resulting pups at some practicable frequency.[6] Immediately on the heels of this revelation was the finding that if the exogenous DNA was injected along with the appropriate regulatory sequences, the injected DNA was not only incorporated, but was also successfully expressed.[7-9] Once transgenesis was accomplished in the mouse, many scientists embraced the new approach and applied it to their own research programs with vigor.

Several years later, progress was made in the related area of targeted germ line modification in mice. Several groups[10,11] were able to modify a specific gene in mouse embryonic stem (ES) cells using homologous recombination, incorporate the modified ES cells into developing blastocysts, and demonstrate that the modifications were heritable. The ability to precisely alter the genome in a fashion that would be retained in the germ line presented a wide range of exciting possibilities.[12] The new technology made it possible to introduce a subtle point mutation or render a gene completely nonfunctional. If combined with transgenic techniques, it would be possible to construct "humanized" mice, where the endogenous mouse genes are turned off and human disease genes introduced. The new techniques offered approaches to explore the potential of gene therapy.

Technological developments in genome manipulation served as necessary prerequisites for subsequent advancements. Alone, they were insufficient to bring about the explosion of induced mutant mouse generation that followed. Both mouse and human genomic sequence information was required to exploit the full potential of the new techniques. This information became accessible under the auspices of the Human Genome Project (HGP).[13] In addition to its primary objective to completely sequence the human genome, the Comparative Genomics component of the project included the sequencing of the mouse genome by 2005. Initially, the HGP generated criticism from scientists who objected to the large diversion of resources to a single program.[14] There were concerns that the products of science would become more important than the process of science and that training for a generation of scientists would suffer, as postdoctoral programs became mundane exercises in nucleotide sequencing. Despite these concerns, work on the HGP proceeded as an international project funded by governmental and private commercial entities. With the benefit of remarkable advances in sequencing technology, both the human[15,16] and mouse sequences[17] were completed ahead of schedule. With sequence information readily available, researchers were free to exercise their imaginations and design a wide variety of novel mutant mice for experimental use.

It is difficult to quantify with any degree of accuracy the number of mutant mice generated by the scientific community. The scientific literature can provide an indication of broad trends in usage. The number of publications listed in PubMed that make reference to transgenic and knock-out mice has, not surprisingly, increased significantly over the last decade (Figure 3.1). A number of large mouse-oriented research projects contributed to this trend. The Gene Expression Nervous System Atlas (GENSAT) project will characterize the tissue expression pattern for thousands of genes active within the central nervous system.[18,19] The GENSAT project employs bacterial artificial chromosome (BAC) transgenic mice whose large mouse genomic inserts have a gene of interest replaced by an enhanced green fluorescent protein reporter gene. The International Gene Trap Consortium offers embryonic cell lines that collectively represent mutations covering two-thirds of genes in mice.[20] The Knockout Mouse Project (KOMP), proposed by the Comprehensive Knockout Mouse Project Consortium, endeavors to generate knock-out mutations for all mouse genes.[21] The academic scientific community and the biopharmaceutical industry view these projects enthusiastically as avenues to characterize gene function and novel drug targets. Thus, the future role of the mutant mouse in research appears secure for the foreseeable future.

THE RESEARCHER/REPOSITORY INTERFACE

Each repository operates with a level of autonomy, and as such, has its own operating procedures. However, some generalizations can be made regarding what a researcher can reasonably expect when interacting with a mouse repository.

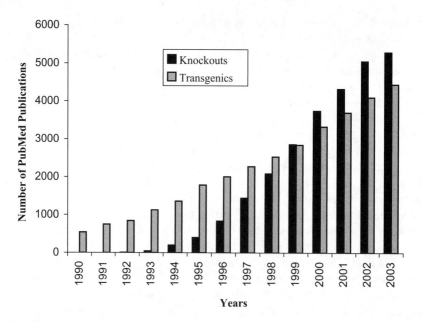

FIGURE 3.1 Literature citations for transgenic and knock-out mutant mice. Values were obtained by querying the National Center for Biotechnology Information PubMed interface (http://www.ncbi.nlm.nih.gov/entrez/query.fcgi?db=PubMed).

SUBMISSION POLICIES

Any repository is only able to accept a finite number of mouse strains as dictated by the resources available to it. Most repositories utilize a scientific review process to select biomedically relevant mouse strains. As an aid to this review process, the prospective donating investigator should be prepared to provide complete information related to the development, husbandry, and phenotype of a submitted mutant mouse line. Acceptance standards are subjective but, in general, desirable mutants should be characterized within a publication from a peer-reviewed journal. If a paper is in press, a strain submission is best coordinated with the publication's progress so that the availability of the strain, a catalogue number, and the distributing repository can be announced in the text of the article. This can save an investigator the trouble of dealing with a deluge of inquiries for the mutant once the paper is published. The submitted strain should exhibit a phenotype that is generally useful to other researchers, not be excessively difficult to maintain, nor should it be encumbered with onerous licensing conditions. Redundancies between repositories are avoided so donating investigators should expect only one repository to distribute their mouse.

DONOR BENEFITS

Although many scientists are motivated solely by altruism, it should be clear that there are a number of benefits realized by the donor who places a mutant mouse strain in a repository. All repositories cryopreserve their holdings as insurance against inadvertent loss. When torrential rainfall in 2001 caused flooding at Baylor College of Medicine in Houston, Texas, an estimated 35,000 mice were tragically lost.[23] On the other hand, unique mutants that had been safely deposited in repositories were systematically recovered, allowing interrupted research projects to resume within a reasonable time frame.

Health levels vary dramatically among research institutions. Upon arrival at a repository, mice are rederived (hysterectomy or embryo transfer) before being admitted to a barrier facility. If the donating investigator is interested in reacquiring mice with a higher health status, he or she should make a request to repository representatives, as such mice, in limited numbers, are frequently provided *gratis*. Such donor requests should be made early in the donation process as the likelihood of successfully obtaining mice free of charge drops precipitously should a strain be removed from the shelf and need to be recovered from cryopreservation.

In addition to distributing mice, repositories are also adept at distributing information. Publicly accessible databases maintained by most repositories supply information that address straightforward questions researchers might have about their mice. Additionally, repositories with technical support representatives address complex questions and help troubleshoot with some procedures such as genotyping.

Investigators from the United States that generate mutant mice with funds provided from the National Institutes of Health (NIH) should be mindful of the NIH policy on sharing mouse resources.[24] In short, the policy encourages timely sharing and distribution of mouse resources to investigators at academic institutions for noncommercial research following characterization and peer-reviewed publication. By placing a mutant mouse in a repository, an investigator fulfills the requirement to share the mouse. Alternatively, scientists may elect to respond directly to requests for mice.

COSTS ASSOCIATED WITH DONATION

With the exception of shipping charges, once a mouse strain is accepted into a repository, all of the charges for importation and cryopreservation are usually borne by the accepting repository. In the event that a repository declines a submitted mouse, a repository may offer to distribute the mutant provided the donating investigator bears the cost of importation and cryopreservation. Policies vary among institutions, so inquiries should be made directly to repository representatives.

FINDING AND OBTAINING MICE

Searching the Internet for the existence and whereabouts of a specific mutant mouse can be time consuming. An excellent place to begin a search is the International Mouse Strain Resource (IMSR) maintained at The Jackson Laboratory (TJL). The IMSR is a searchable online mouse database that includes inbred, mutant, and genetically engineered mice. The goal of the IMSR is to assist the international scientific community in locating and obtaining mouse resources for

TABLE 3.1
Internet Repository Resources

Resource	Web Page
The Jackson Laboratory	http://www.jax.org
Mouse Mutant Regional Resource Centers	http://www.mmrrc.org
Mouse Models of Human Cancers Consortium Mouse Repository	http://web.ncifcrf.gov/researchresources/mmhcc
Center for Animal Resources and Development	http://cardb.cc.kumamoto-u.ac.jp/transgenic/index.jsp
Riken Bioresource Center	http://www.brc.riken.jp/lab/animal/en/
European Mouse Mutant Archive	http://www.emma.rm.cnr.it
International Mouse Strain Resource	http://www.informatics.jax.org/imsr
Bay Genomics Consortium	http://baygenomics.ucsf.edu

research. The IMSR began as a collaboration between the Mouse Genome Informatics (MGI) group at TJL and the Medical Research Council (MRC) Mammalian Genetics Unit at Harwell, United Kingdom. A concerted effort is underway to include the holdings of the major mouse repositories in the IMSR.[25]

REPOSITORIES

THE JACKSON LABORATORY

The Jackson Laboratory (TJL) in Bar Harbor, Maine, was established in 1929 as a nonprofit genetic research facility and has a long history of maintaining laboratory mouse colonies. Current holdings exceed 2800 distinct mouse lines, the largest subset consisting of spontaneous and induced mutants. Although most of the mutant mouse repository activities at TJL are organized under several major programs, information on all publicly available mouse stocks is accessible through a single database interface (JAX® Mice Database) found at the TJL Web site (Table 3.1). Researchers who have a mouse stock that they would like considered for inclusion in the TJL collection may find an online submission form at the TJL Web site. An internal committee consisting of scientists and resource managers reviews submissions, with supplemental reviews provided by numerous external advisory committees, organized by discipline and staffed by prominent scientists.

MOUSE MUTANT RESOURCE (MMR)

As early as the 1930s, Drs. Elizabeth S. Russell and George D. Snell began selecting naturally occurring mutants from among TJL's inbred mouse colonies.[26] This collection of mutant stocks first received public funding from the National Science Foundation in the early 1960s. Approximately a decade later, the MMR obtained additional ongoing support from the National Center for Research Resources

(NCRR) at NIH. The primary functions of the MMR are to: (1) characterize novel spontaneous mutations that arise from the large breeding colonies at TJL; (2) maintain and distribute established mutant stocks; and (3) cryopreserve the mutant stocks accepted into the resource. Although the majority of stocks within the MMR are derived from TJL mouse colonies, submission of spontaneous mutant mice for acceptance into the MMR from the scientific community is encouraged.

THE INDUCED MUTANT RESOURCE (IMR)

The IMR was established at TJL in 1992. Genetically modified mice were being made in increasing numbers and it became clear that a centralized clearinghouse was vital to preserving these research resources. With support from various NIH institutes (NCRR, NIAID, NIAMS, NIA) as well as numerous private foundations and charities, the IMR collection has grown to over 1000 transgenic and targeted mutant mouse lines with a wide variety of research applications. Similar to the MMR, the IMR maintains, distributes, and cryopreserves its holdings. The majority of mutants held by the IMR are derived from sources external to TJL. Newly acquired mutants are posted on the TJL Web site for a period of several months. After several months, if a particular mutant fails to generate interest from the scientific community, live colonies are discontinued and the strain remains available only as cryopreserved embryos or germplasm. In this way, the IMR is able to maintain a dynamic selection of approximately 300 live mutant stocks, with hundreds more available in a cryopreserved state.

TYPE 1 DIABETES RESOURCE (T1DR)

The T1DR was established in 2000 with funding from NCRR and NIDK. It operates in a fashion similar to the MMR and IMR, but with a specialized focus on stocks important to research in type 1 diabetes. The T1DR holdings include stocks that are congenic for chromosomal intervals containing type 1 diabetes susceptibility or resistance loci, as well as stocks bearing transgenes, targeted and spontaneous mutations—many on NOD and NOD-related backgrounds. Eventually, the T1DR plans to maintain approximately 150 individual mouse stocks.

Most of the T1DR stocks are held only in a cryopreserved state. To make these mice more accessible to researchers, a unique management system was established. Several times a year, a number of functionally related strains (all tetracycline-inducible models, for instance) are recovered from cryopreservation and live colonies are temporarily established. Several months notice is given to the scientific community via targeted e-mails and Web site postings in advance of the imminent availability of a particular "functional wave." By offering variously themed functional waves at different times throughout the year, the cost incurred by researchers to acquire these stocks is periodically lowered.

MOUSE MUTANT REGIONAL RESOURCE CENTERS (MMRRC)

In 1998, in response to a meeting of the Trans-NIH Mouse Genomics and Genetic Resources Coordinating Group, the MMRRC network was established to complement the TJL repository and address the increasing need for mouse repository capacity.[27,28] Funded by NCRR, the MMRRC network consists of four centers located at the University of North Carolina at Chapel Hill, the University of California–Davis, Harlan Sprague Dawley, Inc. (Indianapolis) in collaboration with the University of Missouri at Columbia, and Taconic Farms in collaboration with the State University of New York–Albany. In addition to these four repository centers, an Informatics Coordinating Center based at TJL is responsible for strain information management, customer service, and database development and maintenance.

The MMRRC maintains, distributes, and cryopreserves its holdings and offers a single, publicly accessible Web site that displays strain information (Table 3.1). An online strain submission form is available at the Web site and submissions of all mouse strain types are accepted. Generally, each center leverages its expertise in certain areas and further characterizes submitted mice with supplementary phenotyping projects. A unique aspect of the MMRRC network is its collaboration with the Bay Genomics consortium, an organization of research groups in the San Francisco area that has generated a large number of embryonic stem cells that bear gene-trap mutations covering hundreds of genes. Detailed information for each ES cell line can be obtained at the Bay Genomics Web site (Table 3.1). Investigators may contract with the MMRRC to perform blastocyst microinjection with a Bay Genomics ES cell line to generate a mutant mouse. The MMRRC network expects to aggressively expand its ES cell holdings in the next few years.

MOUSE MODELS OF HUMAN CANCERS CONSORTIUM (MMHCC) MOUSE REPOSITORY

In 2001, the National Cancer Institute (NCI) established the MMHCC repository in Frederick, MD, for the purpose of making mouse models of cancer more accessible to the research community. The MMHCC repository Web site displays strain information and an online strain submission form. Strains in high demand are maintained as live colonies; those without significant interest are kept in a cryopreserved state. A repository committee reviews strains submitted to the MMHCC repository on an annual basis. Unlike most repositories, mice distributed from the MMHCC repository are available free of charge, although the requesting researcher is responsible for shipping charges.

RIKEN BIORESOURCE CENTER (BRC)

The mouse resource at the Riken BRC is a nonprofit institute formed in 2001 for the purposes of providing biological materials, technical services, and educational programs to private enterprises and academic organizations worldwide. The BRC imports biomedically important mouse strains (inbred and mutant/genetically engineered), rederives the mice to a specific pathogen-free state, and distributes the mice to the scientific community as live animals and cryopreserved embryos or sperm.

The BRC maintains a Web site both in English and Japanese (Table 3.1) that offers a database of mouse strain information. Scientists interested in depositing mice with the BRC are urged to contact a customer service representative: animal@brc.riken.jp.

CENTER FOR ANIMAL RESOURCES AND DEVELOPMENT (CARD)

The CARD located at Kumamoto University was established in 1998 and maintains Web sites in both English and Japanese (Table 3.1). The CARD repository maintains, cryopreserves, and distributes a growing number of inbred strains, and transgenic and targeted mutations. A limited number of ES cell lines are also available in the CARD repository. Submissions to the CARD repository may be sent to card@kaiju.medic.kumamoto-u.ac.jp.

EUROPEAN MOUSE MUTANT ARCHIVE (EMMA)

The nonprofit EMMA repository network collects, archives, and distributes biomedically important mutant mouse stocks. The network is a partnership of prominent European institutions including CNR Instituto di Biologia Cellulare in Monterotondo, Italy (project coordinator); the CNRS Centre de Distribution, de Typage et d'Archivage Animal in Orleans, France; the MRC Mammalian Genetics Unit in Harwell, United Kingdom; the Karolinska Institutet in Stockholm, Sweden, the Instituto Gulbenkian de Ciência in Oeiras, Portugal; the GSF Institute of Experimental Genetics in Munich, Germany; and the EMBL European Bioinformatics Institute in Hinxton, United Kingdom. EMMA is supported by member institutional funds and the European Commission's Framework Programmes. Strain information for all EMMA strains, a single online submission form, and links to the home page of individual EMMA members may be found at the EMMA Web site (Table 3.1).

REFERENCES

1. Guenet J-L, Bonhomme F. Origin of the laboratory mouse and subspecies. In: Hedrich H, Bullock G, Petrusz P, eds. *The Handbook of Experimental Animals, the Laboratory Mouse*. Elsevier Academic Press, London, U.K., 2004:3–13.
2. Davisson MT, Linder CC. Historical foundations. In: Hedrich H, Bullock G, Petrusz P, eds. *The Handbook of Experimental Animals, the Laboratory Mouse*. Elsevier Academic Press, 2004:15–24.
3. Silver LM. An introduction to mice. In: Silver LM, ed. *Mouse Genetics, Concepts and Applications*. Oxford Press, New York, N.Y., 1995:3–14.
4. Paigen K. One hundred years of mouse genetics: An intellectual history. I. The classical period (1902–1980). *Genetics*. 2003; 163:1–7.
5. Beck JA, Hafezparast M, Lennon-Pierce M, Eppig JT, Festing MF, Fisher EM. Genealogies of mouse inbred strains. *Nat Genet*. 2000; 24:23–25.
6. Gordon JW, Scangos GA, Plotkin DJ, Barbose JA, Ruddle FH. Genetic transformation of mouse embryos by microinjection of purified DNA. *Proc Natl Acad Sci USA*. 1980; 77:7380–7384.

7. Wagner TE, Hoppe PC, Jollick JD, Scholl DR, Hodinka RL, Gault JB. Microinjection of a rabbit beta-globin gene into zygotes and its subsequent expression in adult mice and their offspring. *Proc Natl Acad Sci USA*. 1981; 78:6376–6380.

8. Brinster RL, Chen HY, Messing A, van Dyke T, Levine AJ, Palmiter RD. Transgenic mice harboring SV40 T-antigen genes develop characteristic brain tumors. *Cell*. 1984; 37:367–379.

9. Harbers K, Jahner D, Jaenisch R. Microinjection of cloned retroviral genomes into mouse zygotes: Integration and expression in the animal. *Nature*. 1981; 293:540–542.

10. Thomas KR, Capecchi MR. Site-directed mutagenesis by gene targeting in mouse embryo-derived stem cells. *Cell*. 1987; 5:503–512.

11. Doetschman T, Gregg RG, Maeda N, Hooper ML, Melton DW, Thompson S, Smithies O. Targeted correction of a mutant HPRT gene in mouse embryonic stem cells. *Nature*. 1987; 330:576–578.

12. Capecchi MR. Altering the genome by homologous recombination. *Science*. 1989; 244:1288–1292.

13. Collins FS, Patrinos A, Jordan E, Chakravarti A, Gesteland R, Walters L, et al. New goals for the U.S. human genome project: 1998–2003. *Science*. 1998; 282: 682–689.

14. Rechsteiner MC. The Human Genome Project: Misguided science policy. *TIBS*. 1991; 16:455–458.

15. International Human Genome Sequencing Consortium. Initial sequencing and analysis of the human genome. *Nature*. 2001; 409:860–921.

16. Venter JC, Adams MD, Myers EW, Li PW, et al. The sequence of the human genome. *Science*. 2001; 291:1304–1351.

17. Mouse Genome Sequencing Consortium. Initial sequencing and comparative analysis of the mouse genome. *Nature*. 2002; 420:520–562.

18. Gong S, Zheng C, Doughty ML, Losos K, Didkovsky N, Schambra UB, Nowak NJ, Joyner A, Leblanc G, Hatten ME, Heintz N. A gene expression atlas of the central nervous system based on bacterial artificial chromosomes. *Nature*. 2003; 425:917–925.

19. Heintz N. Gene expression nervous system atlas (GENSAT). *Nat Neurosci*. 2004; 7:483.

20. Skarnes WC, von Melchner H, Wurst W, Hicks G, Nord AS, Cox T, Young SG, Ruiz P, Soriano P, Tessier-Lavigne M, Conklin BR, Stanford WL, Rossant J. et al. A public gene trap resource for mouse functional genomics. *Nat Genet*. 2004; 36:543–544.

21. The Comprehensive Knockout Mouse Project Consortium. The knockout mouse project. *Nat Genet*. 2004; 36:921–924.

22. Zambrowicz BP, Sands AT. Knockouts model the 100 best-selling drugs—Will they model the next 100? *Nat Rev Drug Discov*. 2003; 2:38–51.

23. Sincell M. Houston flood. Research toll is heavy in time and money. *Science*. 2001; 293:589.

24. NIH, Statement on Sharing and Distributing Mouse Resources, http://www.nih.gov/science/models/mouse/sharing/index.html.

25. Eppig JT, Bult CJ, Kadin JA, Richardson JE, Blake JA and the Mouse Genome Database Group. The Mouse Genome Database (MGD) from genes to mice—A community resource for mouse biology. *Nucleic Acids Res*. 2005; 33 Database Issue:D471–475.

26. Davisson MT. The Jackson Laboratory Mouse Mutant Resource. *Lab Animal*. 1990;19: 23–29.
27. Battey J, Jordan E, Cox D, Dove W. An action plan for mouse genomics. *Nat Genet*. 1999; 21:73–75.
28. Grieder FB. Mutant Mouse Regional Resource Center Program: A resource for distribution of mouse models for biomedical research. *Comp Med*. 2002; 52:203.

4 Mouse Genome Informatics: Database Access to Integrated Phenotype Data

Carroll-Ann W. Goldsmith, Martin Ringwald, John P. Sundberg, Carol J. Bult, and Janan T. Eppig

TABLE OF CONTENTS

The Mouse Genome Informatics Database[1] (MGI) provides integrated access to data on the genetics, genomics, and biology of the laboratory mouse. MGI is an integrated and curated database supporting biological queries, hypothesis generation, and knowledge discovery. The database is freely available on the World Wide Web at http://www.informatics.jax.org. Although there are many components of MGI (Table 4.1), this chapter focuses on those aspects specific to phenotypes.

HISTORY OF MGI

The Mouse Genome Informatics Database traces its beginnings to 1989 during the early phases of the Human Genome Initiative. The initial goal was to unify and integrate many small research databases that were developing in response to

TABLE 4.1
Overview of Data and Source Information within MGI[a]

Data Type	Data Examples	Source Examples	External Link Examples
Nomenclature	Official symbols and names for genes, genomic markers, and strains of mice	MGI is the authoritative source for these data; collaborations include HUGO nomenclature committee for human, rat nomenclature committee, and gene family experts	LocusLink, RGD, ILAR
Genes and gene products	Molecular function, biological process, and cellular location	Gene ontology annotations inferred via SwissProt, InterPro, ENZYME, and manual curation of biochemical literature	GO Consortium Web site, SwissProt, ENZYME, InterPro
Molecular and sequence data	DNA and protein sequence data; clone, probe, and antibody data from experiments and from large consortia efforts	GenBank, SwissProt, MGC, RIKEN, IMAGE; probe/PCR/antibody data from literature and data submission	GenBank, SwissProt, MGC, RIKEN, IMAGE, DoTs, TIGR, Unigene
Homology	Orthologue gene sets from mouse/human/rat and other mammals, comparative maps, Oxford grids	Curated data from literature, selected downloads from HomoloGene	Other species databases, such as RGD (rat), LocusLink (human, rat), ArkDB (sheep, cow, pig)
Maps and map data	Chromosome location, experimental data, graphical maps	Mapping panel data submissions, mouse genome sequencing consortium, published literature	Map panel home Web sites, ENSEMBL, UCSC mouse assembly, NCBI
Gene expression	*In situ* hybridizations, RT-PCR, Northern/Western blots, immunohistochemistry, cDNAs, literature index	Published literature, dataloads from large-scale experiments	Mouse Atlas Web site
Tumor biology	Tumor classification, incidence and frequency, host genetics and tumor genetic changes, pathology images and diagnoses	Published literature, Jackson Laboratory animal health department, submissions from mouse pathologists at many institutions	Mammary cancer Web sites, JAX® Mice and NCI mouse repositories
Phenotypes	Allelic and genomic mutations, mouse strains, polymorphisms, phenotypic characterizations, images	ENU mutagenesis centers, Gene Trap Consortium, Complex Traits Consortium, published literature	OMIM (for human disease models), Consortia and Center Web sites

TABLE 4.1 (continued)
Overview of Data and Source Information within MGI[a]

Data Type	Data Examples	Source Examples	External Link Examples
Vocabularies and data standards	Ontologies: Gene ontology, mammalian phenotype ontology, mouse anatomy, and many standard vocabularies such as tissue types, tumor types	Various vocabularies are developed by MGI and as part of collaborative efforts with other database resources	GO Consortium Web site, Mouse Atlas Web site
Literature	Reference data and abstracts		PubMed

ArkDB: Public database of genome mapping data from farm and other animal species (cat, chicken, cow, deer, horse, pig, salmon, sheep, tilapia, turkey), maintained by the Roslin Bioinformatics Group (U.K.).

DoTs: Database of Transcribed Sequences maintained by the Computational Biology and Informatics Laboratory, University of Pennsylvania.

ENSEMBL: Joint project between EMBL-EBI (European Molecular Biology Laboratory–European Bioinformatics Institute) and the Sanger Institute to develop software for automatic annotation of metazoan genomes.

ENZYME: Enzyme nomenclature database following recommendations of the Nomenclature Committee of the International Union of Biochemistry and Molecular Biology.

GenBank: Genetic sequence database, U.S. National Institutes of Health (NIH).

GO: Gene Ontology.

ILAR: Institute for Laboratory Animal Research, U.S. National Academy of Sciences.

IMAGE: Integrated Molecular Analysis of Genome Expression Consortium, maintained by Lawrence Livermore National Laboratory.

InterPro: Database of protein families, domains, and functional sites maintained by the European Bioinformatics Institute.

JAX®: The Jackson Laboratory.

LocusLink: Sequence and descriptive information about genetic loci, maintained by U.S. National Center for Biotechnology Information (NCBI).

MGC: Mouse Genome Consortium, formed by the U.S. National Institutes of Health (NIH), the Wellcome Trust (U.K.), and three private companies (SmithKline Beecham, the Merck Genome Research Institute, and Affymetrix, Inc.).

NCBI: U.S. National Center for Biotechnology Information.

NCI: U.S. National Cancer Institute.

OMIM: Online Mendelian Inheritance in Man, joint project of the Johns Hopkins University and U.S. National Center for Biotechnology Information (NCBI).

TABLE 4.1 (continued)
Overview of Data and Source Information within MGI[a]

PubMed: Bibliographic information maintained by U.S. National Center for Biotechnology Information (NCBI).

RGD: Rat Genome Database maintained by the Bioinformatics Program at the Medical College of Wisconsin.

Riken: Institute of Physical and Chemical Research, Japan.

SwissProt: Protein knowledge database maintained by Swiss Institute of Bioinformatics.

TIGR: The Institute for Genomic Research, U.S.

UCSC: University of California, Santa Cruz, Genome Bioinformatics Group.

Unigene: System for automatically partitioning GenBank sequences into a nonredundant set of gene-oriented clusters, maintained by the U.S. National Center for Biotechnology Information (NCBI).

the growth of mouse genetic data and the advent of molecular technologies. These included databases on experimental genetic mapping, homologies among genes in mammalian species, molecular probes and clones for mice and their polymorphisms, and text descriptions of phenotypes (at that time, mostly spontaneous mutations or induced mutations with visible phenotypes).

The first public access, provided in 1992, was via a stand alone software package featuring interactive graphical displays and accompanying data files that were distributed freely to the community (coined the Encyclopedia of the Mouse Genome[2]). In addition, a gene mapping database, GBASE,[3] was provided via a command line interface on the Internet. Shortly thereafter, the World Wide Web began to provide a new public and easily accessible means of communication and data access, lowering the barrier of using database resources for even naïve computer users. The Mouse Genome Database[4] (MGD), first released in 1994, was developed to coalesce and expand upon the initial mouse database efforts mentioned above. Its rapid development and adoption by the scientific community was accelerated by the rapid dissemination provided by universal Web access. In 1996, the Gene Expression Database[5] (GXD) saw its first release with access through a Web site shared with MGD. The actual integration of GXD with MGD at the infrastructure level gave rise to MGI, with a first release of this combined database resource in early 1998. The Mouse Tumor Biology Database[6] (MTB) was added in late 1998, further increasing the depth and breadth of the Mouse Genome Informatics resources. Since then, MGI has continued to grow, adapt, and evolve with the needs of the scientific community and the progress of the genome initiative.

INTEGRATED DATA WITHIN MGI

As stated previously, the data found within MGI are integrated, meaning that information from multiple sources is organized into a unified structure, common objects are rationalized, and connections among related data defined. This contrasts with the warehousing of data where various data sets are simply housed in one place. As a result of data integration, visitors to the MGI site benefit by being able to ask complex questions, spanning different types of data, from multiple sources.[7,8]

Data in MGI are expertly curated by staff who evaluate and determine what associations are appropriate among data from disparate sources. Data integration presents a number of challenges, including: (1) data from different sources may be of varying quality, include conflicting data, or have missing components; (2) nomenclature used may not be standard; (3) some databases do not maintain unique identifiers or allow these identifiers to change, making updates problematic; and (4) new data may be difficult to integrate if objects disappear or have been split in two. It is imperative that information be analyzed and reviewed by MGI staff to determine the most accurate way to integrate new data into the database.

Genes are an important core data object in MGI. MGI maintains the official symbol and name of all genes for the international mouse research community. Information associated with genes is extensive. MGI contains curated data on gene maps and experimental data supporting these maps, orthology data comparing mouse genes with other mammals, particularly human and rat, sequence links and sequence data, mutant phenotype and strain descriptions, polymorphism data, gene function information, expression data, molecular reagent information, and references. These major data types are summarized on the detail page for a given gene, with links to more extensive information in MGI and to relevant external data resources.

PHENOTYPE INFORMATION WITHIN MGI ON GENETICALLY ENGINEERED AND MUTANT MICE

Information related to phenotype is represented through the allele and phenotype descriptions as well as via gene expression data and mouse tumor biology data. Each perspective looks at a phenotype with different emphasis, but all share the commonality of capturing information about the observed state of a particular mouse, strain, or allele. The following sections describe these areas of MGI in more detail.

GENE EXPRESSION DATA

Gene expression data[5] for mouse development are curated for both normal and mutant mice and placed in context with mouse genetic, genomic, and phenotypic information, facilitating understanding of the underlying molecular biology of mouse development and disease. Both mRNA and protein expression data are represented, with detailed data from experiments using Western and Northern blots, RT-PCR, immunohistochemistry, RNA *in situ* hybridization, and RNase protection assays. Much of these data are derived from published literature and curated into the database manually by MGI staff. However, large-scale data sets are electronically acquired, including large-scale RT-PCR and *in situ* expression screens and expression data from cDNAs and their source tissues. In addition, an index of the literature containing mouse developmental expression data provides users with a quick survey of developmental stage and assay. Currently over 9400 references have been indexed, representing expression data for almost 5900 genes.

Expression information is standardized by employing several standardized vocabularies. Descriptions of anatomical structures are standardized using a mouse anatomical dictionary developed in concert with members of The Edinburgh Mouse Atlas Project.[9] Pattern and level of expression are standardized by utilizing sets of controlled vocabularies developed within MGI. Further standardization is accomplished by using standard gene, strain, and allele nomenclature for all data descriptions. Digitized images of original data, for example, of *in situ* hybridizations or of gel blots, are included in expression records when possible.

THE GENE EXPRESSION DATA QUERY FORM

(http://www.informatics.jax.org/searches/expression_form.shtml) can be used to interrogate expression data in MGI. In response to a submitted query, a summary page of all results matching the search criteria is presented, including links to each specific record. Figure 4.1 shows the detail of a record obtained by searching for RNA *in situ* data for the kit oncogene (Kit). This result page shows the source of the data, the type of assay, a unique accession ID, the official gene symbol and name, the probe used for the assay, how the probe was prepared, and how visualization was conducted. Tables provide specific information for the strain of mouse used, any mutations the mouse carries, the age and sex of the mouse, and specific details of the assay. If images are available, links to these images are also found within the tables. For example, Figure 4.2 shows RNA *in situ* results at embryonic day 13.5 in wild-type mice (+/+), mice heterozygous for the patch allele (*Ph/Ph+*), or mice homozygous for the patch allele (*Ph/Ph*), using a probe for kit oncogene.

MOUSE TUMOR BIOLOGY DATA

Mouse tumor biology data[6] include information on cancer genetics, tumor frequency and incidence, and pathology in genetically defined mice. A primary application of these data is to supply information on the effects of genetic background on tumorigenesis in the mouse that will enable researchers to make better decisions on choices for experimental cancer models and on genetic backgrounds to choose for creating new models. Descriptions of tumors, data on tumor frequency and latency, genes associated with tumors and tumor development, and tumor pathology reports and images can be found at MGI. As with other types of data, integration of information allows investigators to explore complex questions such as the intrinsic cancer-related phenotypes of different mouse strains, the patterns of mutations in specific tumors, and the identification of genes that are commonly mutated across a spectrum of cancers in the mouse.

?	**Gene Expression Data**
	Query Results -- Details

Reference: J:34586 Wehrle-Haller B, Dev Biol 1996 Aug 1;177(2):463-74
Assay type: RNA in situ
MGI Accession ID: MGI:1339505
Gene symbol: Kit
Gene name: kit oncogene
Modification date: 7/ 8/1999

Probe: Kit probe3
Probe preparation: Antisense RNA continuously labelled with Digoxigenin
Visualized with: Alkaline phosphatase

Specimens Used

	6A (a)	6B (a)	6D left embryo	6D middle embryo	6D right embryo	7 left embryo	7 middle embryo	7 right embryo
Strain	Not Specified	involves: BALB/c # C57BL/6	Not Specified	C57BL/6	involves: BALB/c # C57BL/6	Not Specified	C57BL/6	involves: BALB/c # C57BL/6
Mutations		*Ph/Ph*		*Ph/Ph*$^+$	*Ph/Ph*		*Ph/Ph*$^+$	*Ph/Ph*
Age	E10.5 (b)	E10.5 (b)	E11.5 (b)	E11.5 (b)	E11.5 (b)	E13.5 (b)	E13.5 (b)	E13.5 (b)
Sex	Not Specified	Not Specified	Not Specified	Not Specified	Not Specified	Not Specified	Not Specified	Not Specified
Type	whole mount	whole mount	whole mount	whole mount	whole mount	whole mount	whole mount	whole mount
Fixation	Not Specified	Not Specified	Not Specified	Not Specified	Not Specified	Not Specified	Not Specified	Not Specified
Embedding	Not Applicable	Not Applicable	Not Applicable	Not Applicable	Not Applicable	Not Applicable	Not Applicable	Not Applicable

Notes:
(a) Lateral view.
(b) Age of embryo at noon of plug day not specified in reference.

Results: 6A (embryonic day 10.5)

Structure	Level	Pattern	Image
TS17: 1st arch	Present	Not Specified	Figure 6.A
TS17: 2nd arch	Present	Not Specified	Figure 6.A
TS17: limb	Present	Not Specified	Figure 6.A
TS17: trunk mesenchyme: paraxial mesenchyme	Present	Not Specified	Figure 6.A
TS17: future spinal cord	Present	Not Specified	Figure 6.A

FIGURE 4.1 Gene expression detail page from MGI. This screen shot of an MGI Web page describes RNA *in situ* data for the gene, *Kit* (kit oncogene). Data include standard nomenclature for the gene, a unique accession identifier, probe details, reference information, and detailed information about the specimens studied. For each specimen, information on the strain of origin; mutations, if any; and age and sex of the mouse from which the specimen originated is provided. Further details about the methods used to produce the specimen are also noted. If there are images available, links to these images are provided. Note that the specimens in this example include some derived from mice harboring a spontaneous mutation of the patch deletion region, *Ph*.

Mouse tumor data are derived largely from published literature. Photomicrographs of histopathology and immunohistochemistry images are derived from The Jackson Laboratory animal health program and via direct submission from collaborating pathologists. These data can be submitted via a link to the JaxPath submission interface (http://tumor.informatics.jax.org/). As with other areas of

Images
Query Results -- Details

Reference: J:34586 Wehrle-Haller B, Dev Biol 1996 Aug 1;177(2):463-74
Figure: 7.
MGI Accession ID: MGI:1339466
Assays that refer to this image:

- MGI:1339505 Kit

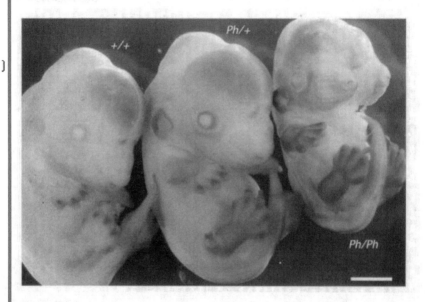

Note: Bar= 2.4mm

Copyright: Reprinted with permission of the author and Academic Press from Wehrle-Haller B, Dev Biol 1996 Aug 1;177(2):463-74. Copyright 1996 by Academic Press.

FIGURE 4.2 Image detail page for expression data from MGI. By choosing the image link from the Gene Expression detail page, the image for the selected expression data is displayed. This example shows RNA *in situ* results for kit oncogene (*Kit*) in wild-type mice (+/+), and mice heterozygous (*Ph+/Ph*) or homozygous (*Ph/Ph*) for the allele patch (*Ph*). Note that these results have a unique identifier and associated reference and copyright information.

MGI, tumor data are standardized by utilizing controlled vocabularies for such parameters as: anatomical system, organ or tissue, tumor classification, treatment, agent, mouse strain, mouse sex, reproductive status, and gene mutation type. In addition, official nomenclature for mouse gene, strain, and allele is followed.

The major data types for tumor biology include: mouse strains; genes and mutations; tumor types/classifications; organs, tissues, or cell types of tumor origin; pathology images; and literature, and can be searched (Figure 4.3), each icon linking to a user query form. For example, a search conducted from the Tumor Search Query Form for papillomas with photomicrographs yields a table of results

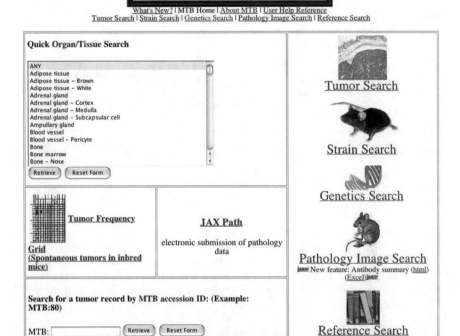

FIGURE 4.3 Entry point into mouse tumor biology data in MGI, with links to search by mouse strains; genes and mutations; tumor types/classifications; organs, tissues, or cell types of tumor origin; pathology images; and literature.

containing information on tumor name, organ/tissue affected, treatment (if any), mouse strain and sex, tumor frequency, images (if any, in this case requested, so only records with images are returned), and metastasis. Links to detailed records are provided, as are links to any photomicrographs, as seen in Figure 4.4. Choosing the link to photomicrographs provides information on the tumor pathology and a further link to a more detailed pathology report (Figure 4.5).

The mouse tumor data in MGI provides the most extensive data on mouse tumor pathobiology on the Web.[10] Currently the database contains over 14,000 tumor records, almost 21,000 tumor frequency reports, representing over 3000 strain/cohort records, and over 3000 pathology images. Providing this information in a standardized, integrated format allows investigators to interrogate the database to find the broadest range of accurate mouse tumorigenesis data available.

ALLELES AND PHENOTYPE DATA

Until a few years ago, phenotypic information was maintained in the Mouse Locus Catalog[11] (MLC), which provided text-based summaries of data on

Mouse Tumor Biology Database (MTB)

MTB Home I About MTB I User Help Reference
Tumor Search I Strain Search I Genetics Search I Pathology Image Search I Reference Search

Tumor Search Results

Displaying records 1 through 3 of 3 records found.

MTB ID	Tumor Name	Organ/ Tissue Affected	Treatment Agents	Strain Name Sex	Tumor Frequency Range (%)	Pathology Images	Metastasizes to the:	Additional Information
MTB:11721	Forestomach squamous cell papilloma	Forestomach	None (Spontaneous)	B6.129S2-Plau^{tm1Mlg}/J male	observed	📷		Show Record Details
MTB:11720	Skin - Epidermis squamous cell papilloma	Skin - Epidermis	None (Spontaneous)	BALB/cByJ male	observed	📷		Show Record Details
MTB:8933	CNS - Brain - Choroid plexus papilloma	CNS - Brain - Choroid plexus	None (Spontaneous)	C57BL/6J-Tg(SV)7Bri/J female	observed	📷		Show Record Details

FIGURE 4.4 Tumor search result page. A search for papillomas with pathology images was conducted and yielded this Web page of results. As seen, each record has a unique identifier, descriptions of the tumor name, information on the organ or tissue affected, information on any treatments, data on the mouse strain and sex, a range of the tumor frequency, and metastasis information. Links to further information, such as the pathology images, or more detailed record results are provided when available.

phenotypic mutations and information on genes and gene products associated with particular phenotypes. Because MLC was text, each entry could include unique terms based on the author's writing style and word choice. Full-text searching of MLC, therefore, returned both unintended results and missed relevant records. For example (Figure 4.6a), if one searches MLC for "hairless", 16 results are returned, listing markers in which the text "hairless" was found in the MLC descriptions. If, however, one searches for "bald" (Figure 4.6b), a different set of 8 results is returned, with some gene markers in common and some unique to each search. Although public access to the combined mouse/human phenotype search is still provided through MGI, the MLC is no longer maintained and its contents should be viewed with this in mind.

Today, descriptions of mutations, whether spontaneous, induced, or created through genetic engineering, are found in the Alleles and Phenotypes section of MGI,[12] and, rather than being primarily text, defined fields and vocabularies are used, with supplementary descriptive notes as needed. The following information is curated for each allele within the database: standardized nomenclature, unique accession identifiers, strain background on which the allele arose, molecular definition of the mutational change, mode of inheritance, phenotype descriptions using controlled vocabularies, associations with human disease syndromes through links to Online Mendelian Inheritance in Man (OMIM), and supporting references. Standard nomenclature and vocabularies support a number of these data types, including mutation type, gene, allele, and strain designations, mode

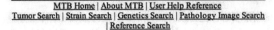

Mouse Tumor Biology Database (MTB)

MTB Home | About MTB | User Help Reference
Tumor Search | Strain Search | Genetics Search | Pathology Image Search
| Reference Search

Pathology Record Details

MTB:11721

Organ/Tissue of Origin:	**Forestomach**
Tumor Classification:	**squamous cell papilloma**
Organ/Tissue Affected:	Forestomach
Treatment:	None (Spontaneous)

Strain Name:	**B6.129S2-Plau^{tm1Mlg}**
Sex:	**male**
Strain Type:	congenic & targeted mutation (knockout)
Strain Notes:	Mice homozygous for a targeted mutation of Plau.

Pathology Report

Description:	Forestomach - papilloma - squamous
Age at Necropsy:	123 days
Source of Pathology Report:	I. Mikaelian and J.P. Sundberg

View Large Image

Image Caption:
Forestomach: a small exophytic papillary epithelial structure is arising from the squamous epithelium of the forestomach. This mass is made of a thick stratified squamous epithelium supported by a small amount of fibrovascular connective tissue. There is no evidence of invasion of the epithelium into the underlying submucosa.

Method:	H&E
Organism:	mouse
Image ID:	152
Source of Image:	I. Mikaelian and J.P. Sundberg

(Copyright: John Sundberg)

FIGURE 4.5 Pathology record detail. From the example in Figure 4.4, links lead to this page, which contains pathology details. Information on the organ/tissue of origin, tumor classification, strain name, sex of the mouse from which the specimen was obtained, the unique identifier, and a detailed pathology report are all found on this page.

of inheritance, and phenotype characteristics. Currently nearly 9500 phenotypic alleles caused by mutations or genetically engineered constructs are catalogued in MGI, representing mutations in almost 4200 genes. In addition, more than 1500 named QTLs are annotated, each with two or more detected variants.

To improve phenotype searches and allow for better analysis and comparison of phenotype data, a standardized phenotype vocabulary[13] is being developed and applied at MGI, the Mammalian Phenotype Ontology (Figure 4.7). Phenotype data are described using this standardized vocabulary, and tools are being developed to fully implement and integrate this vocabulary within the database.

Combined Mouse/Human Phenotypes
Integrated MLC/OMIM Search

This interface allows you to perform simultaneous keyword searches of the Mouse Locus Catalog and McKusick's Online Mendelian Inheritance in Man.

Because MLC and OMIM use different text searching systems, there are some restrictions in the available searching options.

This is a searchable index. Enter keyword(s):

MLC Glimpse Query: hairless

16 row(s) returned.

```
    symbol      chr name
1.  Bda          11 bald arthritic
2.  dep           4 depilated
3.  eb           10 eye blebs
4.  Eda           X ectodysplasin-A
5.  Foxn1        11 forkhead box N1
6.  hr           14 hairless
7.  Nras          3 neuroblastoma ras oncogene
8.  Otc           X ornithine transcarbamylase
9.  Rbpsuh         5 recombining binding protein suppressor of hairless (Drosophila)
10. Rbpsuh-ps1   UN recombining binding protein suppressor of hairless (Drosophila), pseudogene 1
11. Rbpsuh-ps2   UN recombining binding protein suppressor of hairless (Drosophila), pseudogene 2
12. Rbpsuh-rs3    6 recombining binding protein suppressor of hairless (Drosophila), related sequence 3
13. Scd1         19 stearoyl-Coenzyme A desaturase 1
14. st           UN shaker-short
15. Tgfb1         7 transforming growth factor, beta 1
16. wal          14 waved alopecia
```

(a)

FIGURE 4.6A Portion of a result page from a combined mouse/human phenotype search for the term "hairless". Results shown here include 16 mouse gene symbols for which the text string "hairless" was found in an MLC text record.

Combined Mouse/Human Phenotypes
Integrated MLC/OMIM Search

This interface allows you to perform simultaneous keyword searches of the Mouse Locus Catalog and McKusick's Online Mendelian Inheritance in Man.

Because MLC and OMIM use different text searching systems, there are some restrictions in the available searching options.

This is a searchable index. Enter keyword(s):

MLC Glimpse Query: bald

8 row(s) returned.

```
    symbol  chr name
1.  Bda      11 bald arthritic
2.  Dsg3     18 desmoglein 3
3.  Ebp       X phenylalkylamine Ca2+ antagonist (emopamil) binding protein
4.  Eda       X ectodysplasin-A
5.  Edar     10 ectodysplasin-A receptor
6.  Edaradd  13 EDAR (ectodysplasin-A receptor)-associated death domain
7.  hr       14 hairless
8.  Pdcd8     X programmed cell death 8
```

(b)

FIGURE 4.6B Portion of a result page from a combined mouse/human phenotype search for the term "bald". Results shown here include eight mouse gene symbols for which the text string "bald" was found in an MLC text record.

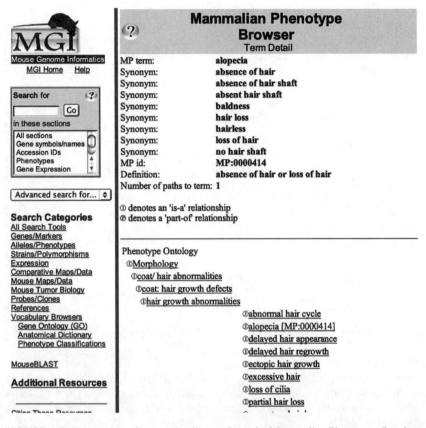

FIGURE 4.7 Screen shot of a term detail page from the Mammalian Phenotype Ontology in MGI. The term "hairless" was searched in the vocabulary. Note that the vocabulary supports multiple synonyms, so that searching on synonyms, such as "bald", results in the same term being returned from the query.

Full implementation of the vocabulary, including linking between the ontology display and mutant alleles associated with specific terms in the Mammalian Phenotype vocabulary is now available and users can browse the vocabulary at http://www.informatics.jax.org/searches/MP_form.shtml. In addition, a broad set of Phenotype Classification Terms is available as standard search terms for mutant alleles at http://www.informatics.jax.org/searches/allele_form.shtml. Finally, from the Allele Query Form, a search for a specific Mammalian Phenotype Ontology term can be conducted, by specifying the term, or a portion of the term (see Figure 4.8).

Figures 4.9 and 4.10 show a Web display of a representative allele record in MGI, a targeted mutation of the gene adenylate cyclase activating polypeptide 1 (*Adcyap1*). Highlights of the information on this allele include: standard nomenclature for the allele symbol and name, and for the gene symbol and name; a

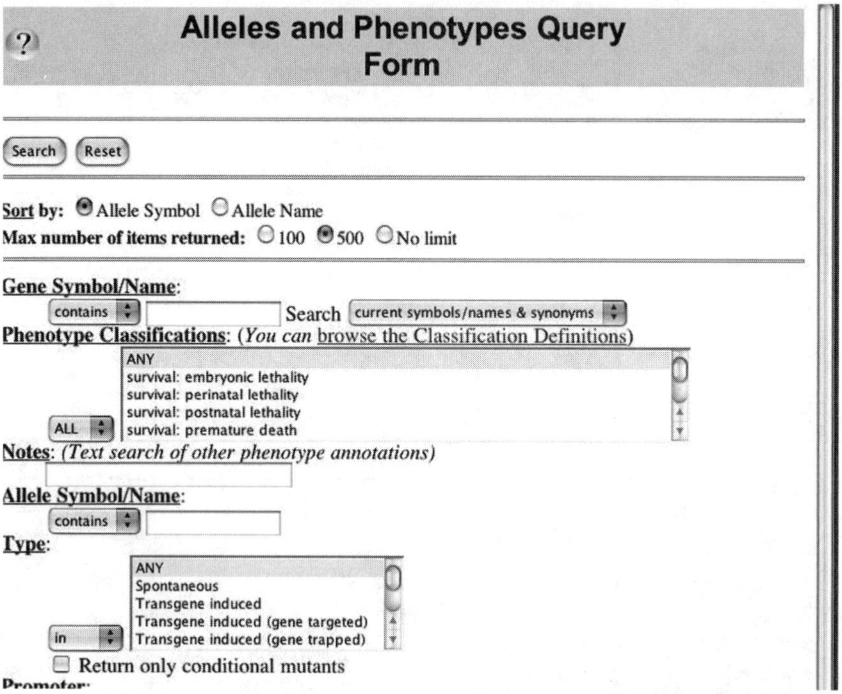

FIGURE 4.8 Alleles and phenotypes query form in MGI. A portion of the whole form is shown here, to highlight the ability to search for specific categories of phenotypes in the Phenotype Classification pick list, or to search the Notes field for specific terms found in the Mammalian Phenotype Ontology and used in the phenotype descriptions of individual alleles.

unique allele accession ID; and a description of the molecular nature of mutation. Information on strain background, references, and orthologous human genes is also provided. The implementation of the phenotype descriptions using the controlled phenotype vocabularies described above can be seen in Figures 4.9 and 4.10. The table in Figure 4.9 contains phenotype information derived from the broad Phenotype Classification Terms, and is fully searchable. The Mammalian Phenotype Ontology terms applicable to this allele are shown in the Associated Phenotype Controlled Terms section of the allele detail page (Figure 4.10).

SUMMARY

Pathology and phenotype related information on mutant and genetically engineered mice can be found in multiple sections of the Mouse Genome Informatics Database. This information includes, but is not limited to, phenotype descriptions, gene expression data, and mouse tumor biology data. Each area of the database emphasizes a different aspect of the phenotype and pathology information, but all share the commonality of capturing information about the observed state of

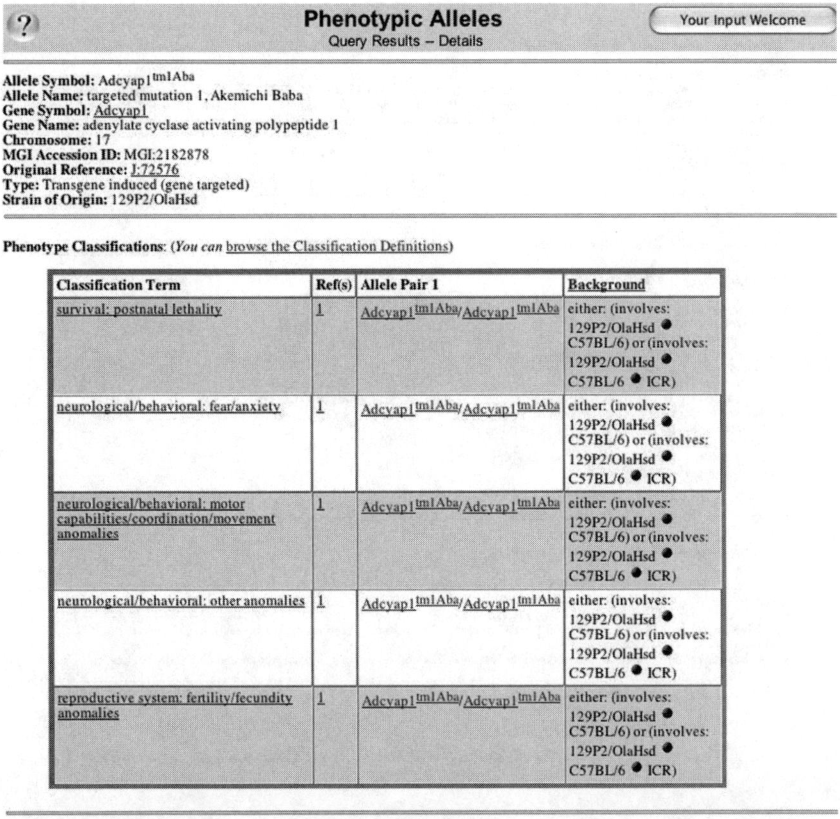

Phenotypic Alleles
Query Results -- Details

Your Input Welcome

Allele Symbol: Adcyap1^{tm1Aba}
Allele Name: targeted mutation 1, Akemichi Baba
Gene Symbol: Adcyap1
Gene Name: adenylate cyclase activating polypeptide 1
Chromosome: 17
MGI Accession ID: MGI:2182878
Original Reference: J:72576
Type: Transgene induced (gene targeted)
Strain of Origin: 129P2/OlaHsd

Phenotype Classifications: (*You can* browse the Classification Definitions)

Classification Term	Ref(s)	Allele Pair 1	Background
survival: postnatal lethality	1	Adcyap1^{tm1Aba}/Adcyap1^{tm1Aba}	either: (involves: 129P2/OlaHsd ● C57BL/6) or (involves: 129P2/OlaHsd ● C57BL/6 ● ICR)
neurological/behavioral: fear/anxiety	1	Adcyap1^{tm1Aba}/Adcyap1^{tm1Aba}	either: (involves: 129P2/OlaHsd ● C57BL/6) or (involves: 129P2/OlaHsd ● C57BL/6 ● ICR)
neurological/behavioral: motor capabilities/coordination/movement anomalies	1	Adcyap1^{tm1Aba}/Adcyap1^{tm1Aba}	either: (involves: 129P2/OlaHsd ● C57BL/6) or (involves: 129P2/OlaHsd ● C57BL/6 ● ICR)
neurological/behavioral: other anomalies	1	Adcyap1^{tm1Aba}/Adcyap1^{tm1Aba}	either: (involves: 129P2/OlaHsd ● C57BL/6) or (involves: 129P2/OlaHsd ● C57BL/6 ● ICR)
reproductive system: fertility/fecundity anomalies	1	Adcyap1^{tm1Aba}/Adcyap1^{tm1Aba}	either: (involves: 129P2/OlaHsd ● C57BL/6) or (involves: 129P2/OlaHsd ● C57BL/6 ● ICR)

FIGURE 4.9 A portion of an allele detail page in MGI. This screen shot shows the top portion of an MGI Web page describing a targeted mutation in the *Adcyap1* (adenylate cyclase activating polypeptide 1) gene, *Adcyap1^{tm1Aba}*, the first targeted mutation created in the laboratory of Akemichi Baba for this gene. Data included are: standard nomenclature for the gene and allele, the allele's unique identifier, and phenotype classifications for genotypes studied using this allele.

a particular mouse, strain, or allele, and the strength of being integrated within MGI as a whole. Information within MGI is freely available on the World Wide Web at: http://www.informatics.jax.org.

REFERENCES

1. Blake, J.A., Eppig, J.T., Bult, C.J., Mouse and Rat Genome Informatics. In *Bioinformatics for Geneticists*, Barnes, M.R. and Gray, I.C., eds., Wiley Press, London, 2003, 119.
2. Eppig, J.T., et al., The Encyclopedia of the Mouse Genome, an update, *Third International Conference on Bioinformatics and Genome Research*, Tallahassee, FL, 1994, 73.

Allele Definition

Mutation(s): Disruption caused by insertion of vector
ES Cell Line: E14TG2a
ES Cell Line Strain: 129P2/OlaHsd
Notes: The gene was disrupted by replacement of the 5' portion of exon 5 with a PGK-neo cassette via homologous recombination. RT-PCR analysis using primers to exons 3-4 and exons 3-5 did not detect gene expression in the midbrain and diencephalons of homozygous mutant animals. In situ hybridization of parasagittal brain sections from homozygous mutants using a probe to the deleted portion of the gene did not detect gene expression. The null allele was further demonstrated by radioimmunoassay (RIA) of brain samples from homozygous mutant animals, which did not detect protein product.
Reference: J:72576

Additional Notes

Associated Phenotype Controlled Terms

Homozygous mice display:

lethality/postnatal
- survival, postnatal lethality:
  - incomplete penetrance

behavior/neurological
- altered anxiety response
  - reduced anxiety in open field emergence, elevated plus maze, and novel-object tests
- increased exploration in new environment
  - open field abnormalities, increased activity
- hyperactivity
- increased vertical activity = rearing and jumping
- decreased grooming behavior

reproductive system
- reduced fertility; homozygous matings resulted in fewer successful pregnanacies than littermate controls

Genetic Background Associated With Phenotype

- Targeting construct was derived from a genomic clone isolated from a mouse 129/SvJ library (source and substrain not specified).
- Chimeric founder mice were outbred to C57BL/6 female.
- Allele was maintained on both a mixed 129P2/OlaHsd x C57BL/6 background and backcrossed five times onto an ICR background.
- Number of generation used for phenotypic analysis was F2 or F3 on the mixed background and N5 for the backcross.

Relationship to Human Genes and Disease

- Orthologous Human Gene: OMIM

Synonyms:

Symbol	Reference
PACAP$^{-/-}$	J:72576

References:

J:72576. Hashimoto H; Shintani N; Tanaka K; Mori W; Hirose M; Matsuda T; Sakaue M; Miyazaki J; Niwa H; Tashiro F; Yamamoto K; Koga K; Tomimoto S; Kunugi A; Suetake S; Baba A. Altered psychomotor behaviors in mice lacking pituitary adenylate cyclase-activating polypeptide (PACAP)., Proc Natl Acad Sci U S A 2001 Nov 6;98(23):13355-60

FIGURE 4.10 A portion of an allele detail page in MGI. This screen shot shows the bottom portion of an MGI Web page describing a targeted mutation in the *Adcyap1* (adenylate cyclase activating polypeptide 1) gene, *Adcyap1^{tm1Aba}*, the first targeted mutation created in the laboratory of Akemichi Baba for this gene. In addition to the data seen in Figure 4.9, data from this portion of the Web page include the molecular details about the allele; phenotype information, using controlled terms from the Mammalian Phentoype Ontology; details on the genetic background associated with the phenotype; relationships to human genes and disease; allele synonyms; and references.

3. Doolittle, D.P., et al., GBASE—The genomic database of the mouse. In *Fifth International Workshop on Mouse Genome Mapping*, Lunteren, Netherlands, 1991, 27.

4. Bult, C.J., et al., The Mouse Genome Database (MGD): Integrating biology with the genome, *Nucleic Acids Res.*, 32, D476, 2004.

5. Hill, D.P., et al., The mouse Gene Expression Database (GXD): Updates and enhancements, *Nucleic Acids Res.*, 32, D568, 2004.

6. Näf, D., et al., The Mouse Tumor Biology database: A public resource for cancer genetics and pathology of the mouse, *Cancer Res.*, 62, 1235, 2002.

7. Bult, C.J., Data integration standards in model organisms: From genotype to phenotype in the laboratory mouse, *Targets*, 1, 163, 2002.

8. Bult, C.J., et al., Mouse Genome Informatics in a new age of biological inquiry. In *Proc. IEEE Int. Symp. Bio-Informatics and Biomedical Engineering*, The Institute of Electrical and Electronics Engineers, Los Alamitos, CA, 2000, 29.

9. Bard, J.B.L., et al., An Internet-accessible database of mouse developmental anatomy based on systemic nomenclature, *Mech Dev.*, 74, 111, 1998.

10. Bult, C.J., et al., A survey of Web resources for basic cancer genetics research, *Genome Res.*, 9, 397, 1999.

11. Snell, G.D., Genes and chromosome mutation. In *Biology of the Laboratory Mouse*, 1st ed., Snell, G.D., ed. McGraw-Hill, New York, 1941, 234.

12. Eppig, J.T., et al., Corralling conditional mutations: A unified resource for mouse phenotypes, *Genesis*, 32, 63, 2002.

13. Eppig, J.T., Algorithms for mutant sorting: The need for phenotype vocabularies, *Mammalian Genome*, 11, 584, 2000.

5 Genetic Resource Databases in Japan

Yukiko Yamazaki

TABLE OF CONTENTS

ESTABLISHMENT OF THE RESOURCE INFORMATION ARCHIVE IN JAPAN

Although we have a long history of development and preservation of mouse genetic resources in Japan, it took quite a long time to establish the resource centers and to construct the database. Recently the Japanese government decided to reinforce the infrastructure of research resource preservation and provision.

In 1997, in order to share genetic resource information, the Center for Genetic Resource Information was opened in the National Institute of Genetics, Mishima, Shizuoka, Japan.

In 1999, the Genetic Resource Committee consisted of researchers in charge of resource preservation for each organism, not only mouse/rat but also other experimental animals, plants, microbes, cells, and DNA clones. This committee was established to exchange and provide information and to discuss current problems and the future direction of Japanese resources.

Since the National BioResource Project (NBRP) was started in 2002, the organization to maintain stocks and information concerning them has become more centralized. Twenty-five groups, nine for animals, eight for plants, and eight for microbes, cells, and DNA clones, and one information center are proceeding with the project. The objectives of NBRP are (1) collection of genetic resources for experimental research, (2) establishment of stock centers for distribution of the genetic resources, and (3) construction of a genetic resource information archive. The Center for Genetic Resource Information is also a member of the National BioResource Project and is expected to play a role as a portal site and an information archive of bioresources in Japan.

MOUSE AND RAT RESOURCE DATABASES IN JAPAN

There are four main centers, Center for Animal Resources and Development (CARD), Kumamoto University; Riken BioResource Center (BRC); Institute of Laboratory Animals, Kyoto University; and National Institute of Genetics (NIG), which are already established. The objectives of each bioresource center are collection, cryopreservation, and distribution of mouse/rat strains, and DNA clones. Each organization provides an online database of its resources and all are available through the Internet.

Current activities of the above resource centers are as follows.

CARD R-DB

The Center for Animal Resources and Development (CARD), Kumamoto University was established in 1998 as one of the core centers for the production, cryopreservation, and supply of genetically engineered mice. The center supplies frozen embryos or sperm and currently has collections of more than 300 strains and stocks of mice. CARD R-DB (Resource Database) provides resource-related information including journal references. The database system was implemented with a direct online data submission and validation system, so that all processes can be done automatically through the Internet. The database is accessible at http://cardb.cc.kumamoto-u.ac.jp/transgenic/. Because nomenclature of the strain, genes, and laboratory code was not always done properly by submitters, all data are checked manually by a specialist before adding data to the database. An online automatic nomenclature system, Godfather, through which submitters can easily find a candidate name according to international nomenclature rules by filling in necessary information, is now being developed.

CARD also encourages submitters to permit their strains and stocks to be part of the IMSR (International Mouse Strain Resources) collections. One third of the CARD colonies are included in the latest version of IMSR. IMSR is available at http://www.informatics.jax.org/imsr/.

Riken GSC/BRC DB

The Riken BioResource Center (BRC) was established in 2001 to provide biological materials, technical services, and educational programs. The Riken BioResource Center is a member of the National BioResource Project (NBRP). Their collection consists of chromosome anomalies/aberrations, congenic, inbred, insertional mutant, and spontaneous mutant mice.

The Riken Genomic Sciences Center (GSC) has advanced the development and establishment of a large number of ENU-induced mutant mice and all resources developed by GSC are distributed through the Riken BRC. MSM mouse BAC clones and their sequences will be available to the public in the near future. Information on Riken GSC and GRC is available from http://www.gsc.riken.go.jp and http://www.brcriken.go.jp/Eng/, respectively.

NIG Mouse DB

The NIG mouse database is more of a knowledge database than a genetic resource database. All NIG collections will be distributed by Riken BRC in the future.

More than 100 laboratory mouse strains, including 27 wild mouse-derived strains and 43 intra-H2-recombinant strains, are maintained and distributed in the National Institute of Genetics, and all information is accessible through the Internet at http://www.shigen.nig.ac.jp/mouse/strain/.

Mouse Polymorphism DB

The Mouse Polymorphism Database is a compilation of microsatellite marker polymorphisms found in inbred mouse strains, most of which were derived from the wild stocks of four *Mus musculus* subspecies, *M.m. domesticus*, *M.m. castaneus*, and *M.m. molossinus*. The inbred strains incorporated in the database are A/J, C57BL/6J, CBA/J, DBA/2J, SM/J, SWR/J, 129Sv/J, MSM/Ms, JF1/Ms, CAST/Ei, NC/Nga, BLG2/Ms, NJL/Ms, PGN2/Ms, SK/CamEi, and SWN/Ms. The number of polymorphic microsatellite loci is over 6000 in all strains and all data are available at http://www.shigen.nig.ac.jp/mouse/mmdbj/. Over 500,000 single nucleotide polymorphism sites (SNPs) were found between MSM and B6 and mapped on the B6 genome. Using the Ensembl system, both microsatellite and SNPs data are integrated into genomic information, and the Ensembl viewer is accessible at the above address.

Rat DB

The National BioResource Project (NBRP) for the Rat, led by Kyoto University, is engaged in the collection, characterization, and distribution of all rat strains.

More than two hundred rat strains, including rat models for human diseases, were collected. The database of all collected strains and acquired data is accessible at http://www.anim.medkyoto-u.ac.jp/nbr/.

FUTURE DIRECTIONS

As an information center of the NBRP, the NIG also provides a one-stop shop (JMSR, Japan Mouse Strain Resource) for mouse genetic resources derived from different organizations in order to help researchers find the appropriate resources quickly and easily. Currently, the JMSR compiles information from four organizations. Each data set consists of minimal items such as (1) strain name, (2) strain group, (3) stock type, (4) genes, (5) model disease name or application field, and (6) the URL address from which full information about the certain strain can be retrieved. We use the free PostgreSQL relational database in order to make the system easily attainable under an Apache open source license. JSMR expects new members to join in the near future. Data distribution is in the XML format with the associated document type definition (DTD) file as an ongoing project. JMSR provides Web access at http://www.shigen.nig.ac.jp/mouse/jmsr/.

6 Computational Pathology: Challenges in the Informatics of Phenotype Description in Mutant Mice

Paul N. Schofield, Jonathan B.L. Bard, Björn Rozell, and John P. Sundberg

TABLE OF CONTENTS

INTRODUCTION:
REPRESENTING MUTANT MOUSE PHENOTYPES

With the completion of its genomic sequence, the mouse has become *par excellence* the prime surrogate organism for the investigation of human biology and disease.[1] The success of the "post-genomic" phase of the mouse genome project now depends on the critical analysis and description of the phenotypes associated with spontaneous and experimentally induced genetic lesions. Through detailed phenotype analysis and compilation of the data obtained it is possible to establish gene function in the intact organism, develop and validate mouse models for specific human diseases, and gain an understanding of common normal developmental and physiological processes (e.g., see Mancuso et al.[2]).

The volume of data being generated from novel genetically engineered mice (GEM), for example, by high-throughput ENU mutagenesis screens,[3] together with the detailed analysis of existing mutant mice, is becoming overwhelming. This means that it has now become crucial to develop computerized methods for storing, recovering, and comparing phenotype data both in text and image formats. This in itself raises the central problem of how phenotype data are recorded. The description of the phenotype of mutant mice has traditionally been presented as a mixture of text (natural language), images, and numerical data, where quantitative assays are applicable. All of these modes present problems when it comes to unambiguous retrieval of data, and the former present rather specific problems for computing/machine processing in that, although natural language is highly expressive, it is very difficult to handle computationally. Furthermore, although images contain a great deal of information, image *files* have no intrinsic semantic content searchable with current technology. Quantitative assays themselves can be problematical, as, without standardization of the assay and assay conditions, it is not possible to compare, for example, two different strains of mice when the assay was carried out by two separate groups using unspecified variants of a given assay technique. This means that phenotype data deposited in a database must have associated with them a full description of the strain, the assay, husbandry conditions, handling, and so forth.

Taking into account all of these issues, it became apparent around the year 2000 that, in order to archive and exploit all of this phenotype data, it was essential to generate a consistent and standardized description framework that was not only appropriate for aspects of mouse phenotype, but for other organisms as well, and a system that would allow semantically unambiguous data to be stored, retrieved, and machine processed.[4]

The approach, which has proved most successful to date, is one based on the construction of ontologies. An ontology can be defined in several ways; however, the most appropriate and useful definition here is that an ontology represents a domain of knowledge structured through formal rules so that it can be interpreted and used by computers.[5] Ontologies are more or less simple hierarchies based on *facts* where a fact is a triad of the form: *Term-Relationship-Term*, and typical relationships for bio-ontologies include: *is-part-of*, *is-a* (*is a member of the set of*), and *descends-from*. Simple ontologies, where a term has only a single parent, can be seen as hierarchies (Linnaen taxonomy is a good example), but, where a term can have more than one parent (and the relationship is directed, or one-way), the ontology is represented by a directed acyclic graph (the Gene Ontology,[6] e.g., includes *is-a* and *part-of* relationships). Thus the mouse developmental anatomy ontology (EMAP[7] [Figure 6.1]) is based on *part-of* relationships whereas the mouse pathology (MPATH) and the cell-type ontologies are based on *is-a* relationships (for further details, see Bard et al.[8]). Key to the use of an ontology is the fact that each of its terms has a unique ID (e.g., orthokeratosis; MPATH:155).

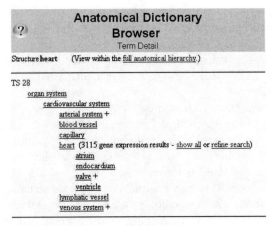

(a)

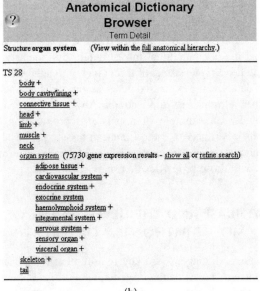

(b)

FIGURE 6.1A The Pathbase Mouse Anatomical Dictionary Browser. The adult anatomical hierarchy may be entered either directly by browsing from the top level which includes topographically and organotypically specified terms (a) and by opening up subterms such as (b) the cardiovascular system (a plus sign next to a term indicates that subterms, i.e., substructures, exist), or by searching for specific terms. Searching for a term in the Anatomical Dictionary, which combines MA (adult) and EMAP (prenatal) anatomy requires a text string and optional specification of Theiler stages. The number of expression results currently annotated to the selected structure is displayed next to the structure name (data from GXD, the mouse gene expression database). Clicking on this link retrieves expression assay result summaries that lead to detailed expression data.

Data associated with ontology terms can be stored independently of that ontology (e.g., in a relational database) provided that the appropriate ID is associated with the data.

The knowledge contained within an ontology and the facts stored in a relational database or knowledge base, are distinct (discussed in Gkoutos et al.[9]): because it contains relationships, an ontology provides a framework through which a knowledge base, such as a data collection of observations relating to individual mice in the real world, can be structured. That is, it contains knowledge about the relationships between defined entities and their attributes. It is clear from this that an ontology can only be successfully applied to a set of data if each of the entities contained in the ontology is fully defined, so that, for example, each aspect of the phenotype of an animal can be coded unambiguously.

Because the terms within an ontology are linked using specific relationships, properties of elements within the ontology can be propagated up or down the hierarchy (depending on the nature of the relationships), this means that ontologies can not only be used for inference and for consistency checking, but as frameworks for guided data retrieval. An obvious example of upwards propagation is provided by the Gene Expression Database (GXD, http://www.informatics.jax.org/mgihome/GXD/aboutGXD.shtml),[10] where searching for the genes expressed in, say, the heart requires that the computational infrastructure first identify in the ontology all the parts of the heart (ventricles, atria, septa, etc.) together with their IDs, then retrieve from the database the expressed genes associated with each ID, and finally add them together, eliminating duplications. Any hierarchy based on the *is-a* relationship provides an example of downwards propagation (known in this specific context as inheritance): in the Linnaean taxonomy, any lower-level term (e.g., species) inherits all the properties of its higher-level terms (e.g., family, genus), together with those properties that mark it as unique.

PATHBASE AND THE DEVELOPMENT OF A MOUSE PATHOLOGY ONTOLOGY

Systematic pathological analysis is a key requirement for the understanding of the response of mice to experimental or genetic manipulation, and it was to meet this need that Pathbase (http://www.pathbase.net) was created[11] to support histopathologic analysis and the dissemination of data on mutant mice. Pathbase stores high-quality images of lesions from genetically engineered and mutant mice, together with spontaneous examples from the background strains, primarily in order to provide a reference and teaching resource, but also to provide a repository for experimental data that could not be published, for example, due to space constraints in journals. For reasons of rapid file transfer, lower-resolution JPGS are currently served from Pathbase, but high-resolution TIFFS, which have greater bit-depth, are available on request.

A key aspect of the development of Pathbase was, of necessity, the generation of a system for the coding of images. Although Pathbase does not aim to hold

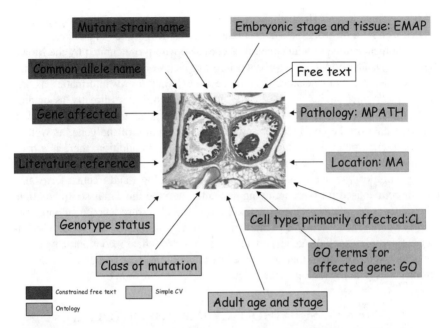

FIGURE 6.2 The Pathbase data model. Images are described by a combination of terms from a set of ontologies, controlled vocabularies, and freetext. Pathbase uses five extant ontologies for the description of images and these allow searches by lesion type, location, embryonic or adult stage, cell type affected, and term from the Gene Ontology (http://www.geneontology.org). Use of these ontologies standardizes queries, facilitates database interoperability, and allows hierarchical searches of the relational database. In addition images are associated with the standardized nomenclature allele name, strain name, and locus ID together with information about the properties of the gene, the mutation, and experimental manipulation.

complete systematic descriptions of the phenotype associated with individual mutations, multiple allelic mutations, or experimental manipulations, it does, nevertheless, allow a user to retrieve images on the basis of, for example, the underlying gene mutation, strain, anatomical location, or pathologic lesion. To allow this, each image in Pathbase has an associated datafile containing the relevant terms from a set of orthogonal ontologies and short controlled vocabularies (CVs) that were recently developed by us and others.[5] These ontologies are shown diagrammatically in Figure 6.2. The combined data make Pathbase a powerful tool for discovery, and increase its value as a specialized repository from which other gene-centric or phenotype-centric databases may obtain information.

In Pathbase the anatomic attributes of the image are coded by using either the time-dependent embryo anatomy ontology developed by the EMAGE consortium (http://genex.hgu.mrc.ac.uk)[12,13] or the mouse adult anatomy (MA: http://www.informatics.jax.org/searches/anatdict_form.shtml) developed for the gene-expression database (GXD) initiated by and curated at The Jackson Laboratory. Other attributes are coded using cell type (CL: Bard et al.[8]), and pathology

(MPATH) ontologies. The pathologic phenotypes, as determined histologically, are coded using a term from the MPATH ontology (see next section). Allele names and strain nomenclature are maintained in accordance with those curated by the Mouse Genome Database (MGD: http://www.informatics.jax.org), using the International Mouse Genetic Nomenclature Committee rules[14] (http://www.informatics.jax.org/mgihome//nomen/strains.shtml). In addition, the records are directly associated with the Gene Ontology (GO) terms[15] of the mutant gene, assigned by MGD[16] to provide data on the cell location and molecular function of the gene, as well as the biological process in which the gene is involved. In addition, there is a brief plaintext file associated with each image. Each image is therefore defined by the intersection of a series of controlled vocabulary, ontology, and freetext terms that are of course used for both archiving data and querying the database. In addition, the concept of a "constrained freetext" term was also introduced to incorporate terms where it is not feasible to maintain a defined ontology or controlled vocabulary. Examples include gene and allele names, which are continuously under review, and those names that are defined by an authority (e.g., the Mouse Genome Informatics group which provides a curated reference resource).

THE MOUSE PATHOLOGY (MPATH) ONTOLOGY

MPATH was developed by a group of veterinary and medical pathologists and anatomists who work extensively with laboratory mice as a description ontology for histological images of tissue lesions generated in response to underlying genetic or extrinsic damage (it is not intended to be used as a complete phenotypic disease description). Images of gross lesions are specifically excluded, as images of whole or parts of animals/organs are considered unsuitable for a freely publicly accessible database. Release 2 of MPATH contains full definitions and contains terms covering all the major classes (580 to date) of pathological lesions, with specific reference to the mouse. These classes are arranged as a hierarchy within a directed acyclic graph (DAG), six levels deep using the is-a relationship with each item having an MPATH ID that can be used for database interoperability and analysis (Figure 6.3). The ontology is accessible both from Pathbase and from the Gene Ontology Consortium's OBO (Open Biological Ontologies) site[15,17] where bio-ontologies are archived (http://www.geneontology.org, http://obo.sourceforge.net/). The structure of the ontology is such that it is easily expanded and modified as new concepts are developed and old ones change, in line with the approach taken by the GO consortium.

The top level of the hierarchy is arranged in categories of general pathology and covers cell and tissue damage, circulatory disorders, developmental and structural defects, growth and differentiation defects, healing and repair, immunopathology, inflammation, and neoplasia. Within these classes, each of the subsidiary terms represents an instance of the parent class and the broad arrangement is designed to be familiar to trained pathologists so as to make its use as intuitive as possible. Many tissue responses are common to multiple anatomical sites and as far as possible the redundancy of specifying a particular

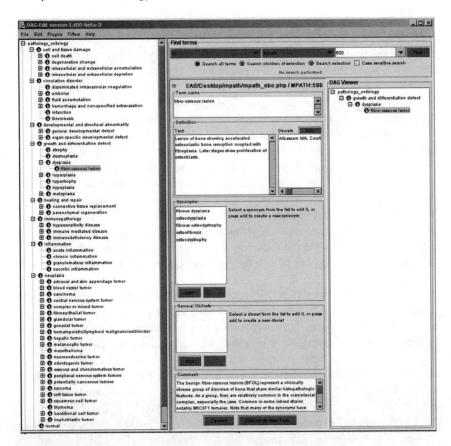

FIGURE 6.3 The MPATH ontology, as displayed in the DAGEDIT ontology editor (see http://www.geneontology.org), currently contains more than 570 pathology terms arranged in a simple tree format that extends to a depth of six levels. Each class can be viewed as a leaf attached to a higher-level node by being "an instance of" that higher level. The top levels of the ontology are arranged as general pathology and terms can be searched by ID or text using an ontology browser on the Pathbase site (http://eulep.anat.cam.ac.uk/Pathology_Ontology/index.php).

response in multiple tissues has been avoided, the additional topographical or anatomical information for each image coming from other orthogonal ontologies, and the coding for each image is therefore combinatorial. In other instances, however, there is either an intrinsic anatomical element embedded in the term or traditional pathology terms include information about the cell type or tissue of origin. This is most frequent with the neoplasias and we felt that such terms were best included in their familiar form.

All the terms available in MPATH have definitions and associated synonyms; the only other resources that achieve this are the NCI CTRM/NCI Thesaurus (http://nciterms.nci.nih.gov/NCIBrowser/Startup.do), where automatic matching pulls together the definition fields from sources such as *Stedman's Medical*

Dictionary,[18] MeSH (http://www.nlm.nih.gov/mesh/meshhome.html), and the NCI glossary. More recently a taxonomy for the developmental lineage classification of neoplasms has been developed by Berman,[19,20] and this contains 122,632 different terms encompassing 5376 neoplasm concepts that also have synonyms listed and are defined from the UMLS and SNOMED-CT which itself contains ICD-O. The primary purpose of this is the indexing of textual information rather than phenotype description, and this is reflected in the rather broad range of expressions included as synonyms. In the case of MPATH, definitions were drawn from these sources and the primary literature, especially where classification has been under recent review. Care was taken to include all of the known synonyms for each term from different nomenclatures and pathology traditions and, in the case of the neoplasias where there is an equivalent human tumor, terms are mapped onto ICD-O (version 3).[21] It is important to note that a panel of 14 European and North American pathologists (The Pathbase Consortium)[11] was gathered to review and debate these terms and definitions to arrive at a consensus during annual meetings, and the ontology will be kept under continuous review by this group.

Because of the importance of cancer to human health, the analysis of mouse tumors has received particular attention. The development of mouse models for human cancers, and the use of direct genetic analysis and molecular profiling have in recent years led to extensive revision of the classification of mouse tumors. Recent work on the rationalization and evidence-based classification of mouse cancers by the MMHC consortium (http://emice.nci.nih.gov/emice/mouse_models) has yielded a new set of diagnostic terms and definitions that are proving to be useful in the classification of mouse lesions and underpin MPATH.[22–30] The Natinal Toxicology Program (NTP) pathology tables on which the *MPHASYS* system (see below) is based (http:// ntp-server.niehs.nih.gov/Main_Pages/NTP_PATH_TBL_PG.html) do however, only give four classes of lymphoma: histiocytic, lymphoblastic, mixed, and undifferentiated. By contrast, the MMHC classification,[22] as reflected in MPATH (Figure 6.4), gives ten fully defined terms for the B cell lymphomas alone, which differ markedly from each other, are defined by molecular and morphological criteria, and are also correlated with specific human lesions. Although the NTP tables are designed for rapid coding of disease states and lesions in response to toxicological challenge, they are not really intended as a system aimed at discovering gene function and producing new models for human diseases.

There is a further point here. Any pathological ontology that aims to be helpful in this wider context must first be kept under continuous expert curation and, second, have each term precisely defined. The MPATH ontology is maintained in accordance with reviews on other mouse cancer types from this body as they are published. One problem faced when developing a pathology ontology for an experimental organism is that new lesions may appear which need to be incorporated within the ontology. A number of lesions in mutant mice appear to have no equivalent in humans. Adrenal cortical dysplasia (*acd*)[31] is one that did not fit easily into a model for a human disease. Others could not be accurately compared until the mutated genes were known in both species. Examples include the nude

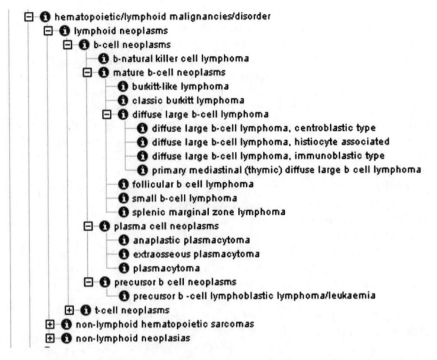

FIGURE 6.4 Segment of the MPATH ontology giving the various categories of lymphoid malignancies.

mouse and human (mutations in *Foxn1* and *FOXN1*, respectively)[32] and hairless (*Hr* and *HR*, respectively).[33–36] Likewise, mutations that seem to be unique in the mouse, such as lanceolate hair (*lah*)[37,38] were later found in the human and a novel gene identified.[39] Contrary examples exist where the creation in mice of a mutation found in humans generates novel lesions that mimic the human disease. This is exemplified by mice in which specific point mutations in p53 (the structural mutant $p53^{R172H}$ and the contact mutant $p53^{R270H}$) give rise to a tumor spectrum similar to Li-Fraumeni syndrome, unlike that found in simple $p53^{-/-}$ mice.[40]

OTHER APPROACHES TO A PATHOLOGY DESCRIPTION FRAMEWORK

The knowledge held within MPATH is formalized within an ontology based on classes (with *is-a* relationships) as described earlier, and so differs from other extant approaches to descriptive pathology in a computer-readable form. These include the "ontologization" of ICD-9 (the *Disease Ontology*) by the Center for Genetic Medicine at Northwestern University (http://diseaseontology.sourceforge.net), and the CTRM knowledge base and NCI thesaurus (http://nciterms.nci.nih.gov/NCIBrowser/Startup.do) that are used to index the NCI Cancer Images Database (http://cancerimages.nci.nih.gov/caIMAGE/index.jsp).

These other approaches, which often include many terms that are etiologically or anatomically predicated, demonstrate how difficult it is to describe lesions and diseases in mice in a complete and universally applicable way. The *Disease Ontology* (DO) is a description framework for human disease containing many terms that are either inappropriate to describe lesions in mutant mice or are redundant for rodents. There are plans to increase the semantic richness of the ontology and to increase granularity. Together with the proposed mapping of the *DO* onto the *Systematized Nomenclature of Medicine* (SNOMED) and *Current Procedural Terminology* (CPT) codes, these will increase the application of this useful development. The collection of controlled vocabularies implemented by the NCI Cancer Images database are related at a high level, but some of its lower-level terms are not related to each other in an intuitive way, and have highly variable levels of granularity (the measure of the fineness of divisions between recognizable subclasses of a term), accuracy, and completeness in different segments.

The NCI Thesaurus Taxonomy has one of the most complete sets of terms and concept relationships of any extant disease description framework, but is largely focused on human disease. Definitions and database cross-references to other classifications are extensive; however, its aim is far broader than describing abnormal histology and it therefore fails to provide an intuitive means of accessing pathological situations.

An alternative approach, known as eVOC, has been produced by the South African National Bioinformatics Institute (http://www.sanbi.ac.za/evoc)[41] and is composed of a set of orthogonal controlled vocabularies, which includes a pathology ontology designed for use in humans. This was developed pragmatically as a tool for the indexing of, for example, cDNA and SAGE libraries and contains 174 pathology terms organized at the higher levels as congenital anomalies, genetic, infectious, inflammatory, neoplastic, metabolic, degenerative, and other disorders. More than 8000 cDNA libraries have so far been annotated using eVOC, but there is currently insufficient coverage of pathology to use the eVOC ontology generally for the description of mutant mice.

More recently the Mouse Phenotype Analysis System (MPHASYS) (http://mphasys.info/), which aims to integrate phenotypic, pathological, and anatomical data from experiments on mutant mice, developed a rich pathology ontology adapted from the National Institutes of Environmental Health Sciences (NIEHS) National Toxicology Program's (NTP) Pathology Code Tables (PCT; http://ntp-server.niehs.nih.gov/Main_Pages/NTP_PATH_TBL_PG.html). This pathology ontology is, as discussed below, helpful as a local recording and data-retrieval tool, and contains a very useful segment describing clinical observations, and short segments for severity qualifiers and sample condition. However, it lacks the granularity and organization required for its use as a discovery tool. For example, all 85 neoplastic processes are listed alphabetically as children of the class: neoplastic processes by morphology, "MORPH_NEOPLASTIC_Neoplasm_NOS", without further classification or relation.

THE DEVELOPMENT OF MPATH: PROBLEMS AND PITFALLS IN THE CLASSIFICATION OF LESIONS

Lesions, both neoplastic and nonneoplastic, have traditionally been classified primarily by morphology and secondarily by anatomic location (hence the term anatomic pathology), leading to what is usually expressed as a hierarchical classification or taxonomic system that emphasizes the similarities between classes of lesion. In the case of neoplastic lesions, these also often reflect the presumed cell type of origin and major variant classes based on observable differentiated features. A hierarchical classification such as this invites the inference of properties of one class or instance of, say, a tumor from its larger parent class. Although we might intuitively expect that it should be possible to produce a logically consistent taxonomy for neoplasia, there is still no overall classification of neoplasia in which the relations between instances and classes are consistent and biologically validated.

The point is well made by Berman[20] when he points out that most "classifications" are generated for an individual organ or organ system, and this is exemplified by the recent work of the NCI MMHC consortium which worked through the mouse system by organ system producing coherent, biologically meaningful taxonomies that for the first time take into account the molecular characteristics of individual tumor classes, such as the expression of particular patterns of genes or cell surface markers, and the presence of specific chromosomal rearrangements or mutations.

Any attempt to merge the classifications produced by the various groups discussed above into a standard ontology for neoplasia that could be used computationally would fail for an interesting reason: most terms within the MPHASYS classification have at least one hidden qualifier associated with them, that of anatomic location.

Adenosquamous carcinoma, for example, appears as class 1.2.3.3 in the classification of lung neoplasia in the mouse; it also appears in the classification of mammary neoplasia, but with the same class identifier. Although it might therefore seem that the two adenosquamous carcinomas are the same tumors, this is not the case as each arises from a different cell of origin.

There are two possible alternative approaches here. Adenosquamous carcinoma of each tissue could either be coded into the ontology directly so that either the two terms become "adenosquamous carcinoma of the mammary gland" and "adenosquamous carcinoma of the lung", or the ontology could be expanded to include anatomic sites so that the term becomes related to the anatomic concept "lung" through an *originated-in-the* relationship and the second has the equivalent relationship with the mammary gland. Neither approach is desirable. (This dilemma arose from the traditional methods used for morphological diagnoses by pathologists where the organ system is defined in conjunction with a series of modifiers that have broad interpretations unless used combinatorially.) The former will generate a huge number of terms, as many lesions, both neoplastic and nonneoplastic, can arise in many tissues, and the latter will generate a deeply tangled ontology[42] with intrinsically different logical types of class and term mixed in together. Both approaches will generate very hard ontologies on a scale that will be difficult to handle at both the computational and end-user levels. There may also be a high degree of redundancy, although whether or not that redundancy is meaningful will remain unclear until we integrate a molecular taxonomy of pathology with pathobiological evidence-based terminology and relations.

There is an underlying issue that this case exemplifies. The same term is used to describe two lesions that, although they originate from similar cell types and share similar morphologic properties, are clearly distinct; as a result, we cannot assume that they will behave similarly, respond similarly to therapeutic intervention, or have the same prognosis. On balance, we therefore believe that the most effective strategy is to strip as much of the anatomical or topographical component from each class or term and to provide this through a second, orthogonal ontology, the combination of the two then generating an accurate code for each lesion. This is the approach that we have implemented in Pathbase, and it works easily, both computationally and for the user. Inevitably there are some lesions that are intrinsically tissue-specific, such as odontoma, nephroblastoma, and so forth, and in MPATH these have been left within the ontology (usually at the bottom level) as they are familiar terms to practicing pathologists and already encapsulate the element of tissue specificity with the accepted term, but this is an uneasy compromise.

It is possible that, in the not too distant future, combinations of large-scale gene expression profiling, both at the genomic (massive scale) and functional (focused scale) levels will provide genomic profiles or signatures of very specific diseases that will further focus these ontologies. Using relatively large-scale quantitative real-time RT PCR methods and focusing on immune regulatory gene sets, we are now able to generate molecular signatures for a variety of immune-mediated diseases in the mouse.[43] These types of technologies will refine the definitions of disease in mice and humans, thereby defining far more precisely their similarities and differences.

DISEASES AND THE PATHOLOGIC RESPONSES

So far we have discussed formalizing pathology within an ontology. Histopathology is of course only one aspect of a disease, and a full description of a disease entity must include, for example, anatomic and temporal qualifiers as well as those of severity and possibly subtype. In many cases, the lesions seen in novel genetically engineered mutant animals differ from spontaneous diseases,[22] whereas many models of human disease have never been seen before in mice, or at least the correlation has not been made. Providing an ontology framework for this is a serious challenge. Although it might seem that incorporating all these descriptors within a single ontology based on existing spontaneous disease or toxicological paradigms might be appropriate, it seems to us that this approach is not only unwieldy but may actually hide potentially important aspects of the novel disease process. We therefore think disaggregation of "disease phenotype" in the sense in which pathologists traditionally make a "diagnosis" (morphologic diagnosis) is highly desirable from the point of view of gene function discovery using mutant mice.

As yet there have been few attempts at a truly disaggregated approach to "disease" description, the closest applicable formalism being that proposed by Gkoutos et al.[9,44] for describing the overall phenotype. The advantage in having general pathology terms available in a description framework, which is based on morphology, is that it becomes possible to associate diseases or lesions with active pathological processes without implying any specific underlying molecular defect. We thus believe that the mixture of both general and specific pathology terms in MPATH will make it a useful tool as part of a much broader description framework for mouse phenotypes or "diseases."

DISCUSSION

The use of ontologies for handling pathology description frameworks within a computational environment is now generally accepted, and the important problems now are those of standardization, acceptability, and use both for coding individual data items and for providing a core component of a framework for handling mutant phenotypes.

MORPHOLOGICAL AND MOLECULAR DESCRIPTIONS FOR PATHOLOGY

A key question is whether the field should use a single ontology for anatomy and lesion morphology or use separate ontologies. In our view, and as we discussed earlier, archiving, searching, and software are all simpler if these aspects are separated. What is less clear is whether it is still appropriate to design an ontology solely on a histological basis. This sort of question is now arising as the pathology of mutant mice is seen as not simply being based on a gene mutation, but on the downstream effects of that mutation (i.e., the tissue response to the underlying biochemical dysregulation), and this may affect growth, apoptosis, and differentiation pathways. The net result is that the relatively crude morphological phenotype displayed in a tissue may be generated by several molecular phenotypes. Furthermore, no gene works independently and genetic polymorphisms between inbred strains result in changes of phenotype when mutated genes are moved to new inbred genetic backgrounds during the creation of congenic strains. This further complicates interpretation and classification but conversely helps explain phenotypic variation with single gene mutations in wild, outbred populations, which best represent most human disease situations.

The ideal tumor classification system appropriate here would be one based on patterns of gene expression or on chromosomal rearrangements combined with histopathology, and, although this is not currently possible, it probably will be in the future. There have been some notable successes in using DNA microarrays for the classification of both mammary and lymphoid neoplasms in humans, and marker gene expression patterns have been linked to pathways of tumorigenesis in the mouse.[45–48] In some cases patterns of gene expression common to all the tumors in a morphologic class have been demonstrated, making it perhaps more likely that terms linked in the ontology through morphologic criteria will also show close biological relatedness. However, exceptions are emerging, as is also the case with histopathology, and the concept of a holy grail of an entirely molecular diagnosis and classification of neoplasias thus still seems some way off. Adequate molecular diagnoses might be thought simply to require larger arrays and more sophisticated analysis; however, there will be an intrinsic problem with the use of gross tumor material for probe preparation in that each sample will contain normal host material, abnormal host material, and, in all likelihood, inflammatory infiltrates of some sort. The response of each host may well be different; here, the genetic profile of the host may limit our ability to find common patterns of gene expression within a tumor class.

An alternative approach is to use immunohistochemistry or *in situ* hybridization to characterize the lesion: the development of panels of antibodies for defining lesion subclasses does provide some promise for building another layer onto the existing morphological criteria for classification, and may avoid the problem of tumor heterogeneity at least with regard to the normal tissue component. This approach proved particularly useful for haematopoietic lesions and those of the mammary gland.[49] However, because the specificity of reagents varies and the numbers and diversity of reagents remain limited, this remains an incomplete and expensive approach. Combining large-scale focused quantitative real-time RT PCR with immunohistochemical and *in situ* approaches provides stronger diagnostic criteria when combined with traditional histopathology, although these remain expensive methods used by relatively few laboratories. As prices decline and availability improves, these will be used more effectively, ultimately defining diseases more accurately and providing better prognostic tools. Integration of these findings into this and other database systems will ultimately expand and validate these approaches.

A richer approach is provided by the use of cluster analysis for tumor classification based on patterns of gene expression,[50] and this raises the possibility of a nonhierarchical taxonomy where related neoplasias may be assigned to the pathology equivalent of clades (see Reference 51 for a discussion concerning the impact of a cladistic approach on the relationship between classification and phylogeny); this approach has the advantage that it makes no assumptions as to tumor ontogeny or anatomical distribution. Such a classification would be theoretically fascinating, but probably extremely hard to implement and use, at least for the next few years. As a result, the use of a classical framework based on morphologic definition will be the most pragmatic way of proceeding until we have a new method for establishing the relationships between tumors consistently at a molecular level. From a more practical perspective, these powerful technologies are currently too expensive and labor intensive for wide-scale use. Until that changes, they will be applicable only to limited and clinically serious disease processes that can be difficult to diagnose accurately and predict reliable treatments for. In short, we have no choice, in practice, but to use morphological criteria for differentiating among pathological phenotypes. This has been the classical approach for nearly 200 years and, as such, a great deal of collective experience translates to practical application at the patient level.

We would naturally believe that the MPATH ontology provides the most appropriate formalism for handling mouse pathology data, but, as discussed above, there are other approaches and, within the limits of each, it will be up to those working in the field to decide which system best meets the needs of the individual group or research community and for which purpose.

HANDLING MUTANT PHENOTYPES

How best to code mutant phenotypes is a problem that is concerning everyone in the field of bio-ontologies, and various solutions are being considered that range from using a single core ontology[52] to requiring sets of orthogonal ontologies.[5,44] Whichever approach is taken, the coding will have to include some aspects of anatomic pathology using morphological diagnoses, if only for the reason that the field will want to search mutant phenotypes using abnormal histology. Our interest here is to consider the extent to which MPATH is adequate in this context, at least for considering abnormal mice.

It is useful here to examine a specific example, technically known as a use-case, that would be part of any phenotypic description of a mutant mouse. Suppose we were to consider tyrosine kinases (GO:0004716), and ask in which genetically engineered mice did *adenosquamous carcinomas* occur. A simple return for this query would, as MPATH now stands, display all the instances in the skin, mammary gland, and lung, and there would be few enough of these for each record to be scanned and evaluated for relevance. But when the search term is *adenocarcinoma,* the return would be far greater as all 18 of the "children" of adenocarcinoma would be displayed (because the *is-a* relationship holds here, each child inherits the properties of the parent). This example demonstrates the power of an ontology, and on the other hand, the importance of having a fairly fine granularity within an ontology if the user is not to be overwhelmed by the size of the response to a query.

Although the responses to such queries give us confidence in the informatics of MPATH, there are nevertheless important caveats that need to be observed, and perhaps the most obvious is the extent to which the lesion in an individual mouse is properly described by the term in the ontology. Here, it is usual for pathologists to provide a single diagnostic term to describe a lesion, but this term is then qualified by a series of adjectives that describe the appearance or behavior of the particular lesion in that individual. Consider the *adenoacanthoma*: the example in a particular mouse is an instance of the class of tumors represented by the term *adenoacanthoma* in the ontology (strictly, all the terms within MPATH or indeed any pathology ontology should be regarded as classes, not instances of occurrences in the real world). When assigning a phenotype to a particular genotype, several individuals are always scored and irrespective of the parameter measured there will be individual-to-individual differences, despite common genetic backgrounds. This means that the assignment of a phenotype—this mouse gets this kind of tumor—is in fact a class property in itself where the class is the "population" of mice studied. It also means that the ontology term is an umbrella for a range of tumor phenotypes, and there really needs to be a means of discriminating among them.

This point is well discussed in Gkoutos et al.[44] who point out that any description of lesions based on ontologies will have to have some qualifier by which individual instances of a closely related class of lesion can be scored uniquely. There are already many formal frameworks for scoring variants within specific classes of tumor, by behavior (Has the basement membrane been breached? Are there signs of local invasiveness? Is there evidence of metastasis?), to nuclear shape, the presence of metaplasia, and so forth. These frameworks are always specific to the particular type of tumor, and attempts to produce ontologies to describe anything more than severity or evidence of spread would immediately run into problems with the range of qualifiers used, notwithstanding the frequent revision of grading on a domain-by-domain basis. This remains a problem, yet the qualifiers and indeed the range of qualifier scores seen for a lesion found on one background might be extremely important in establishing the relationship of the gene to its function. This remains the classic dilemma of merging the classical system of morphological diagnoses, developed over nearly two centuries, with an ontological approach that is practically identical but functionally quite different.

It is hard to see how this problem might be solved. One approach is to adopt a rather broad range of qualifiers such as are used in the NTP code tables, although these are all observational rather than based on known pathobiologically relevant properties and so are unlikely to be of much use in the discovery of novel gene function or new pathways. The other approach is broad brush and not to include any qualifiers, but to give an (it is hoped, typical) image of each instance and leave it to the user to make a decision, perhaps against a standard image. This latter approach has been taken by Pathbase for two reasons: first, it is easy for both archiver and user, a more complex system would be too unwieldy for easy use; and second, in all informatics resources, the coding involves an element of subjectivity (which sequence for a specific gene in Genbank should be viewed as correct), and it is simply impossible to be precise. The informatics field does its best, but, in the limits, its motto has to be *caveat emptor* (let the buyer beware)!

More realistically, pathologists are visually oriented medical specialists. The traditional approach to diagnosis of lesions that the pathologist has never seen before is to read descriptions (expansions of the morphological diagnoses) and evaluate images for similarities and differences (pattern recognition) to the case under investigation. Pathbase addresses all of these needs. The ontology approach addresses the morphological diagnosis. As each term is defined, it addresses, at least superficially, the description. Representative images provide the direct comparisons in a format with which a pathologist is comfortable. These images are each annotated with brief or detailed descriptions, which again meets the text description issues. Curation by the volunteer pathologists is not, of course, always accurate and this has to be moderated by continuous peer reviewing (permitting users to critique images and text and provide feedback to curators who will correct inaccuracies). The key point, of course, is the need for the involvement of the wider scientific community and the pooling of knowledge and experience that will ultimately control the quality of any data resource or description framework adopted.

ACKNOWLEDGMENTS

The authors gratefully acknowledge the enormous contribution of the members of the Pathbase Consortium to the development of MPATH and the ideas articulated in this chapter. The contribution of Professor Wolfgang Goessner will be sorely missed, and this chapter is dedicated to his memory.

This work was supported in part by grants from the European Union (Contract# QLRI-1999-CT-00320, PNS), North American Hair Research Society (JPS, PNS), and the National Institutes of Health (RR173, JPS). BR was supported by grants from the Wallenberg Consortium North for Functional Genomics and by funds from the Karolinska Institute. JBLB acknowledges support from the UK BBSRC.

REFERENCES

1. Sands, A. T., The master mammal, *Nat Biotechnol* 21 (1), 31–32, 2003.
2. Mancuso, M., Pazzaglia, S., Tanori, M., Hahn, H., Merola, P., Rebessi, S., Atkinson, M. J., Di Majo, V., Covelli, V., and Saran, A., Basal cell carcinoma and its development: Insights from radiation-induced tumors in *Ptch1*-deficient mice, *Cancer Res* 64 (3), 934–941, 2004.
3. Auwerx, J., Avner, P., Baldock, R., Ballabio, A., Balling, R., Barbacid, M., Berns, A., Bradley, A., Brown, S., Carmeliet, P., Chambon, P., Cox, R., Davidson, D., Davies, K., Duboule, D., Forejt, J., Granucci, F., Hastie, N., de Angelis, M. H., Jackson, I., Kioussis, D., Kollias, G., Lathrop, M., Lendahl, U., Malumbres, M., von Melchner, H., Muller, W., Partanen, J., Ricciardi-Castagnoli, P., Rigby, P., Rosen, B., Rosenthal, N., Skarnes, B., Stewart, A. F., Thornton, J., Tocchini-Valentini, G., Wagner, E., Wahli, W., and Wurst, W., The European dimension for the mouse genome mutagenesis program, *Nat Genet* 36 (9), 925–927, 2004.
4. Eppig, J. T., Algorithms for mutant sorting: The need for phenotype vocabularies, *Mamm Genome* 11 (7), 584–589, 2000.
5. Bard, J. B. and Rhee, S. Y., Ontologies in biology: Design, applications and future challenges, *Nat Rev Genet* 5 (3), 213–222, 2004.

6. Harris, M. A., Clark, J., Ireland, A., Lomax, J., Ashburner, M., Foulger, R., Eilbeck, K., Lewis, S., Marshall, B., Mungall, C., Richter, J., Rubin, G. M., Blake, J. A., Bult, C., Dolan, M., Drabkin, H., Eppig, J. T., Hill, D. P., Ni, L., Ringwald, M., Balakrishnan, R., Cherry, J. M., Christie, K. R., Costanzo, M. C., Dwight, S. S., Engel, S., Fisk, D. G., Hirschman, J. E., Hong, E. L., Nash, R. S., Sethuraman, A., Theesfeld, C. L., Botstein, D., Dolinski, K., Feierbach, B., Berardini, T., Mundodi, S., Rhee, S. Y., Apweiler, R., Barrell, D., Camon, E., Dimmer, E., Lee, V., Chisholm, R., Gaudet, P., Kibbe, W., Kishore, R., Schwarz, E. M., Sternberg, P., Gwinn, M., Hannick, L., Wortman, J., Berriman, M., Wood, V., de la Cruz, N., Tonellato, P., Jaiswal, P., Seigfried, T., and White, R., The Gene Ontology (GO) database and informatics resource, *Nucleic Acids Res* 32 Database issue, D258–261, 2004.

7. Baldock, R. A., Bard, J. B., Burger, A., Burton, N., Christiansen, J., Feng, G., Hill, B., Houghton, D., Kaufman, M., Rao, J., Sharpe, J., Ross, A., Stevenson, P., Venkataraman, S., Waterhouse, A., Yang, Y., and Davidson, D. R., EMAP and EMAGE: A framework for understanding spatially organized data, *Neuroinformatics* 1 (4), 309–325, 2003.

8. Bard, J., Rhee, S.Y., and Ashburner, M., An ontology for cell types, *Genome Biol* 6, R21, 2005.

9. Gkoutos, G. V., Green, E. C., Mallon, A. M., Hancock, J. M., and Davidson, D., Using ontologies to describe mouse phenotypes, *Genome Biol* 6 (1), R8, 2005.

10. Hill, D. P., Begley, D. A., Finger, J. H., Hayamizu, T. F., McCright, I. J., Smith, C. M., Beal, J. S., Corbani, L. E., Blake, J. A., Eppig, J. T., Kadin, J. A., Richardson, J. E., and Ringwald, M., The mouse Gene Expression Database (GXD): Updates and enhancements, *Nucleic Acids Res* 32 Database issue, D568–5671, 2004.

11. Schofield, P. N., Bard, J. B., Booth, C., Boniver, J., Covelli, V., Delvenne, P., Ellender, M., Engstrom, W., Goessner, W., Gruenberger, M., Hoefler, H., Hopewell, J., Mancuso, M., Mothersill, C., Potten, C. S., Quintanilla-Fend, L., Rozell, B., Sariola, H., Sundberg, J. P., and Ward, A., Pathbase: A database of mutant mouse pathology, *Nucleic Acids Res* 32 Database issue, D512–515, 2004.

12. Bard, J. L., Kaufman, M. H., Dubreuil, C., Brune, R. M., Burger, A., Baldock, R. A., and Davidson, D. R., An Internet-accessible database of mouse developmental anatomy based on a systematic nomenclature, *Mech Dev* 74 (1–2), 111–120, 1998.

13. Bard, J. and Winter, R., Ontologies of developmental anatomy: Their current and future roles, *Brief Bioinform* 2 (3), 289–299, 2001.

14. Mice, C. o. S. G. N. f., Rules for nomenclature of inbred strains, in *Genetic Variants and Strains of the Laboratory Mouse*, 3rd edition. Ed., Lyon MF, R. S., Brown SDM Oxford University Press, Oxford, 1996, pp. 1532–1536.

15. Ashburner, M., Ball, C. A., Blake, J. A., Botstein, D., Butler, H., Cherry, J. M., Davis, A. P., Dolinski, K., Dwight, S. S., Eppig, J. T., Harris, M. A., Hill, D. P., Issel-Tarver, L., Kasarskis, A., Lewis, S., Matese, J. C., Richardson, J. E., Ringwald, M., Rubin, G. M., and Sherlock, G., Gene ontology: Tool for the unification of biology. The Gene Ontology Consortium, *Nat Genet* 25 (1), 25–29, 2000.

16. Hill, D. P., Davis, A. P., Richardson, J. E., Corradi, J. P., Ringwald, M., Eppig, J. T., and Blake, J. A., Program description: Strategies for biological annotation of mammalian systems: Implementing gene ontologies in mouse genome informatics, *Genomics* 74 (1), 121–128, 2001.

17. Consortium, T. G. O., Creating the gene ontology resource: Design and implementation, *Genome Res* 11 (8), 1425–1433, 2001.

18. *Stedman's Medical Dictionary for the Health Professions and Nursing*, 5th ed. Lippincott, Williams and Wilkins, 2005.
19. Berman, J. J., Tumor taxonomy for the developmental lineage classification of neoplasms, *BMC Cancer* 4 (1), 88, 2004.
20. Berman, J. J., Tumor classification: Molecular analysis meets Aristotle, *BMC Cancer* 4 (1), 10, 2004.
21. Fritz, A.C. P., Jack, A., Shanmugaratnam, K., Sobin, L., Parkin, D. M., and Whelan, S., *International Classification of Diseases for Oncology (ICD-O)*, 3rd ed. World Health Organisation, Geneva, 2000.
22. Morse, H. C., 3rd, Anver, M. R., Fredrickson, T. N., Haines, D. C., Harris, A. W., Harris, N. L., Jaffe, E. S., Kogan, S. C., MacLennan, I. C., Pattengale, P. K., and Ward, J. M., Bethesda proposals for classification of lymphoid neoplasms in mice, *Blood* 100 (1), 246–258, 2002.
23. Kogan, S. C., Ward, J. M., Anver, M. R., Berman, J. J., Brayton, C., Cardiff, R. D., Carter, J. S., de Coronado, S., Downing, J. R., Fredrickson, T. N., Haines, D. C., Harris, A. W., Harris, N. L., Hiai, H., Jaffe, E. S., MacLennan, I. C., Pandolfi, P. P., Pattengale, P. K., Perkins, A. S., Simpson, R. M., Tuttle, M. S., Wong, J. F., and Morse, H. C., 3rd, Bethesda proposals for classification of nonlymphoid hematopoietic neoplasms in mice, *Blood* 100 (1), 238–245, 2002.
24. Shappell, S. B., Thomas, G. V., Roberts, R. L., Herbert, R., Ittmann, M. M., Rubin, M. A., Humphrey, P. A., Sundberg, J. P., Rozengurt, N., Barrios, R., Ward, J. M., and Cardiff, R. D., Prostate pathology of genetically engineered mice: Definitions and classification. The consensus report from the Bar Harbor meeting of the Mouse Models of Human Cancer Consortium Prostate Pathology Committee, *Cancer Res* 64 (6), 2270–2305, 2004.
25. Cardiff, R. D., Moghanaki, D., and Jensen, R. A., Genetically engineered mouse models of mammary intraepithelial neoplasia, *J Mammary Gland Biol Neoplasia* 5 (4), 421–437, 2000.
26. Boivin, G. P. and Groden, J., Mouse models of intestinal cancer, *Comp Med* 54 (1), 15–18, 2004.
27. Nikitin, A. Y., Alcaraz, A., Anver, M. R., Bronson, R. T., Cardiff, R. D., Dixon, D., Fraire, A. E., Gabrielson, E. W., Gunning, W. T., Haines, D. C., Kaufman, M. H., Linnoila, R. I., Maronpot, R. R., Rabson, A. S., Reddick, R. L., Rehm, S., Rozengurt, N., Schuller, H. M., Shmidt, E. N., Travis, W. D., Ward, J. M., and Jacks, T., Classification of proliferative pulmonary lesions of the mouse: Recommendations of the Mouse Models of Human Cancers Consortium, *Cancer Res* 64 (7), 2307–2316, 2004.
28. Weiss, W. A., Israel, M., Cobbs, C., Holland, E., James, C. D., Louis, D. N., Marks, C., McClatchey, A. I., Roberts, T., Van Dyke, T., Wetmore, C., Chiu, I. M., Giovannini, M., Guha, A., Higgins, R. J., Marino, S., Radovanovic, I., Reilly, K., and Aldape, K., Neuropathology of genetically engineered mice: Consensus report and recommendations from an international forum, *Oncogene* 21 (49), 7453–7463, 2002.
29. Nikitin, A. Y., Connolly, D. C., and Hamilton, T. C., Pathology of ovarian neoplasms in genetically modified mice, *Comp Med* 54 (1), 26–28, 2004.
30. Holland, E. C., *Mouse Models of Human Cancer*, John Wiley & Sons, Hoboken, NJ, 2004.
31. Beamer, W. G., Sweet, H. O., Bronson, R. T., Shire, J. G., Orth, D. N., and Davisson, M. T., Adrenocortical dysplasia: A mouse model system for adrenocortical insufficiency, *J Endocrinol* 141 (1), 33–43, 1994.

32. Frank, J., Pignata, C., Panteleyev, A. A., Prowse, D. M., Baden, H., Weiner, L., Gaetaniello, L., Ahmad, W., Pozzi, N., Cserhalmi-Friedman, P. B., Aita, V. M., Uyttendaele, H., Gordon, D., Ott, J., Brissette, J. L., and Christiano, A. M., Exposing the human nude phenotype, *Nature* 398 (6727), 473–474, 1999.

33. Panteleyev, A. A., Paus, R., Ahmad, W., Sundberg, J. P., and Christiano, A. M., Molecular and functional aspects of the hairless (*hr*) gene in laboratory rodents and humans, *Exp Dermatol* 7 (5), 249–267, 1998.

34. Panteleyev, A. A., Ahmad, W., Malashenko, A. M., Ignatieva, E. L., Paus, R., Sundberg, J. P., and Christiano, A. M., Molecular basis for the rhino Yurlovo (*hr^{rhY}*) phenotype: Severe skin abnormalities and female reproductive defects associated with an insertion in the hairless gene, *Exp Dermatol* 7 (5), 281–288, 1998.

35. Ahmad, W., Faiyaz µl Haque, M., Brancolini, V., Tsou, H. C., µl Haque, S., Lam, H., Aita, V. M., Owen, J., deBlaquiere, M., Frank, J., Cserhalmi-Friedman, P. B., Leask, A., McGrath, J. A., Peacocke, M., Ahmad, M., Ott, J., and Christiano, A. M., Alopecia universalis associated with a mutation in the human hairless gene, *Science* 279 (5351), 720–724, 1998.

36. Ahmad, W., Panteleyev, A. A., Henson-Apollonio, V., Sundberg, J. P., and Christiano, A. M., Molecular basis of a novel rhino (*hr^{rhChr}*) phenotype: A nonsense mutation in the mouse hairless gene, *Exp Dermatol* 7 (5), 298–301, 1998.

37. Sundberg, J. P., Boggess, D., Bascom, C., Limberg, B. J., Shultz, L. D., Sundberg, B. A., King, L. E., Jr., and Montagutelli, X., Lanceolate hair-J (*lah^J*): A mouse model for human hair disorders, *Exp Dermatol* 9 (3), 206–218, 2000.

38. Montagutelli, X., Hogan, M. E., Aubin, G., Lalouette, A., Guenet, J. L., King, L. E., Jr., and Sundberg, J. P., Lanceolate hair (*lah*): A recessive mouse mutation with alopecia and abnormal hair, *J Invest Dermatol* 107 (1), 20–25, 1996.

39. Kljuic, A., Bazzi, H., Sundberg, J. P., Martinez-Mir, A., O'Shaughnessy, R., Mahoney, M. G., Levy, M., Montagutelli, X., Ahmad, W., Aita, V. M., Gordon, D., Uitto, J., Whiting, D., Ott, J., Fischer, S., Gilliam, T. C., Jahoda, C. A., Morris, R. J., Panteleyev, A. A., Nguyen, V. T., and Christiano, A. M., Desmoglein 4 in hair follicle differentiation and epidermal adhesion: Evidence from inherited hypotrichosis and acquired pemphigus vulgaris, *Cell* 113 (2), 249–260, 2003.

40. Olive, K. P., Tuveson, D. A., Ruhe, Z. C., Yin, B., Willis, N. A., Bronson, R. T., Crowley, D., and Jacks, T., Mutant p53 gain of function in two mouse models of Li-Fraumeni syndrome, *Cell* 119 (6), 847–860, 2004.

41. Kelso, J., Visagie, J., Theiler, G., Christoffels, A., Bardien, S., Smedley, D., Otgaar, D., Greyling, G., Jongeneel, C. V., McCarthy, M. I., Hide, T., and Hide, W., eVOC: A controlled vocabulary for unifying gene expression data, *Genome Res* 13 (6A), 1222–1230, 2003.

42. Rector, A., Wroe, C., Rogers, J., and Roberts, A., Untangling taxonomies and relationships: Personal and practical problems in loosely coupled development of large ontologies, in *First International Conference on Knowledge Capture (K-CAP 2001)* ACM, Victoria, BC, Canada, 2001, pp. 139–146.

43. Akilesh, S., Shaffer, D. J., and Roopenian, D., Customized molecular phenotyping by quantitative gene expression and pattern recognition analysis, *Genome Res* 13, 1719–1727, 2003.

44. Gkoutos, G. V., Green, E. C., Mallon, A. M., Hancock, J. M., and Davidson, D., Building mouse phenotype ontologies, *Pac Symp Biocomput*, 178–189, 2004.

45. Alizadeh, A. A., Ross, D. T., Perou, C. M., and van de Rijn, M., Towards a novel classification of human malignancies based on gene expression patterns, *J Pathol* 195 (1), 41–52, 2001.

46. Dyrskjot, L., Classification of bladder cancer by microarray expression profiling: Towards a general clinical use of microarrays in cancer diagnostics, *Expert Rev Mol Diagn* 3 (5), 635–647, 2003.

47. Lyons-Weiler, J., Patel, S., Becich, M. J., and Godfrey, T. E., Tests for finding complex patterns of differential expression in cancers: Towards individualized medicine, *BMC Bioinformatics* 5 (1), 110, 2004.

48. Macgregor, P. F. and Squire, J. A., Application of microarrays to the analysis of gene expression in cancer, *Clin Chem* 48 (8), 1170–1177, 2002.

49. Mikaelian, I., Blades, N., Churchill, G. A., Fancher, K., Knowles, B. B., Eppig, J. T., and Sundberg, J. P., Proteotypic classification of spontaneous and transgenic mammary neoplasms, *Breast Cancer Res* 6 (6), R668–679, 2004.

50. Wang, J., Bo, T. H., Jonassen, I., Myklebost, O., and Hovig, E., Tumor classification and marker gene prediction by feature selection and fuzzy c-means clustering using microarray data, *BMC Bioinformatics* 4 (1), 60, 2003.

51. Benton, M. J., Stems, nodes, crown clades, and rank-free lists: Is Linnaeus dead? *Biol Rev Camb Philos Soc* 75 (4), 633–648, 2000.

52. Smith, C. L., Goldsmith, C. A., and Eppig, J. T., The Mammalian Phenotype Ontology as a tool for annotating, analyzing and comparing phenotypic information, *Genome Biol* 6 (1), R7, 2005.

7 Biological Methods for Archiving and Maintaining Mutant Laboratory Mice

Martin D. Fray, Peter H. Glenister,
Steven Rockwood, Takehito Kaneko, and
Naomi Nakagata

TABLE OF CONTENTS

INTRODUCTION

Mouse genetics plays a pivotal role in the study of mammalian gene function. However, the explosion of new mouse models from transgenic and mutagenesis programmes threatens to overwhelm existing animal facility space and place an intolerable burden on the resources needed to maintain these animals as conventional breeding colonies.

Several factors have contributed to the burgeoning wealth of genetic information. For example, despite the apparent size of the mouse mutant resource, mutations are known for only 1 to 2% of the total number of mammalian genes. Consequently, several institutions have set out to fill this "phenotype gap"[1] by means of large systematic ENU (N-ethyl N-nitrosourea) mutagenesis pro-grammes.[2] Another factor that is enlarging the mouse genetic resource is the increasingly sophisticated ways the mouse genome can be manipulated, resulting in a flood of transgenic, knock-out, and knock-in strains, all of which require maintenance of some form so they are not lost to future researchers. This situation is due to be compounded by consortia that are proposing to systematically mutagenise the entire mouse genome.[3,4] Continuous breeding is costly, labor intensive, and carries the risk of loss due to impaired reproductive performance, death, or disease. Most laboratories do not have the resources to maintain all the new models they generate as breeding colonies, as well as carry out fundamental research. Therefore alternative strategies such as cryopreservation and freeze-drying need to be considered.

The mouse is the only mammalian species for which there are highly effective cryopreservation protocols for all stages of preimplantation embryos, as well as immature and mature oocytes, primordial follicles, ovarian tissue, and sperm. In contrast, freeze-drying has more limited application and at the present time is restricted to the archiving of spermatozoa. However, despite the range of options now open to the mouse biologist it should be emphasised that the most reliable strategy for archiving a mouse line is still embryo or sperm freezing.

This chapter describes the background and the methodologies most com-monly used in our laboratories for the archiving and long-term storage of embryos and gametes collected from genetically altered mice.

CRYOPRESERVING THE MOUSE

The first methods for the successful cryopreservation of mouse embryos were reported independently by Whittingham et al.[5] and Wilmut.[6] Until that time the only way to maintain a line was through continuous breeding. The potential benefits of archiving mouse stocks for future research were immediately recog-nized by geneticists.[7] Not only could nonessential stocks be economically pre-served and removed from the shelf, but critical mouse models could also be frozen and thus safeguarded against disease, fire, and genetic contamination. Embryo banks were established shortly afterwards at several major laboratories, notably at the MRC Mammalian Genetics Unit, Harwell, United Kingdom; The Jackson Laboratory, Maine, United States; the National Institutes of Health, Maryland, United States; and the Centre for Animal Resources & Development (CARD), Kumamoto University, Kumamoto, Japan

In 1977, Whittingham[8] published a successful protocol for the cryopreserva-tion of unfertilised mouse oocytes. Cryopreservation of mouse ovarian tissue and recovery of fertile mice following the grafting of frozen/thawed tissue had earlier been reported by Parrott in 1958[9] and 1960.[10] These investigations showed that, although some live young were produced, the number of surviving oocytes was low and the reproductive life of the host females was shortened. Further progress on the freezing of mouse ovaries and ovarian tissue remained limited until the 1990s, but the technique is now well documented.[11–14]

Cryopreservation of spermatozoa from a wide variety of species has been commonplace for many years. However, for reasons that remain unclear, mouse spermatozoa proved far more difficult to cryopreserve, and it was only during the 1990s that successful methods appeared in the literature.[15–22] Although various alternative protocols have been published using differing cryoprotective agents, the combination of 18% raffinose and 3% skim milk[17] has been adopted with success by the majority of the cryopreservation laboratories.

DISSEMINATION OF MOUSE MODELS AS FROZEN MATERIALS

Shipping live animals around the world is inconvenient, expensive, and fraught with problems such as incorrect handling and complex customs regulations. In addition, verification of the health status of the donor animals is of great importance because most facilities are extremely reluctant to accept live mice that may carry undesirable microorganisms. Historically, lengthy quarantine procedures coupled with Caesarean rederivation were used to circumvent these problems. However, some pathogens such as mouse hepatitis virus (MHV) may be transmitted via the placenta and are therefore not removed by Caesarean rederivation.[23] In contrast, embryo transfer has been shown to eliminate many viruses including MHV, Reovirus3, Sendai virus, Theiler's encephalomyelitis virus, mouse rotavirus, and mouse adenovirus providing careful washing of the embryos in a sterile medium is carried out before embryo transfer.[24–26]

The availability of liquid nitrogen dry-shipping containers has made the dissemination of mouse stocks as frozen embryos and gametes a relatively simple and safe procedure. These containers absorb liquid nitrogen into a heavily insulated outer layer so the contents remain at liquid nitrogen temperature for periods in excess of two weeks, thus providing a time buffer in the event of transport delays. As the shipment does not contain nitrogen in liquid form it is not classed as "dangerous goods" by the International Air Transport Association (IATA), thus minimising the cost and bureaucracy associated with the shipping process.

EMBRYO CRYOPRESERVATION: AN OVERVIEW

The original protocol described by Whittingham, Leibo, and Mazur in 1972[5] involved slow cooling (<1°C/min) to –80°C with the embryos suspended in a cryoprotectant solution of 1 M dimethylsulphoxide (DMSO). Slow warming (~10°C/min) was found to be necessary for survival together with a stepwise dilution of the DMSO. This protocol is essentially similar to the freezing methods followed by many cryolaboratories, but at the MRC-Harwell we follow a method originally proposed by Renard and Babinet.[27] This protocol uses 1.5 M propylene glycol as cryoprotectant, 1 M sucrose as diluent, and a slow cool at 0.3°C/min to –30°C. The embryos are contained in 0.25 ml plastic insemination straws and require rapid warming (>1000°C/min) for survival.

The major advantages of these slow-cooling methods is that they are very reliable and the relative nontoxicity of the cryoprotectants used allows a certain amount of tolerance when collecting large numbers of embryos for cryopreservation. At the MRC-Harwell we find that in many instances it is possible to recover sufficient live-born offspring to reestablish a breeding colony from one straw of approximately 20 frozen embryos. One disadvantage of this procedure is that precise control of the cooling rate generally requires costly programmable freezing machines, although it is possible to construct cheaper alternatives in most laboratories.[28]

CONTROLLED RATE FREEZING

Methods for controlled rate freezing of mouse oocytes and embryos are usually based around 1–2 M solutions of permeating cryprotectants such as propylene glycol, dimethylsulphoxide DMSO, glycerol, or ethylene glycol. In these protocols the embryos are equilibrated with the cryoprotectant before being cooled, and then frozen under precisely controlled conditions. During this process the cells undergo a characteristic series of volumetric changes brought about by the changing osmotic environment to which they are exposed. For a more detailed explanation of these changes, see Glenister and Rall.[29]

Successful cryopreservation of cells using controlled rate cooling is achieved by the gradual dehydration of the cell as water is drawn from the intracellular compartments by extracellular ice crystallisation combined with ingress of a permeable cryoprotectant. However, the degree of dehydration achieved during freezing is determined by the duration of the controlled cooling phase, that is, the temperature reached before the cells are plunged into liquid nitrogen. What is more, the degrees of dehydration determine how the cells should be thawed. In essence, cells cooled to –70 to –80°C are highly dehydrated and need to be thawed slowly to allow adequate rehydration. On the other hand, embryos cooled to –30°C to –35°C are only moderately dehydrated and can be thawed more rapidly.

METHODS FOR CONTROLLED RATE FREEZING

This method describes the procedure followed at the MRC-Harwell and is based on the method of Renard and Babinet.[27]

Equipment and Reagents

Instruments for Embryo Collection
   133-mm (0.25-ml) plastic semen straws (Planar UK; A201)
   An automatic freezer, for example, Kryo 10 (Planer UK)
Analar grade propylene glycol (Sigma; P-1009)
Analar grade sucrose (Merck; 10247)
35-mm culture dishes (Falcon; 351005)
M2 medium (Sigma; M7167)
Cristaseal or similar for sealing the straws (Hawksley)

Solutions

Embryos are frozen in medium M2 containing 1.5 M propylene glycol (PROH). To prepare this solution, pipette 4.4 ml of M2 into a culture tube and add 0.6 ml PROH. Filter sterilize into a 35-mm culture dish.

A solution of 1.0 M sucrose in M2 is used for dilution. To prepare this solution dissolve 1.71 g sucrose in 5 ml M2 and filter sterilise into a 35-mm culture dish.

FREEZING PROCEDURE

1. Start the programmable freezer.
2. Collect the embryos in M2 and screen carefully for abnormalities at room temperature.
3. When all the embryos have been collected, wash them three times in M2 (2 ml of M2/wash). If more than one stock is being frozen on the same day ensure that each stock is handled with a unique pipette and has its own set of labeled wash dishes.
4. Prepare the straws as shown in Figure 7.1. Take a 133-mm straw and using a metal rod with a stop, push the plug from A to B, that is, to within 75 mm of the open end of the straw. Attach the identifying label to location A.
5. With a marker pen make three marks on each straw (see Figure 7.2). The first mark should be 20 mm from the plug; the second mark should be 7 mm from mark 1, and the third mark 5 mm from mark 2. Use graph paper as a guide, or make the marks on a straw rack.
6. To fill the straws, aspirate sucrose to mark 3 using a 1-ml syringe, then aspirate air so that the sucrose meniscus reaches mark 2. Then aspirate 1.5 M PROH so that the sucrose meniscus reaches mark 1. Finally, aspirate air until the column of sucrose reaches halfway up to the plug and forms a seal with the polyvinyl alcohol.
   Pipette the embryos into a 150-µl drop of 1.5 M PROH and equilibrate for 15 min.

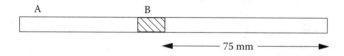

FIGURE 7.1

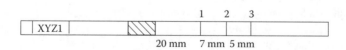

FIGURE 7.2

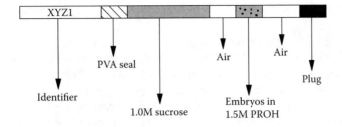

FIGURE 7.3

7. Using a finely drawn pipette, load the embryos into the 1.5 M PROH fraction of each straw (Figure 7.3).
8. Seal the straws using Cristaseal or a heat sealer.
9. Place the straws into the cooling apparatus and cool to –6°C (cooling rate not critical).
10. Wait 5 min to equilibrate.
11. Seed the sucrose fraction by touching it near the plug with a cotton wool bud or the tips of forceps cooled in liquid nitrogen.
12. Wait 5 min; then check that the ice has migrated to the embryo fraction.
13. Cool at 0.3°C/min to –30°C.
14. Plunge the straws directly into liquid nitrogen.
15. Transfer the straws to the appropriate liquid nitrogen storage vessel, taking care at all times to keep the embryo fraction submerged in liquid nitrogen.

THAWING

1. Transfer the straw from the storage vessel to a smaller container of liquid nitrogen.
2. Using forceps, hold the straw near the label for 30 sec in air and then place in water at room temperature until the contents of the straw thaw completely.
3. Wipe the straw with a clean paper tissue.
4. Cut off the plug and also cut through the cotton/PVA seal, leaving about half the seal in place to act as a plunger.
5. Using a metal rod, expel the entire liquid contents of the straw into a 35-mm Falcon dish.
6. Wait 5 min, during which time the embryos will shrink considerably.
7. Transfer the embryos to a drop of M2. They will rapidly take up water and assume normal appearance.
8. Wash the embryos in M2 and then either transfer to the oviducts of a 0.5-day pseudopregnant recipient, or culture to the blastocyst stage in M16 or KSOM and transfer to the uterus of a 2.5-day pseudopregnant recipient.

ADDITIONAL CONSIDERATIONS WHEN FREEZING EMBRYOS AND GAMETES

1. A permanent method of identifying the cryocontainers needs to be considered. At the MRC-Harwell we use self-adhesive printed cryolabels (Brady UK) to label the straws, but other possibilities exist including the use of a fine indelible marker pen.
2. Most samples are lost through the inappropriate handling of samples when they are being removed from secure archives. In order to avoid premature thawing of your samples always place the cryocontainers in liquid nitrogen holding vessels immediately after they have been removed from the archive. The sample transfer should be done as quickly as possible.
3. Liquid nitrogen will enter poorly sealed straws and vials. If this occurs there is a risk that the cryocontainer will explode as the liquid nitrogen rapidly converts to its gaseous phase. In order to prevent injury and the destruction of valuable samples it is important to ensure that the cryocontainers are properly sealed before freezing. Plastic semen straws should be hermetically sealed at both ends.
4. If screw-topped cryotubes are tightened up properly before they are placed in the freezing apparatus they do not present a problem. However, on occasion liquid nitrogen does enter these vials. This problem is easily identified because cryotubes filled with fluid sink when placed in a liquid nitrogen holding vessel. However, the contents of these vials can still be recovered if the liquid nitrogen is allowed to escape by loosening the top of the vial and placing it on the bench top before the designated thawing protocol is started.

FREEZE-DRYING: AN OVERVIEW

Advances in technology have demonstrated that it is possible to obtain live born pups from freeze-dried mouse sperm or even to freeze sperm in the absence of cryoprotectants.[30–33] Unfortunately, the sperm is immotile and it is necessary to use intracytoplasmic sperm injection (ICSI) to fertilise the oocytes and recover live offspring. However, 50% of embryos after transfer were developed to normal live offspring.[32] Moreover, freeze-dried spermatozoa can be stored long-term in a refrigerator (4C), and also normal live offspring were born from samples stored for 1.5 years.[34] The authors are improving their freeze-drying method to increase its efficiency and believe that the freeze-drying of spermatozoa will become more widespread in the future because it circumvents the expense and inconvenience of liquid nitrogen storage and the shipment of frozen samples.

FREEZE-DRYING METHOD

This is the method described by Kaneko et al.[32]

Equipment and Reagents

1.5-ml polypropylene microcentrifuge tube (Fisher Scientific, Pittsburgh, PA)
2-ml long-neck glass ampoules (Wheaton, Millville, NJ)
Freeze-dry systems (Labconco Co., Kansas City, MO)
Liquid nitrogen
Refrigerator

Solution

Sperm are freeze-dried in a solution of 50 mM EGTA, 50 mM NaCl, and 10 mM Tris-HCl in sterile distilled water. Adjust to pH 8.0 by adding 1 M HCl, dropwise. Filter sterilise freeze-drying solution before use.

Preparation of Epididymal Spermatozoa

1. Warm 1 ml of freeze-drying solution to 37°C in a 1.5-ml polypropylene microcentrifuge tube.
2. Remove two caudae epididymides from a male using a pair of small scissors.
3. Puncture the distal end of each epididymis with a pair of small forceps.
4. Put the sperm mass at the bottom of a microcentrifuge tube containing freeze-drying solution.
5. Keep the tube at 37°C for 10 min, allowing spermatozoa to disperse into the buffer solution.
6. Transfer the upper 800 μl of the sperm suspension to another tube.
7. Transfer a 100-μl aliquot of sperm suspension into a long-necked glass ampoule.

Freeze-Drying of Spermatozoa

1. Dip the glass ampoules with sperm suspension in liquid nitrogen.
2. Connect ampoules to the freeze-dry systems and lyophilise for 4 h.
3. Flame-seal each ampoule using a gas burner.
4. Store ampoules at 4°C until use.

Rehydration of Freeze-Dried Spermatozoa and ICSI

1. Open an ampoule and rehydrate spermatozoa by adding 100 μl of sterile distilled water.
2. Collect a small volume of the sperm suspension and mix it with Hepes–CZB medium with 12% PVP.
3. Collect intact-looking spermatozoa with heads still attached to the tails.
4. Draw a single spermatozoon, tail first, into the injection pipette in such a way that its neck (the junction between the head and tail) is at the opening of the pipette.

5. Separate the head from the tail by applying a few piezo-pulses to the neck region.
6. Inject sperm head only into each oocyte.
7. Culture injected oocytes in CZB medium in 5% CO_2 and 95% air at 37°C and allow them to develop to the two-cell stage.

Anticipated Results

In experienced hands more than 70% of the oocytes will survive sperm injection and more than 90% of surviving oocytes should fertilise normally. Most of the fertilised oocytes will develop to the two-cell stage. Again in experienced hands more than 80% of transferred two-cell embryos will implant and 50% of transferred embryos develop to live offspring.

VITRIFICATION: AN OVERVIEW

Numerous protocols exist in the literature that do not require the expensive equipment associated with controlled rate cooling regimes. Such methods include ultra-rapid freezing[35] and vitrification.[36,37] These techniques have the advantage that the only equipment needed is a vessel suitable for holding liquid nitrogen.

Vitrification refers to the physiochemical process by which solutions solidify into a glass state during cooling without crystallisation of the solvent or solute solutions.[29] Successful vitrification is achieved by carefully exposing the embryos to high concentrations of cryoprotectant combined with very rapid cooling (>1000°C/min). In contrast to controlled rate freezing, vitrification is a nonequilibration procedure; that is, the embryos do not equilibrate with the cryoprotectant but are partially dehydrated by it.

The principal advantage of vitrification over controlled rate freezing is that the embryos are not subject to chilling injury or blastomere damage resulting from intra- or extracellular ice crystal formation. However, precise control of the time the embryos are exposed to the vitrification solutions and the temperature of those solutions is required to ensure proper equilibration and to avoid toxicity effects. For a more detailed overview of the biophysics of vitrification, see Glenister and Rall.[29]

Methods for Vitrification of Mouse Embryos

This simple vitrification method describes the procedure followed in the CARD and is commonly used throughout Japan.

Equipment and Reagents

Culture dishes (Nunc; 153066)
Freezing tubes (Nunc; 366656)
Cooling block (IWAKI Co. Ltd; CHT-100, or NALGENE; 1556-0012)

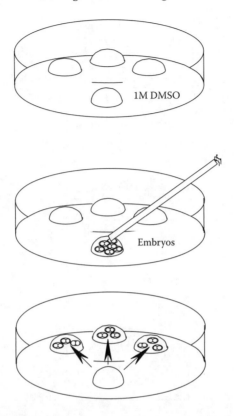

FIGURE 7.4–7.6

Solutions

The cryoprotectant (DAP213) solution used is prepared as a mixture of $2\,M$ DMSO, $1\,M$ acetamide, and $3\,M$ propylene glycol made up in PB1 medium.

PB1 medium is prepared as a solution of NaCL (800.0 mg/100 ml); KCL (20.0 mg/100 ml); $CaCl_2.2H_2O$ (13.2 mg/100 ml); KH_2PO_4 (20.0 mg/100 ml); $MgCL_2.6H_2O$ (10.0 mg/100 ml); $Na_2HPO_4.12H_2O$ (289.8 mg/100 ml); glucose (100.0 mg/100 ml); Na-pyruvate (3.6 mg/100 ml); and penicillin G (100 U/100 ml).

Embryo Cooling

1. Put four drops (100 l/drop) of PB1 containing $1\,M$ DMSO ($1\,M$ DMSO solution) into a dish; see Figure 7.4.
2. Place up to 120 embryos in one of four drops and divide them into three groups. This will be sufficient to freeze up to 40 embryos in each of three freezing tubes (Figure 7.5).
3. Place each group of embryos into a separate drop (Figure 7.6) and then transfer the embryos, in 5 l of $1\,M$ DMSO solution, into a freezing tube see (Figure 7.7).

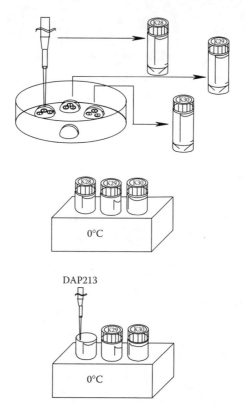

FIGURE 7.7–7.9

4. Put the freezing tube into the block cooler at 0°C and wait 5 min (Figure 7.8).
5. Add 45 l of cryoprotective solution (DAP213) at 0°C into the freezing tube and wait 5 min (Figure 7.9).
6. Immediately fix the freezing tubes to a cane and plunge the samples directly into liquid nitrogen (Figure 7.10).

Thawing

1. After removing the sample from the liquid nitrogen (Figure 7.11), open the cap of the freezing tube and let it stand at room temperature for 30 sec (Figure 7.12).
2. Put 0.9 ml of PB1 containing 0.25 M sucrose into the freezing tube and thaw the sample quickly by pipetting (Figure 7.13).
3. Pour the contents of the freezing tube into a culture dish (Figure 7.14).
4. Recover the embryos (Figure 7.15), and transfer them in a drop of HTF medium (Figure 7.16).
5. After 5 min, take the embryos through two washes of fresh HTF medium. In the case of blastocysts, wash the embryos after 60 min.

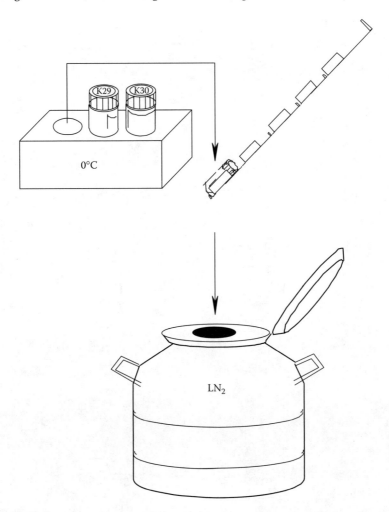

FIGURE 7.10

CRYOPRESERVATION OF MOUSE SPERMATOZOA

A major advantage of sperm freezing is that it is essentially very simple and requires little, if any, expensive equipment. As with the vitrification and ultra-rapid cooling methods for embryos, all that is required is a suitable container to provide a gradient of liquid nitrogen vapor such as a polystyrene box with a lid. A further benefit is that spermatozoa from numerous mice (30 to 50) can be frozen in a single day by a small team of technicians. These factors introduce one of the primary benefits of mouse sperm cryopreservation, which is its ability to economically archive the prodigious numbers of mutants that arise from large-scale ENU mutagenesis programmes and then recover large numbers of progeny by *in vitro* fertilisation.[22] It is now commonplace to cryopreserve spermatozoa

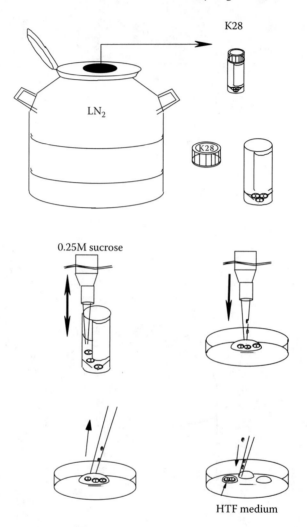

FIGURE 7.11–7.16

and tissue samples for DNA extraction during mutagenesis screen so that mutations identified retrospectively in "gene driven" screens can be recovered by *in vitro* fertilisation (IVF).

The potential to produce large numbers of mice by IVF also allows the rapid production of substantial numbers of back-cross progeny for genetic mapping. Fresh spermatozoa from males carrying new mutations of interest can be collected and one aliquot used for IVF. The remainder can be frozen in order to archive the mutant or, if required, used in further IVF sessions. In this way regular production of >100 or more back-cross progeny can be achieved from individual IVF sessions. A further aspect of these "speed-back-crosses" [22,38] are the economies of scale that result from the simultaneous generation of the back-cross progeny. For example, there are considerable savings in time and animal house space, progeny are weaned and classified on the same day, and DNA samples for mapping can be rapidly and efficiently accessed.

One drawback of sperm cryopreservation is the marked differences in the fertility of frozen/thawed spermatozoa according to the genetic background of the donor mouse.[39,40] C57BL/6 sperm in particular is notoriously difficult to cryopreserve. However, with the judicious use of assisted fertilisation techniques such as partial zona dissection,[41] partial zona digestion (Doyle A, personal communication), and sperm selection,[42] reasonable fertilisation rates can be achieved. For a more detailed overview of cryopreservation of mouse spermatozoa, see Nakagata.[43]

FREEZING PROCEDURE

The method outlined below describes the procedure followed at the MRC-Harwell and is based on the work originally published by Nakagata et al.[18]

Equipment and Reagents

1.8-ml cryotubes (Nunc; Cat No 363401K)
Cryotube rack
1.5-ml microfuge tubes
Raffinose (Sigma; R-7630)
H_2O (Sigma; W-1503)
Skim milk (Difco; 232100)
35-mm culture dishes (Falcon; Cat No 351008)
15-ml tubes (Falcon; Cat No 2097)
Fine watchmakers' forceps
30-g needle/insulin syringe (Beckton Dickinson)
Deep polystyrene box with lid suitable for holding liquid nitrogen
Small dewar of liquid nitrogen
Hot block held at 37°C
Sexually mature male mouse at least 8 weeks old that has not recently mated

Solutions

Cryoprotective agent (CPA) comprising 18% raffinose and 3% skim milk. Place 9 ml H_2O in a screw-top 15-ml Falcon tube and equilibrate to 60°C in a water bath. Add 1.8 g raffinose and dissolve by gentle inversion. Add 0.3 g skim milk and dissolve by gentle inversion. Make up to 10 ml if necessary. Aliquot into microfuge tubes and centrifuge at 14,000 rpm for 10 min. Tip off supernatant and millipore (0.45 μm) filter into cryotubes. Store at –20°C in 1.1-ml aliquots.

CRYOPRESERVATION METHOD

1. To prepare the cooling apparatus, place a platform (e.g., the insert from a Gilson yellow tip box) into the polystyrene box. This acts as a support for the cryotube rack. Carefully pour liquid nitrogen into the polystyrene box to just cover the platform. Place a cryotube rack on top of the platform so that it is suspended in liquid nitrogen vapor. Replace the lid on the polystyrene box and allow it to fill with vapor. Replenish the liquid nitrogen as necessary during the freezing session, but do not allow the level to rise above the platform.

2. Thaw an aliquot of CPA solution for each male and bring to 37°C in the incubator or hot block. Mix by inversion if there is any precipitation.

3. Pipette 1.0 ml CPA into a small culture dish and place on a heated plate at 37°C. Dissect the vasa deferentia and cauda epididymides from the mouse and clean off any excess adipose or vascular tissue. To expel the sperm from the cauda epididymides make 6 or 7 cuts across the structure with a 30-g needle/insulin syringe. Then using fine watchmakers' forceps gently expel the sperm from the vasa deferentia.

4. To disperse the sperm, shake the dish gently for approximately 30 sec and then incubate for 10 min on a hot plate at 37°C.

5. Using a wide-bore pipette tip, aliquot 100 µl into each of 8 cryotubes. Replace the screw cap and tighten to seal the cryotube.

6. Place the cryotubes in the liquid nitrogen vapor phase in the precooled freezing apparatus and leave for 10 min.

7. After 10 min has elapsed plunge cryotubes into liquid nitrogen. Store in a liquid nitrogen refrigerator until required.

Thawing

1. Using forceps, hold the cryotube in air for 30 sec, and then thaw rapidly by placing in a 37°C water bath. Take special care that the tube has not filled with liquid nitrogen before plunging into the water bath (such tubes may explode). If liquid nitrogen is present in the tube, loosen the lid and wait for it to evaporate.

2. Gently agitate the cryotube to mix the sperm. Using a wide-bore pipette tip, take 5- to 10-µl aliquots and pipette gently into 500-µl fertilisation drops (see IVF protocol below). There should be enough sperm for 9×500 µl drops (five dishes).

METHOD USED FOR *IN VITRO* FERTILISATION

A number of protocols have been published for performing IVF with mouse gametes. Although the type of culture medium (MEM, HTF, T6) used may vary among these different methods the basic principles remain constant. The IVF system employed at the MRC-Harwell is as follows.

Equipment and Reagents

30-g needle/insulin syringe (Beckton Dickinson)
37°C incubator gassed with 5% CO_2 in air
35-mm dishes for oviduct collection (Falcon; 351008)
60-mm dish for IVF and embryo culture (Falcon; 353004)
Instruments for dissection
Embryo tested mineral oil (Sigma; M84100)
BSA (Sigma; A-3311)
Penicillin (Sigma; P4687)
Streptomycin (Sigma; S1277)
Mineral oil (Sigma; M8410)
HTF media (Cambrex Bio Services; BE02-021F)

Solutions

HTF media supplemented with BSA at 4.0 g/liter, penicillin at 0.75 g/liter, and streptomycin at 0.5 g/liter.

IVF Setup Procedure

Note: The culture dishes should be prepared on the day before the IVF session is set up.

1. Prepare one 1×35 mm oviduct collection dish for every five females of the donor stock. Each dish should contain 2 ml of the HTF. Incubate the dishes overnight at 37°C, in 5% CO_2 in air.
2. Prepare one 1×60 mm dish/five females of the donor stock. Each dish should contain 1×500 μl drop of HTF for the fertilisation step, plus 4×150 μl drops of HTF for the culturing/washing steps (see Figure 7.17). Overlay the drops with mineral oil and incubate overnight at 37°C, in 5% CO_2 in air.
3. If the freshly harvested sperm is to be used, prepare a sperm dispersal dish by placing a 500-μl drop of HTF into the center of a 60-mm culture dish. Overlay with mineral oil and equilibrate overnight at 37°C, in 5% CO_2 in air (this step can be omitted when using frozen/thawed sperm).

Sperm Sample Preparation—Freshly Harvested Sperm

1. The male should be at least 8 weeks old, and not have been mated within 3 days of sperm collection.
2. Sacrifice the male and dissect out the cauda epididymides and vasa deferentia.

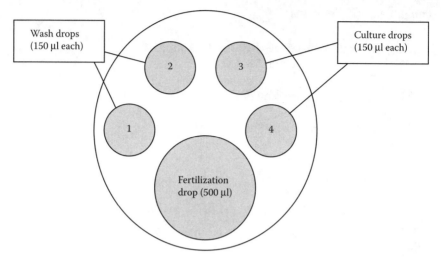

FIGURE 7.17

3. Place the cauda epididymides and vasa deferentia on a clean tissue, and remove as much adipose and vascular tissue as possible, using fine watchmakers' forceps. Work quickly under a dissecting microscope to prevent the material from desiccating.
4. Place the cauda epididymides and vasa deferentia into the 500-µl sperm dispersal drop and extrude the sperm from the vasa deferentia by "walking" your forceps along it.
5. Then, nick the surface of the cauda epididymides 5 or 6 times with the 30-g needle/insulin syringe and return the dish to the incubator for approximately 10 min to disperse the sperm.
6. After the sperm has been dispersed, pipette 5 µl to 10 µl sperm into each fertilisation drop using a wide-bore tip and return the dishes to the incubator.

SPERM SAMPLE PREPARATION—CRYOPRESERVED SPERM

Thaw the sperm sample in accordance with the relevant freezing/thawing protocol and add the recommended volume of sperm to each fertilisation drop. However, if the sperm has been frozen down using the method outlined in this chapter, proceed as follows.

1. Using forceps, hold the cryotube in air for 30 sec or roll the cryotube around on the bench.
2. Thaw the sample by rapidly placing in a 37°C water bath. Take special care that the tube has not filled with liquid nitrogen before plunging into the water bath (such tubes may explode). If liquid nitrogen is present in the tube, loosen the cap and wait for it to evaporate.

3. Once thawed, gently mix the sample by agitating the cryotube and then, using a wide-bore tip, pipette 10 μl of the sperm sample directly into each fertilisation drop and return the dishes to the incubator.

HARVESTING OOCYTES

1. Dissect the oviducts from superovulated females ~14 h after the hCG injection and place them into a dish of HTF (no more than 10 oviducts/dish).
2. Under a dissecting microscope and fine forceps, gently release the cumulus masses into the HTF.
3. Gently draw them up into a wide-bore pipette tip in the minimum amount of HTF and transfer the cumulus masses to one of the fertilisation drops containing spermatozoa.
4. Incubate the dish at 37°C, in 5% CO_2 in air for approximately 5 h to allow fertilisation to occur.
5. Repeat steps 1 to 4 for each fertilisation dish in succession so that the minimum time elapses between killing the oocyte donors and returning the fertilisation dishes to the incubator.

WASHING AND CULTURING THE FERTILISED OOCYTES

1. After incubating the fertilisation drop for ~5 hours remove all the presumptive zygoytes from the fertilisation drop and place them in wash drop 1. Then move the oocytes to wash drop 2 (see Figure 7.17). During this process transfer as little cell debris as possible.
2. Divide the washed oocytes approximately equally between the culture drops 3 and 4.
3. Incubate overnight at 37°C, in 5% CO_2 in air and then harvest the two-cell embryos.
4. Either transfer the two-cell embryos to the oviducts of 0.5-day pseudopregnant foster mothers, or prepare the two-cell embryos for cryopreservation according to a standard protocol, or transfer the embryos into KSOM for prolonged *in vitro* culture.

CRYOPRESERVATION OF MOUSE OOCYTES

Mouse metaphase II oocytes were first cryopreserved by Whittingham[8] by suspending the oocytes in a solution of 1.5 M DMSO and slowly cooling them to –80°C. This method was subsequently improved by the addition of 10% foetal calf serum to the cryoprotectant solution which was thought to prevent changes to the zona pellucida that occur during freezing and thawing.[44] During the intervening years Nakagata[45] published a method for the successful vitrification of mouse oocytes.

Methods for Oocyte Freezing

This procedure can be used to freeze immature and mature mouse oocytes and is based on the method by Carroll et al.[44]

Equipment and Reagents

133-mm (0.25-ml) plastic semen straws (Planar UK; A201)
Cristaseal or similar for sealing the straws (Hawksley)
An automatic freezer, for example, Kryo 10 (Planar UK)
M2 medium (Sigma; M7167)
Analar grade DMSO (BDH; 103234L, MW 78.13, d = 1.099 – 1.101 g/ml)
Foetal calf serum (ICN Flow)
Hyaluronidase (Sigma; H3506)

Solutions

M2, plus 10% FCS. The pH of this solution needs to be adjusted to pH 7.2 to 7.4 with 10 µl 2 M HCl/10 ml medium.
Cryoprotectant: 1.5 M dimethylsulphoxide (DMSO) made up in M2, plus FCS. Make up fresh each day and do not filter sterilise.
Hyaluronidase (300 µg/ml) made up in M2.

Freezing Procedure

1. Prepare the cooling apparatus and set the automatic freezer to 0°C.
2. Using crushed ice and water, prepare an ice slurry in a low-sided container. Cover the ice with foil and lay two glass boiling tubes and a glass embryo dish on top of the foil. The embryo dish should contain 3.0 ml of 1.5 M DMSO in M2, plus FCS. Leave the container in the refrigerator until needed.
3. Collect the oocytes 15–16 hours after the donor females were injected with hCG. Remove cumulus with hyaluronidase and hold at 37°C in M2, plus FCS. It is important not to let the oocytes cool until they are introduced to DMSO.
4. Prepare the straws as follows. First label the straws; then using a marker pen make three marks on each straw: at 17 mm from the plug (mark 1), 87 mm from the plug (mark 2), and 95 mm from the plug (mark 3); see Figure 7.18.

FIGURE 7.18

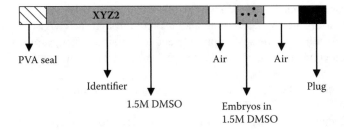

FIGURE 7.19

5. Aspirate 1.5 *M* DMSO to mark 3. Aspirate air so that the meniscus reaches mark 2 and then aspirate 1.5 *M* DMSO so that the meniscus reaches mark 1. Aspirate air until the column of cryoprotectant wets the PVA plug. Place the straws horizontally in the boiling tubes on ice.
6. Place the remaining 1.5 *M* DMSO solution on ice.
7. Place the oocytes in the precooled 1.5 *M* DMSO contained in the cold embryo dish taking care to transfer only a small amount of M2. Leave for 2 min on ice before loading the oocytes into the straws. The oocytes are not fully equilibrated at this stage and appear shrunken.
8. Using a finely drawn pipette, load the oocytes into the large fraction of 1.5 *M* DMSO in each straw; see Figure 7.19. Seal the end of the straw. Return the straw to the glass boiling tube, on ice. Work quickly to minimise the oocytes' exposure to room temperature. Note: Return the watch glass to ice while loading straw.
9. Place the straws containing the oocytes in the cooling apparatus, again trying to minimise the oocytes' exposure to room temperature.
10. Total exposure to 1.5 *M* DMSO at 0–4°C is 12 min.
11. Cool to –8°C at 2°C/min.
12. Wait 5 min to allow the sample to reach –8°C. Then using the tips of fine forceps or cotton wool bud, previously cooled in liquid nitrogen, seed the fraction of 1.5 *M* DMSO near the meniscus.
13. Wait 5 min to allow the ice to migrate to the fraction containing the oocytes; then cool at 0.3°C/min to –40°C. Then plunge the straws into liquid nitrogen.
14. Transfer the samples to a liquid nitrogen storage vessel taking care to keep the straws submerged at all times.

THAWING

1. Transfer the selected straw from the storage vessel to a small dewar filled with liquid nitrogen.
2. Using a pair of forceps, pick the straw up at the end farthest away from the oocytes and hold it in air, at room temperature, for 40 sec. Then place the straw in a water bath set at 30°C until the straw's contents thaw completely.

3. Dry the outside of the straw with a clean paper tissue.
4. Cut off the Cristaseal plug and also cut through the PVA seal leaving about half of the seal in place to act as a plunger. Then carefully expel its contents by pushing the seal down the straw with a metal rod into a 35-mm dish.
5. Incubate the oocytes at room temperature for 5 min; then add 0.8 ml M2, plus FCS to the dish and gently mix.
6. Incubate the oocytes for a further 10 min at room temperature.
7. Transfer the oocytes to a fresh dish of M2, plus FCS, and incubate at room temperature for a further 10 min.
8. Wash the oocytes in a fresh dish of M2, plus FCS and incubate at 37°C until ready to carry out the IVF. This incubation period should not last longer than 30 min.
9. The oocytes should be washed through the appropriate IVF medium before being transferred to the fertilisation drops.

CRYOPRESERVATION OF MOUSE OVARIAN TISSUE

The cryopreservation of ovarian tissue has been possible for many years.[9,10,46] However, the transplantation of frozen/thawed ovarian tissue is not totally reliable and the number of litters and animals produced after transplantation is often reduced compared with control animals. Another complication with ovarian transplantation is the need to transfer the explants into a histocompatible or immunodeficient recipient female. Nevertheless, this technique can be used to cryopreserve female gametes collected at various stages of postnatal development.[12–14, 47]

METHODS FOR OVARY FREEZING

Whole small ovaries or pieces of ovary (1–2 mm^3) can be successfully frozen in 1.5 M DMSO.

Equipment and Reagents

Plastic 1.80-ml cryotubes (Nunc 363401)
An automatic freezer, for example, Kryo 10 (Planar UK)
M2 medium (Sigma; M7167)
Analar grade DMSO (BDH; 103234L, MW 78.13, d = 1.099 – 1.101 g/ml)
Foetal calf serum (ICN Flow)

Solutions

M2, plus 10% foetal calf serum. The pH of this solution needs to be adjusted to pH 7.2 to 7.4 with 10 µl 2 M HCl/10 ml medium.
Cryoprotectant: 1.5 M DMSO made up in M2, plus 10% FCS. Add 1.07 ml DMSO to 8.93 ml M2, plus FCS. Make up fresh each day and do not filter sterilise.

Freezing Procedure

1. Prepare the cooling apparatus and set the automatic freezer to cool and hold at 0°C.
2. Collect the ovaries in M2, plus FCS. Dissect away any attached fat or mesentery. 10-day mouse ovaries can be frozen whole. Adult ovaries should be cut into quarters.
3. Prepare an ice slurry with crushed ice and water in a low-sided container. Cover the ice with foil. Lay a glass embryo dish on top of the foil. Leave the container in the refrigerator until needed.
4. Prepare the cryotubes as follows. Label the cryotubes. Place 0.3 ml 1.5 M DMSO into each cryotube. Embed the cryotubes into the ice slurry.
5. Place 3.0 ml of 1.5 M DMSO solution in the chilled embryo dish and leave on ice.
6. Transfer the ovaries into the 1.5 M DMSO solution in a cold embryo dish taking care to transfer only a small amount of M2. Leave for 2 min on ice before loading the ovaries into the cryotubes.
7. Using fine forceps, load the ovaries into the cryotube (1–4 ovaries/tube). Seal the tube using the screw cap. Place the cryotube back on ice.
8. Total exposure to 1.5 M DMSO at 0–4°C is 20 min.
9. Cool to –8°C at 2°C/min.
10. Wait 5 min to allow the sample to reach –8°C.
11. Using the tips of fine forceps or cotton wool bud that has been pre-cooled in liquid nitrogen, touch the meniscus of the solution in the cryotube to induce ice crystal formation.
12. Wait 5 min, confirm that ice crystals are visible in all the cryotubes, and then cool at 0.3°C/min to
13. –40°C.
14. Cool at 10°C/min to –150°C; then plunge the cyrotubes into liquid nitrogen.
15. Transfer the samples to a liquid nitrogen storage vessel.

Thawing

1. Transfer the frozen cryotubes to a small dewar containing liquid nitrogen. Make sure no liquid nitrogen is trapped in the cryotube (such tubes may explode). If liquid nitrogen is present in the tube, loosen the lid and wait for it to evaporate.
2. Using precooled forceps, hold the cryotube in air at room temperature for 40 sec; then place the vial in water at 30°C until the contents completely thaw.
3. Using fine forceps remove the ovaries from the cryotube. Place into 1–2 ml M2, plus FCS, at room temperature.

4. Gently swirl the dish to mix the solutions. Wait 5 min at room temperature.

5. Transfer the ovaries to M2, plus FCS at room temperature; leave for 5 min. Wash the ovaries in M2, plus FCS, at 37°C. Transplant to a suitable recipient female as soon as possible after thawing.

HOW DO YOU DEFINE AN ADEQUATELY ARCHIVED STOCK?

When setting up an archive it is important to prepare for all eventualities and to this extent it is advisable to split the frozen sample between two separate storage vessels. What is more, these vessels should be housed in separate locations where they are not going to be subject to the same risk factors, such as fire, flood, or loss of liquid nitrogen supply and the like.

Regardless of the method chosen for cryopreservation of embryos and gametes, it is vital to establish that the material has been archived successfully, particularly if the stock is to be killed off. When archiving spermatozoa it is usually sufficient to freeze down around 50 aliquots from four to six males/strain. To confirm the stock can be recovered, a test IVF needs to be performed.

When considering embryo freezing, the archivist should ensure there are enough embryos in the bank to allow for several rederivations before the stock has to be reestablished in order to replenish the bank. At the MRC-Harwell we apply rules that loosely allow for a stock to be rederived ten times or more before embryo replacement becomes essential. For example, where a mutant or transgenic line can be frozen in the homozygous state, then 200+ frozen embryos would normally be considered sufficient to safeguard the line. This rule is also applied to inbred strains because all surviving embryos will be of the required genotype.

Where a gene of interest is segregating in the cross setup to provide embryos for cryopreservation then an appropriate adjustment in the number of embryos to be frozen is calculated. However, it should be stressed that genetic background exerts an effect on the ability of frozen/thawed embryos to develop *in vivo*.[48] Unfortunately, the mechanism for these apparent genotype-specific differences is unknown. Consequently a viability test which compares the number of embryos thawed to the number of pups born with the desired genotype is the only true test of whether sufficient embryos are stored in the bank.

LONG-TERM SURVIVAL OF CRYOPRESERVED GERMPLASM

Several observations suggest that frozen mouse embryos will probably remain viable indefinitely. For example, studies investigating the cumulative effects of background radiation demonstrated that low-dose exposure equivalent to 2000 years of normal background radiation had no detectable effect on embryonic survival, development *in vitro* or *in vivo*, or breeding performance, nor did it increase the incidence of mutations among the offspring.[49–51] What is more, numerous stocks have been reestablished from the Harwell, Jackson, and NIH archives after more than 20 years of storage and none have shown any detrimental effects of extended periods in liquid nitrogen.

Similar studies have not been conducted on cryopreserved spermatozoa, but there is no reason to believe that long-term exposure to liquid nitrogen will have an adverse effect on the fertility of frozen/thawed mouse sperm. Tentative support for this statement is provided by the observation that cattle spermatozoa can retain their fertilising capacity after 37 years of cryopreservation.[52]

MOUSE MUTANT RESOURCES

To enable the scientific community to extract the maximum benefit from cryopreserved embryos and gametes a number of laboratories have developed public access databases (see the appendix). These organisations offer not-for-profit archiving services to the global scientific community, although nominal charges are levied to withdraw materials from the archives. In addition, many laboratories are working together to expand the scope of the International Mutant Strain Resource network (http://www.informatics.jax.org/imsr/index.jsp) in order to establish a one-stop genetic resource for scientists looking for a particular mouse line.

CONCLUDING REMARKS

Cryopreservation of embryos, gametes, and ovarian tissue clearly has a major role to play in mouse genetics research. Although the preservation of mutants by way of oocytes and ovarian tissue is not commonly practiced, embryo freezing is an established procedure that is still the most reliable means of banking mouse strains of major importance and offers the advantage that the entire genotype can be archived. The chief disadvantage is that it is a rather skilled process and may require large numbers of mice to provide enough embryos to securely bank the stock.

Spermatozoa freezing, on the other hand, is relatively simple, rapid, and cheap and has the added advantage that large numbers of animals can be produced from one aliquot of frozen spermatozoa. Consequently, dissemination of mutations as cryopreserved spermatozoa is becoming more widespread, particularly now that more laboratories have developed the necessary skills to deal with frozen spermatozoa and perform IVF. However, the fertility of frozen/thawed spermatozoa is very dependent upon genotype. A further disadvantage of sperm cryopreservation is that only the haploid genotype is conserved and if the original genetic background is required in the future, then appropriate oocytes would also have to be available.

REFERENCES

1. Brown, S.D.M. and Peters, J. Combining mutagenesis and genomics in the mouse—Closing the phenotype gap. *Trends Genet.* 12, 433, 1996.
2. Brown, S.D.M. and Nolan, P.M. Mouse mutagenesis—Systematic studies of mammalian gene function. *Human Mol. Genet.* 7, 10 Review, 1627, 1998.
3. Austin, C.P. et al. The knockout mouse project. *Nat. Genet.* 36, 921, 2004.
4. Auwerx, J. et al. The European dimension for the mouse genome mutagenesis program. *Nat. Genet.* 36, 925, 2004.
5. Whittingham, D.G., Leibo, S.P., and Mazur, P. Survival of mouse embryos frozen to −196°C and −269°C. *Science* 178, 411, 1972.
6. Wilmut, I. The effect of cooling rate, warming rate, cryoprotective agent and stage of development on survival of mouse embryos during freezing and thawing. *Life Sci.* 11, 1071, 1972.
7. Whittingham, D.G. Embryo banks in the future of developmental genetics. *Genetics* 78, 395, 1974.
8. Whittingham, D.G. Fertilization *in vitro* and development to term of unfertilised mouse oocytes previously stored at −196°C. *J. Reprod. Fert.* 49, 89, 1977.
9. Parrott, D.M.V. Fertility of orthotopic ovarian grafts. *Stud. Fert.* 9, 137, 1958.
10. Parrott, D.M.V. The fertility of mice with orthotopic ovarian grafts derived from frozen tissue. *J. Reprod. Fert.* 1, 230, 1960.
11. Carroll, J. and Gosden, R.G. Transplantation of frozen-thawed mouse primordial follicles. *Human Reprod.* 8, 1163, 1993.
12. Candy, C.J., Wood, M.J., and Whittingham, D.G. Long-term fertility of recipients of cryopreserved mouse ovaries. *J. Reprod. Fert. Abstr.* Ser 19, 66, 1997.
13. Gunasena, K.T. et al. Live births after autologous transplant of cryopreserved mouse ovaries. *Human Reprod.* 12, 101, 1997.
14. Sztein, J.M. et al. Cryopreservation and orthotopic transplantation of mouse ovaries: New approach in gamete banking. *Biol. Reprod.* 58, 1071, 1998.
15. Tada, N. et al. Cryopreservation of mouse spermatozoa in the presence of raffinose and glycerol. *J. Reprod. Fert.* 89, 511, 1990.
16. Okuyama, M. et al. *In vitro* fertilization (IVF) and artificial insemination (AI) by cryopreserved spermatozoa in mouse. *J. Fertil. Implant* 7, 116, 1990.
17. Takeshima, T., Nakagata, N., and Ogawa, S. Cryopreservation of mouse spermatozoa. *Exp. Anim.* 40, 493, 1991.
18. Nakagata, N. Cryopreservation of embryos and gametes in mice. *Exp. Anim.* 43, 11, 1994.
19. Songsasen, N., Betteridge, K.J., and Leibo, S.P. Birth of live mice resulting from oocytes fertilized *in vitro* with cryopreserved spermatozoa. *Biol. Reprod.* 56, 143, 1997.
20. Sztein, J.M. et al. Motility of cryopreserved mouse spermatozoa affected by temperature of collection and rate of thawing. *Cryobiology* 35, 46, 1997.
21. Marschall, S. et al. Reliable recovery of inbred mouse lines using cryopreserved spermatozoa. *Mammalian Genome* 10, 773, 1999.
22. Thornton, C.E., Brown, S.D.M., and Glenister, P.H. Large numbers of mice established by *in vitro* fertilization with cryopreserved spermatozoa: Implications and applications for genetic resource banks, mutagenesis screens and mouse backcrosses. *Mammalian Genome* 10, 987, 1999.

23. Harkness, J.E. and Wagner, J.E. In: *Biology and Medicine of Rabbits and Rodents*, 3d ed. London: Lea and Febiger. 1983.

24. Carthew, P., Wood, M.J., and Kirby, C. Elimination of Sendai (parainfluenza type 1) virus infection from mice by embryo transfer. *J. Reprod. Fert.* 69, 253, 1983.

25. Carthew, P., Wood, M.J., and Kirby, C. Pathogenicity of mouse hepatitis virus for preimplantation mouse embryos. *J. Reprod. Fert.* 73, 207, 1985.

26. Morrell, J.M. Techniques of embryo transfer and facility decontamination used to improve the health and welfare of transgenic mice. *Lab. Anim.* 33, 201, 1999.

27. Renard, J.P. and Babinet, C. High survival of mouse embryos after rapid freezing and thawing inside plastic straws with 1-2 propanediol as cryoprotectant. *J. Exp. Zool.* 230, 443, 1984.

28. Wood, M.J., Whittingham, D.G., and Rall, W.F. The low temperature preservation of mouse oocytes and embryos. In: *Mammalian Development, a Practical Approach*, Ed. Monk, M. Oxford: IRL Press, 1987, 255.

29. Glenister, P.H. and Rall, W.F. Cryopreservation of embryos and gametes. In: *Mouse Genetics and Transgenics, a Practical Approach*. Eds. Abbott, C. and Jackson, I. J. Oxford University Press, 2000, Chap. 2.

30. Wakayama, T. and Yanagimachi, R. Development of normal mice from oocytes injected with freeze-dried spermatozoa. *Nat. Biotechnol.* 16, 639, 1998.

31. Kusakabe, H. et al. Maintenance of genetic integrity in frozen and freeze-dried mouse spermatozoa. *Proc. Natl. Acad. Sci. USA* 98, 13501, 2001.

32. Kaneko, T., Whittingham, D.G., and Yanagimachi, R. Effect of pH value of freeze-drying solution on the chromosome integrity and developmental ability of mouse spermatozoa. *Biol. Reprod.* 68, 136, 2003.

33. Kaneko, T. et al. Tolerance of the mouse sperm nuclei to freeze-drying depends on their disulfide status. *Biol. Reprod.* 69, 1859, 2003.

34. Ward, M.A. et al. Long-term preservation of mouse spermatozoa after freeze-drying and freezing without cryoprotection. *Biol. Reprod.* 69, 2100, 2003.

35. Trounson, A., Peura, A., and Kirby, C. Ultra-rapid freezing: A new low-cost and effective method of embryo cryopreservation. *Fert. Steril.* 48, 843, 1987.

36. Rall, W.F. and Wood, M.J. High *in vitro* and *in vivo* survival of day 3 mouse embryos vitrified or frozen in a non-toxic solution of glycerol and albumin. *J. Reprod. Fert.* 101, 681, 1994.

37. Kasai, M. Cryopreservation of mammalian embryos. *Mol. Biotechnol.* 7, 173, 1997.

38. Marschall, S. and Hrabe de Angelis, M. Cryopreservation of mouse spermatozoa, double your mouse space. *Trends Genetics* 15, 128, 1999.

39. Nakagata, N. and Takeshima, T. Cryopreservation of mouse spermatozoa from inbred and F_1 hybrid strains. *Exp. Anim.* 42, 317, 1993.

40. Songsasen, N. and Leibo, S.P. Cryopreservation of mouse spermatozoa II. Relationship between survival after cryopreservation and osmotic tolerance of spermatozoa from three strains of mice. *Cryobiology* 35, 255, 1997.

41. Nakagata, N. et al. The positive effect of partial zona dissection on the *in vitro* fertilizing capacity of cryopreserved C57BL/6J transgenic mouse spermatozoa of low motility. *Biol. Reprod.* 57, 1050, 1997.

42. Bath, M.L. Simple and efficient *in vitro* fertilization with cryopreserved C57BL/6J mouse sperm. *Biol. Reprod.* 68, 19, 2003.

43. Nakagata, N. Cryopreservation of mouse spermatozoa. *Mammalian Genome*, 11, 572, 2000.
44. Carroll, J., Wood, M.J., and Whittingham, D.G. Normal fertilization and development of frozen-thawed mouse oocytes: Protective action of certain macromolecules. *Biol. Reprod.* 48, 606, 1993.
45. Nakagata, N. High survival rate of unfertilized mouse oocytes after vitrification. *J. Reprod. Fert.* 73, 479, 1989.
46. Parkes, A.S. Viability of ovarian tissue after freezing. *Proc. Roy. Soc., B*, 147, 520, 1957.
47. Cox, S.L., Shaw, J., and Jenkin, G. Transplantation of cryopreserved fetal ovarian tissue to adult recipients in mice. *J. Reprod. Fert.* 107, 315, 1996.
48. Dinnyes, A., Wallace, G.A., and Rall, W.F. Effect of genotype on the efficiency of mouse embryo cryopreservation by vitrification or slow freezing methods. *Mol. Repro. Devel.* 40, 429, 1995.
49. Whittingham, D.G., Lyon, M.F., and Glenister, P.H. Long-term storage of mouse embryos at –196°C: The effect of background radiation. *Genet. Res. Camb.* 29, 171, 1977.
50. Glenister, P.H., Whittingham, D.G., and Lyon, M.F. Further studies on the effect of radiation during the storage of frozen 8-cell mouse embryos at –196°C. *J. Reprod. Fert.* 70, 229, 1984.
51. Glenister, P.H., Whittingham, D.G., and Wood, M.J. Genome cryopreservation: A valuable contribution to mammalian genetic research. *Genet. Res. Camb.* 56, 253, 1990.
52. Liebo, S.P., Semple, M.E., and Kroetsch, T.G. *In vitro* fertilisation of oocytes by 37-year-old cryopreserved bovine spermatozoa. *Theriogenology*, 42, 429, 1994.

APPENDIX

Resource centers that can be used to locate a wide variety of mouse strains.

The Centre for Animal Resources Development (CARD)
http://cardb.cc.kumamoto-u.ac.jp/transgenic/index.jsp

The Centre for Modelling Human Disease (CMHD)
http://www.cmhd.ca

The European Mouse Mutant Archive (EMMA)
http://www.emmanet.org

The German ENU-Mouse Mutagenesis Screening Project
http://www.gsf.de/ieg/groups/enu/mutants/index.html

The Frozen Embryo and Sperm Archive (FESA)
http://www.mgu.har.mrc.ac.uk

The International Mouse Strain Resource (IMSR)
http://www.informatics.jax.org/imsr/index.jsp

The Jackson Laboratory
http://www.jax.org/

The National Institutes of Health (NIH)
http://www.nih.gov/science/models/mouse/resources/index.html

The Oak Ridge National Laboratory
http://bio.lsd.ornl.gov/mouse

The Mutant Mouse Regional Resource Centers (MMRRC)
http://www.mmrrc.org

The RIKEN Bioresource Centre
http://www.brc.riken.jp/lab/animal/en

8 Mouse Genetic Resources without Germ Cells: Somatic Cell Nuclear Transfer and ES Technology

Teruhiko Wakayama

TABLE OF CONTENTS

INTRODUCTION

Since first reported in 1997,[1] somatic cell cloning has been demonstrated in several mammalian species. Cloning efficiencies can range from 0 to 20%; however, rates of just 1 to 2% are typical (i.e., one or two live offspring are produced per one hundred initial embryos). Recently, abnormalities in mice cloned from somatic cells have been reported, such as abnormal gene expression in embryos,[2,3] abnormal placenta,[4] obesity,[5,6] or early death.[7] Such abnormalities notwithstanding, success in generating cloned offspring has opened new avenues of investigation, and it provides a valuable tool that basic research scientists have employed to study complex processes such as genomic reprogramming, imprinting, and embryonic development. A number of potential agricultural and clinical applications are also being explored, including therapeutic cloning for human cells, tissues, and organ replacement, and the reproductive cloning of farm animals.

Another such application is the preservation of the genetic resource of mouse strains. Before the success of mouse cloning, the maintenance of mutant mice with severe sterility, such as no germ cells, was impossible even if they showed a number of interesting phenotypes. Unfortunately, the current success rate of mouse cloning from adult somatic cells remains very low and is usually less than 2% of the cloned embryos that develop to term.[4,8] Moreover, the success rate depends on the mouse strain.[9,10] Usually, a hybrid strain is more suitable for mouse cloning than an inbred strain. C57BL/6 and C3H/He are popular mouse strains in mouse genetics but have never been used successfully for producing cloned mice.[9] Therefore, even if somatic cell cloning techniques are available, it remains difficult in practice to apply those techniques to the preservation or rescue of an infertile mouse. Furthermore, even if we succeed in cloning a sterile mouse to preserve its genetic resources, we have to continue cloning to make the next generation because of its sterile phenotype. However, as we have shown previously,[11] the success rate of cloning from a cloned mouse decreases from generation to generation. Thus, there is a limitation to the preservation of the sterile mutant mouse as a genetic resource by the mouse cloning technique.

On the other hand, others and we have shown that nuclear transfer techniques are applicable to the production of embryonic stem (ES) cells from somatic cells.[12-14] These cell lines are considered to harbor the same special and highly desirable properties as those of "conventional" ES cell lines derived from normal embryos produced by fertilization. To distinguish the two, the former are referred to as nuclear transfer ES (ntES) cell lines.[15] The ntES cell lines are expected to be genetically identical to the donor nuclei and promise no immune rejection when used in regenerative medicine.[16-18] Our previous data suggest that the success rate of an ntES cell line establishment from a cloned blastocyst is significantly higher than that of the full-term development of cloned mice.[14] Those ntES cell lines can differentiate into all three germ layers during *in vitro* and sperm or oocyte *in vivo* in chimera mice.[14] Moreover, cloned mice can be obtained from the ntES cell lines by a second nuclear transfer.[14] Taken together, the ntES technique may be a powerful tool not only for regenerative medicine,[15-18] but also for the production of offspring from an infertile mouse.[14]

This review attempts to describe features of cloning and ntES cell technology for the basic study and preservation of genetic resources.

HISTORY OF SOMATIC CELL CLONING AND APPLICATION

The first report of cloned mammals was of mice,[19] which reported and described the production of three mice (one male and two females) by coordinated microinjection of inner cell mass (ICM) cell nuclei into zygotes and immediate removal of the preexisting pronuclei. However, this report was controversial. All subsequent investigators were unable to reproduce the experiment.[20,21] Until the end of the 1980s, only the nuclei of one- or two-cell embryos had been shown to be capable of programming full mouse development following the transfer into an enucleated zygote or two-cell embryo blastomere.[20-23]

The first demonstrably duplicable cloning of a mammal (sheep) was achieved by transferring the eight-cell stage embryonic nuclei into the recipient *unfertilized* eggs (oocytes), instead of one- or two-cell embryos.[24] It is probable that in mammalian cloning the oocyte represents a more viable recipient than the zygote.[25] Thereafter, the techniques of mammalian cloning were developed and more mature cells began to be used as donors. Ten years later, Campbell et al. reported cloned sheep derived from an established embryonic cell line.[26] Using these techniques, the same group later reported the first somatic cell cloned animal.[1]

Cumulina, the first cloned mouse derived from adult somatic cell nuclei, was achieved using a piezo-actuated injection system in 1997.[8] Works on Cumulina and contemporaneously cloned mice demonstrated, for the first time, that mammalian clones were fertile and could produce normal offspring.[8] It was later demonstrated that clones could be obtained from males, using other cell types, such as adult mouse tail-tip cells[4] and newborn Sertoli cells,[27] and that the cloning was not female-specific. It was also shown that clones could be used to generate additional clones.[8,11] Moreover, cloning was performed using cultures from established ES cell lines that had completed G0/G1-phase and were at G2/M-phase.[28] Of the remainder of the acutely isolated cell types tested, all supported embryonic development to the implantation stage.[9] However, development in these cases is typically arrested around embryonic days 6–7.

Recently, adult-derived somatic cell nuclear transfer has been used to produce cloned embryos from which ES cell lines can be derived.[14] These ntES cells exhibit full pluripotency in that they can be made to differentiate along prescribed pathways *in vitro* (e.g., to produce dopaminergic neurons), contribute to the germ line following injection into blastocysts, and support full development following the nuclear transfer.[14] Reports of human ES cell-like cell lines[29,30] coupled to this work on the mouse raise the hope that ntES cells will provide a source of differentiated cells for human autologous transplant therapy, that is, for therapeutic cloning.[16] In 2004, Hwang et al. produced the first human ntES cell line, which was generated from 242 human enucleated oocytes and cumulus cells.[31] Although this in itself does not prove that such cells can be used effectively in regenerative medicine (including the dilemma of persistent rerejection in cases of autoimmune diseases), it does provide an important first proof-of-principle of the feasibility of creating ntES cells using a patient's own ES cells as a source.

TECHNICAL DETERMINANTS IN CLONING

The extent to which the outcome of a nuclear transfer reflects the technical details of the methodologies used versus properties inherent to the biological material is not known. It is probable that many of the differences observed are due to relative differences in operators' skills. The speed of manipulation is an obvious variable in this respect, which may have profound consequences for developmental outcome by reducing trauma to cells and cell fractions during microsurgery. Workers performing experiments side by side with shared samples can produce irreconcilable data.[32] Technical considerations should therefore be

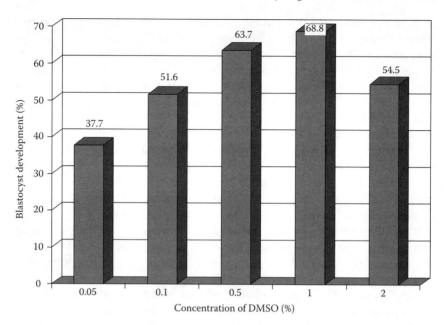

FIGURE 8.1 The effect of DMSO on cloned embryo development. The developmental rate of a cloned embryo to the blastocyst stage was improved when 1% of DMSO was added in the culture medium during the oocyte activation period.

strongly accounted for when interpreting data. This is especially true where a slightly modified method has been employed, such as oocyte activation,[33] donor nucleus introduction,[34,35] and embryo culture medium. The DMSO can exert a negative influence on embryo development, but we discovered that it may enhance the reprogramming (Figure 8.1) of the donor nucleus after the nuclear transfer.[36] Data on nuclear transfer embryos is biologically meaningless without a description of the subsequent developmental potential of the embryos thus generated. The mouse, as a model, lends itself to determining the contribution of technical constraints to the outcome of this development.

THE ABNORMALITY OF CLONED MICE

One striking and possibly diagnostic effect of mouse cloning is that it generates offspring whose placentas are enlarged two- to threefold relative to those of their noncloned counterparts[4,37] (Figure 8.2). The effect is, to our knowledge, universal and occurs irrespective of the nucleus donor source and method of nuclear transfer. This placental enlargement is predominantly due to the exaggerated development of the basal layer: spongiotrophoblasts, giant trophoblasts, and especially glycogen cells.[37] Moreover, placental zonation seems to be disrupted by the apparent invasion of spongiotrophoblasts and glycogen cells into the labyrinthine layer. These abnormalities presumably reflect a dysregulation of gene

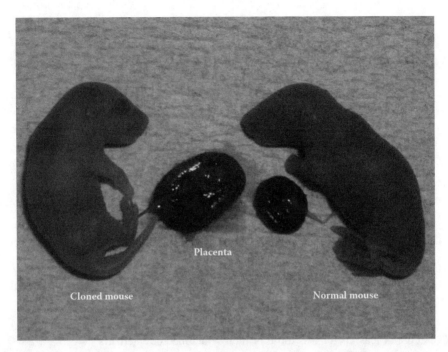

FIGURE 8.2 A comparison of placenta between a cloned and control newborn. All cloned mice were born with two to three times larger placenta (left) than control mice (right).

function in cells of the trophectodermal lineage, which may be accounted for by aberrant reprogramming—the failure to completely reset the differentiative state of the host cell.[37,38]

Abnormalities in cloned offspring have also been reported: when an inbred mouse strain was used as the donor, most cloned mice died soon after birth by respiratory catastrophe,[4,8,9,28] whereas those from a hybrid mouse strain did not.[7] However, with significant variance across strains the postpubertal body mass of clones becomes significantly greater than that of the noncloned controls from approximately eight to ten weeks of age,[5] and those clones have elevated plasma leptin and insulin.[6] The increase in body mass is not a consequence of low activity or increased dietary intake, as behavioral assays do not show any significant difference between the cloned and the controls. Hence, the difference in body mass may be due to lower basal energy expenditure in clones.[6] Furthermore, analysis of the cleavage patterns generated by the methylation-sensitive restriction endonuclease *Not*I in a sample of ~1500 fragments showed that subsets (~0.5%) were differentially methylated in comparisons of samples derived from clones and controls, as well as those derived from different clones.[39] The fact that cloned embryos exhibit such abnormal methylation and yet are able to develop into pups indicates that the success of mammalian cloning is not critically sensitive to the perfect maintenance of the methylation state. Mutually, these findings indicate that a cloned mouse is not a perfect copy of the original mouse in terms of placental development, body weight, or the methylation status of genomic DNA.[39]

However, obesity and enlarged placenta are not heritable and are absent from the progeny of a clone mated with a clone and a clone mated with a wild type.[6] This implies that an imprinting or reprogramming phenomenon that is corrected during gametogenesis causes the nuclear transfer, clone-associated obesity, and enlarged placenta. It is probable that these clone-specific abnormalities are not inherited even after recloning, in which a clone is produced from a clone, because we have previously succeeded in producing six generations of clones.[8,11] The process of aging in animals generated by reiterative cloning is another interesting issue for study. However, at the level of organisms, there is no evidence of age-related attrition of learning ability (as judged by the Morris water maze and Krushinsky tests) or of strength, agility, coordination, or home cage activity compared to the noncloned controls.[5] At the cellular level, we assessed telomere length in the peripheral blood lymphocytes of clones and found no evidence of shortened telomeres in cloned mice.[11] On the contrary, telomeres may exhibit a modest lengthening with each successive generation of clones. Because the mice in this study were sampled simultaneously, an age-related contribution to this increase—in which younger mice have longer telomeres—cannot be excluded.[11] In contrast, these studies on the mouse suggest that cloned animals have telomeres that are at least as long as those of their noncloned counterparts.

Unfortunately, there are not yet any data on the longevity of reiteratively cloned mice compared to age-matched controls. Recently, Ogonuki et al. have reported that many cloned mice die prematurely due to liver necrosis, tumors, and pneumonia.[7] However, some clones survive as long as nonclones, and in fact the first cloned mouse, Cumulina, died aged two years and seven months, which is slightly longer than the lifespan of an average mouse. This indicates that more research into the life expectancy of cloned animals is needed.

GENERATION OF NTES CELL LINES FROM ADULT SOMATIC CELLS

As described above, cloned mice do exhibit some abnormalities. Notwithstanding these abnormalities, cloning has the potential for a number of immediate applications even if the technique itself remains imperfect. One direct application of mouse cloning to biomedicine is exemplified by the recent fusion of stem cell biology with nuclear transfer, with the ultimate goal of therapeutic cloning.[12–14,16,40] Adult-derived somatic cell nuclear transfer can be used to produce cloned embryos from which ES cell lines can be derived. These ntES cells exhibit full pluripotency in that they can be made to differentiate along prescribed pathways *in vitro* (e.g., to produce dopaminergic neurons) and contribute to the germ line following injection into the blastocysts (Table 8.1).[14] The success rate for the establishment of ntES cell lines is much higher than the success rate for the full-term development of cloned mice (Figure 8.3). For example, usually less than

TABLE 8.1
Establishment of ntES Cell Lines From Different Mouse Strains and Mutant Mouse

Mouse Strain	Type of Donor Cell	Sex	No. of Established ntES Cell Lines	% From Morula/Blastocyst	No. Germ Line Transmitting Cell Lines
B6D2F1	Tail	Male	11	26.8	3
	Cumulus	Female	8	16.3	1
B6C3F1	Tail	Male	5	21.7	1
	Tail	Female	1	33.3	1
	Cumulus	Female	27	19.3	ND
BD129F1	Tail	Male	31	58.5	ND
	Cumulus	Female	12	75.0	ND
C3H/He	Tail	Male	1	33.3	1
	Tail	Female	2	16.7	1
	Cumulus	Female	1	5.9	1
C57BL/6	Tail	Male	2	14.3	2
	Cumulus	Female	5	9.8	2
DBA/2	Tail	Male	2	18.2	1
	Cumulus	Female	2	10.0	1
129/Sv	Tail	Male	1	2.7	1
	Cumulus	Female	1	7.7	1
C57BL/6$^{nu/nu}$	Tail	Male	4	4.5	ND
	Tail	Female	5	6.7	2
FVB	Tail	Male	3	23.1	ND
ICR	Tail	Male	1	9.1	ND
	Tail	Female	3	23.1	1
Tg-BDF1	Tail	Male	3	20.0	1
	Cumulus	Female	3	12.0	1
129B6F1	Sertoli	Male	4	26.7	2
Total			138	23.3	

2% of cloned blastocysts (derived from cumulus cells) develop to full term;[8] however, using the same donor cell, about 16% of these cloned blastocysts can give rise to an ntES cell line.[14]

The success rate for the establishment of ntES cell lines is much higher than that for the full-term development of cloned mice, particularly in the inbred strains. For instance, C57BL/6 and C3H/He, which are the major inbred strains in mouse experiments, have never been successfully used for producing cloned mice,[9] but the establishment of ntES cell lines was possible (Wakayama et al., unpublished observation) and the success rate is almost the same as that of fertilized ES cells.[41] This poses the question as to why the success rate for the establishment of an ntES cell line is higher than that of the full-term development

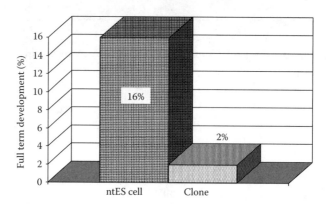

FIGURE 8.3 A comparison between the establishment rate of the ntES cell line and success rate of cloned mice. The establishment rate of ntES cell lines is usually 10 to 20% either in a mouse strain, gender, or cell type. However, the success rate of cloned mice is less than 2%. The most popular mouse strains have never produced cloned mice.

of a cloned mouse. On average, 30 to 50% of nuclear transfer embryos develop to the blastocyst stage; conversely, most of the cloned embryos die immediately after implantation,[42] probably due to incomplete reprogramming. Therefore, even when reprogramming is incomplete, a high percentage of cloned embryos readily develops to the blastocyst stage, suggesting that complete reprogramming in cloned embryos is not required for ntES cell establishment. Once ntES cell lines are established from the tail-tip cells of infertile mutant mice, those cells become immortal and grow infinitely, similar to an ES cell. These cells are easy to preserve under LN2. When the mutant mice are required for an experiment, they will be produced either by cloning of ntES cell nuclei or by germ line transmission of chimera mice.[14]

To further complicate the picture, it appears that even cloned blastocysts destined to die following implantation[42] can contribute to the establishment of ntES cell lines (Figure 8.4). Given this marked discrepancy in developmental competency, it is plausible that the ntES cells derived from cloned embryos doomed to postimplantation lethality are constitutionally different in important ways from ES cells derived from viable embryos produced by natural fertilization.[43] Even the possibility of the existence of these differences makes the application of ntES cells dramatically more problematic in regenerative medicine. Ethical issues aside, many fundamental biological questions must be answered before the clinical use of these cells can be justified even for a purely scientific perspective.[43] Nevertheless, the technology does hold great promise over a long time period. We have already been published demonstrating that the offspring of cloned animals do not share in the abnormalities specific to cloned individuals; the creation of germline cells as part of a regenerative medical protocol or through the germline of chimera mouse should not present any insurmountable biological obstacles.[6,14]

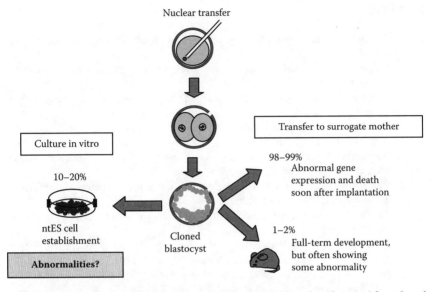

FIGURE 8.4 The abnormalities of ntES cell lines and cloned embryos. After cloned embryos are transferred into surrogate mothers, less than 2% of embryos develop to term; the remainder die soon after implantation. Even the cloned mice that developed to adulthood show some abnormalities. In contrast, when cloned embryos are maintained for more than one month *in vitro*, 10 to 20% of the embryos are able to give rise to ntES cell lines. The extent to which ntES cells harbor abnormalities is unknown.

CONCLUSION

The genetically modified mouse is a very important tool for research in medicine and basic biology, and every year many mutant mice are produced by the ENU mutagenesis method.[44,45] To save space for mouse-breeding facilities, the embryo, oocyte, and spermatozoon of several mouse strains are stored under LN_2 by means of cryopreservation techniques.[46] However, some mutant mice show the infertile phenotype without germ cells. As we have shown herein, the combination of cloning and ntES techniques serves as an alternative method for preserving the genetic resources of invaluable mice, particularly sterile mice, instead of embryo or oocyte/spermatozoon. Recently, oocytes and spermatozoa/haploid cells have been made from ES cells by *in vitro* differentiation.[47–49] Although those studies did not realize fertility and full-term development, it may be possible to make oocyte or spermatozoon from somatic cells via the ntES technique in the near future.

ACKNOWLEDGMENT

We thank D. Sipp for critical and useful comments on the manuscript.

REFERENCES

1. Wilmut, I. et al. Viable offspring derived from fetal and adult mammalian cell. *Nature,* 385, 810, 1997.
2. Boiani, M. et al. Oct4 distribution and level in mouse clones: Consequences for pluripotency. *Genes Dev.,* 16, 1209, 2002.
3. Bortvin, A. et al. Incomplete reactivation of Oct4-related genes in mouse embryos cloned from somatic nuclei. *Development,* 130, 1673, 2003.
4. Wakayama, T. and Yanagimachi, R. Cloning of male mice from adult tail-tip cells. *Nat. Genet.,* 22, 127, 1999.
5. Tamashiro, K.L.K. et al. Postnatal growth and behavioral development of mice cloned from adult cumulus cells. *Biol. Reprod.,* 328, 2000.
6. Tamashiro, K.L. et al. Cloned mice have an obese phenotype not transmitted to their offspring. *Nat. Med.,* 8, 262, 2002.
7. Ogonuki, N. et al. Early death of mice cloned from somatic cells. *Nat. Genet.,* 30, 253, 2002.
8. Wakayama, T. et al. Full-term development of mice from enucleated oocytes injected with cumulus cell nuclei. *Nature,* 394, 369, 1998.
9. Wakayama, T. and Yanagimachi, R. Mouse cloning with nucleus donor cells of different age and type. *Mol. Reprod. Dev.* 58, 376, 2001.
10. Inoue, K. et al. Effects of donor cell type and genotype on the efficiency of mouse somatic cell cloning. *Biol. Reprod.* 69, 1394, 2003.
11. Wakayama, T. et al. Cloning of mice to six generations. *Nature,* 407, 318, 2000.
12. Cibelli, J.B. et al. Transgenic bovine chimeric offspring produced from somatic cell-derived stem-like cells. *Nat. Biotechnol.,* 16, 642, 1998.
13. Munsie, M.J. et al. Isolation of pluripotent embryonic stem cells from reprogrammed adult mouse somatic cell nuclei. *Curr. Biol.,* 10, 989, 2000.
14. Wakayama, T. et al. Differentiation of embryonic stem cell lines generated from adult somatic cells by nuclear transfer. *Science,* 292, 740, 2001.
15. Mombaerts, P. Therapeutic cloning in the mouse. *Proc. Natl. Acad. Sci. U.S.A.,* 100, 11924, 2003.
16. Gurdon, J.B. and Colman, A. The future of cloning. *Nature,* 402, 743, 1999.
17. Rideout, W.M. 3rd et al. Correction of a genetic defect by nuclear transplantation and combined cell and gene therapy. *Cell,* 109, 17, 2002.
18. Barberi, T. et al. Neural subtype specification of fertilization and nuclear transfer embryonic stem cells and application in Parkinsonian mice. *Nat. Biotechnol.,* 10, 1200, 2003.
19. Illmensee, K. and Hoppe, P.C. Nuclear transplantation in *Mus musculus*: Developmental potential of nuclei from preimplantation embryos. *Cell,* 23, 9, 1981.
20. McGrath, J. and Solter, D. Nuclear transplantation in the mouse embryo by microsurgery and cell fusion. *Science,* 220, 1300, 1983.
21. McGrath, J. and Solter, D. Inability of mouse blastomere nuclei transferred to enucleated zygotes to support development *in vitro. Science,* 226, 1317, 1984.
22. Tsunoda, Y. et al. Full-term development of mouse blastomere nuclei transplanted into enucleated two-cell embryos. *J. Exp. Zool.,* 242, 147, 1987.
23. Howlett, S.K., Barton, S.C., and Surani, M.A. Nuclear cytoplasmic interactions following nuclear transplantation in mouse embryos. *Development,* 101, 915, 1987.

24. Willadsen, S.M. Nuclear transplantation in sheep embryos. *Nature,* 320, 63, 1986.
25. Wakayama, T. et al. Nuclear transfer into mouse zygotes. *Nat. Genet.,* 24, 108, 2000.
26. Campbell, K.H.S. et al. Sheep cloned by nuclear transfer from a cultured cell line. *Nature,* 380, 64, 1996.
27. Ogura, A. et al. Production of male cloned mice from fresh, cultured, and cryo-preserved immature Sertoli cells. *Biol. Reprod.,* 62, 1579, 2000.
28. Wakayama, T. et al. Mice cloned from embryonic stem cells. *Proc. Natl. Acad. Sci. U.S.A.,* 96, 14984, 1999.
29. Thomson, J.A. et al. Embryonic stem cell lines derived from human blastocysts. *Science,* 282, 1145, 1998.
30. Shamblott, M.J. et al. Derivation of pluripotent stem cells from cultured human primordial germ cells. *Proc. Natl. Acad. Sci. U.S.A.,* 95, 13726, 1998.
31. Hwang, W.S. et al. Evidence of a pluripotent human embryonic stem cell line derived from a cloned blastocyst. *Science,* 109, 4515, 2004.
32. Perry, A.C.F. and Wakayama, T. Untimely ends and new beginnings in mouse cloning. *Nat. Genet.,* 30, 243, 2002.
33. Kishikawa, H., Wakayama, T., and Yanagimachi, R. Comparison of oocyte-activating agents for mouse cloning. *Cloning,* 1, 153, 1999.
34. Ogura, A. et al. Birth of mice after nuclear transfer by electrofusion using tail tip cells. *Mol. Reprod. Dev.* 57, 55, 2000.
35. Wakayama, S., Cibelli, J.B., and Wakayama, T. Cloning and stem cells. 5, 181, 2003.
36. Wakayama, T. and Yanagimachi, R. Effect of the timing of oocyte activation, cytokinesis inhibitor and DMSO on mouse cloning using cumulus cell nuclei. *Reproduction,* 122, 49, 2001.
37. Tanaka, S. et al. Placentomegaly in cloned mouse concepti caused by expansion of the spongiotrophoblast layer. *Biol. Reprod.,* 65, 1813, 2001.
38. Inoue, K. et al. Faithful expression of imprinted genes in cloned mice. *Science,* 295, 297, 2002.
39. Ohgane, J. et al. DNA methylation variation in cloned mice. *Genesis,* 30, 45, 2001.
40. Kawase, E. et al. Mouse embryonic stem (ES) cell lines established from neuronal cell-derived cloned blastocysts. *Genesis,* 28, 156, 2000.
41. Shoonjans, L. et al. Improved generation of germline-competent embryonic stem cell lines from inbred mouse strains. *Stem Cells,* 21, 90, 2003.
42. Wakayama, T. and Yanagimachi, R. Cloning the laboratory mouse. *Semin. Cell Dev. Biol.,* 10, 253, 1999.
43. Wakayama, T. On the road to therapeutic cloning. *Nat. Biotechnol.,* 22, 399, 2004.
44. Hrabe de Angelis, M.H. et al. Genome-wide, large-scale production of mutant mice by ENU mutagenesis. *Nat. Genet.,* 25, 444, 2000.
45. Nolan, P.M. et al. Systematic, genome-wide, phenotype-driven mutagenesis pro-gramme for gene function studies in the mouse. *Nat. Genet.,* 25, 440, 2000.
46. Nakagata, N. Cryopreservation of mouse spermatozoa. *Mamm. Genome,* 11, 572, 2000.
47. Toyooka, Y. et al. Embryonic stem cells can form germ cells *in vitro. Proc. Natl. Acad. Sci. U.S.A.,* 100, 11457, 2003.
48. Hubner, K. et al. Derivation of oocytes from mouse embryonic stem cells. *Science,* 300, 1251, 2003.
49. Geijsen, N. et al. Derivation of embryonic germ cells and male gametes from embryonic stem cells. *Nature,* 427, 148, 2004.

9 The Present Status of Somatic Cell Cloning

Atsuo Ogura, Kimiko Inoue, Narumi Ogonuki, and Hiromi Miki

TABLE OF CONTENTS

INTRODUCTION

Nuclear transfer cloning of animal individuals began in the 1950s using frogs. Adult cloned frogs were produced from larva (tadpole) cells, but only larva could be produced from adult cells. One reason for this limitation in frog cloning was that adult cell nuclei cannot keep up with the schedule of chromosomal DNA synthesis and division in a limited time (about 90 minutes).[1] In mammals, viable offspring were successfully produced using blastomeres from dividing embryos (blastomere cloning) in the 1980s, and using fetal and adult somatic cells (nuclear transfer cloning) in the 1990s.[2] This demonstrated that the genome of at least some differentiated somatic cells could be almost completely reprogrammed (initialized) in unfertilized oocytes. However, production efficiency and live birth rates differ considerably between blastomere cloning and somatic cell cloning. The reason for this difference has attracted attention on reprogramming abnormalities and limits on plasticity of the somatic cell genome. The general public and researchers alike often equate "somatic cell cloning" with "abnormalities." This is due in large part to unorganized (or indiscriminate) information published in magazines and newspapers. Therefore, this review attempts to present an organized overview of the current status of somatic cell nuclear transfer cloning in mammals. For a more detailed description, the reader is referred to other excellent articles and reviews.

WHY DO ABNORMALITIES OCCUR?

The concept of "genomic reprogramming" is understood insofar as the phenomenon exists, but little is really known about the underlying mechanisms. The reprogramming mechanism in essence exists so that the genome of sperm and oocytes acting as gametes can return to the status of the genome of the fertilized oocyte. Nuclear transfer cloning technology attempts to utilize this mechanism. Thus, one would expect problems to exist in direct reprogramming of the genome of somatic cells, which of course are not germ cells. These often manifest as demethylation abnormalities in reconstructed embryos by nuclear transfer. One would further expect these abnormalities to have long-term effects on abnormal gene expression and phenotypes in cloned fetuses and individuals. Evidence is mounting that errors in genomic reprogramming occur in nuclear transfer cloning, but an understanding of this phenomenon remains a long way off. In pursuing this understanding, researchers are not relying on mammalian oocytes, in which protein analysis can be difficult. Rather, they are using frog oocytes, which have been historically used in nuclear transfer cloning studies, to provide much useful information.

ABNORMALITIES ASSOCIATED WITH SOMATIC CELL NUCLEAR TRANSFER CLONING

Many reports have described abnormalities in cloned individuals. Although much remains unknown, organization and classification of these abnormalities are important to further our understanding of somatic cell nuclear transfer and to set the direction for future research.

STAGES WHEN ABNORMALITIES DEVELOP

Stages of development when abnormalities occur can broadly be divided into the pronuclear, embryonic gene activation (EGA), implantation, placentation, fetal, perinatal, and postnatal periods. Major abnormalities at these stages generally manifest as arrested or delayed development. During pronuclear formation, abnormalities in DNA demethylation, chromatin structure, and nuclear lamina remodeling have been reported, but rarely do these lead to developmental arrest. In earlier experiments with somatic cell cloning, many embryos arrested at the EGA stage, but because of improvements in embryo manipulation technology, this is no longer a problem. Most clone-specific abnormalities occur during the placentation, fetal, perinatal, or postnatal periods. For example, abnormal placentation in mice, cattle, and sheep accounts for many deaths during the placentation–fetal period (discussed later). However, pigs generally have no placental anomalies. There are no reports (to the author's knowledge) about cloned placentas in rats and rabbits. Postnatal abnormalities are often influenced by genetic factors and include obesity and immune dysfunction.[3,4] Obesity has been reported only in mice, whereas immune dysfunction has been found in mice, cattle, pigs, and goats. This poses an interesting question on whether or not the immune dysfunction may be due to some common underlying factor.

CLONE-SPECIFIC ABNORMALITIES

Somatic cell nuclear transfer cloning, like other reproductive engineering technology (e.g., *in vitro* fertilization, microinsemination), involves oocyte manipulation and embryo *in vitro* culture. These procedures themselves have been reported to cause abnormal expression of genes required for development. For example, the "large offspring syndrome" seen in cattle and sheep, associated with decreased expression of the imprinted gene *IGF2R*, is due to embryo *in vitro* culture, which is used in both *in vitro* fertilization and cloning.[5] Genomic imprinting memory in donor cells is basically maintained even after nuclear transfer,[6] so if imprinting memory in the donor cell is already abnormal, this will lead to abnormal gene expression in the cloned embryo. Therefore, to further our understanding of nuclear transfer cloning, one must carefully examine whether gene expression and phenotypic abnormalities are indeed clone-specific.

Heteroplasmy (a mixture of different mitochondrial DNA in cells), although rarely expressed phenotypically, is a clone-specific abnormality. In normal fertilization, the sperm midpiece mitochondria are degraded in the oocyte by the ubiquitin system, whereas in cloning, the donor cell mitochondria are transferred to the embryo. When somatic cell cloning was first performed, donor cell mitochondria were not thought to be transmitted to cloned individuals due to the overwhelming amount of oocyte mitochondria, but with improved DNA detection methods, it is now clear that some, albeit a small amount, of mitochondria are transmitted. Our laboratory has confirmed this phenomenon in mice. Of further interest is what appears to be organ-specific distribution, with significant accumulation of donor mitochondrial DNA in the liver.[7]

TYPICAL OR VARIED ABNORMALITIES

Even though an abnormality is clone-specific, it may also be a "typical" or "varied" abnormal finding. The most well-known typical abnormalities are those related to placental morphology. In mice, there is expansion of the spongiotrophoblast layer, which is derived from diploid trophoblast cells, leading to placentomegaly.[8] In cattle, a reduction in size of the allantochorion leads to placental dysfunction. Obesity is also a typical finding in cloned agouti mice (especially females).[3] These placental anomaly and obesity phenotypes are not transmitted to offspring and thus have been shown to be epigenetic.[3,9] Varied abnormalities are typically associated with various gene expression levels. These can now be readily identified using DNA arrays and have been discussed in many other reports. Telomere length variations have been found among cloned individuals as well as in different organs within the same individual. Even in cloned embryos, telomere length increases during transition from a morula to blastocyst.[10] Variations may also occur in subsequent stages of differentiation and development. Of interest is the normal telomere length found in germ cells (spermatozoa).[11]

What is the significance of these typical and varied abnormalities? Probably, when the somatic cell genome is reprogrammed in the oocyte, some areas always remain with errors, whereas other areas, at a certain rate, are accurately reprogrammed. The combination of both types of abnormalities leads to a decreased efficiency of somatic cell cloning. Interestingly, when a chimeric embryo is produced from two cloned embryos, the birth rate improves significantly.[12] This finding suggests that the two embryos may act to compensate for any variations in gene expression in the chimeric embryo. On the other hand, chimeras with tetraploid embryos for placental rescue do not improve the production efficiency of viable clones. Thus, developmental arrest of cloned embryos cannot be explained by placental insufficiency alone (unpublished data).

Placental abnormalities, particularly in mice, are being investigated in detail with respect to gene expression, DNA methylation, and histopathological features. As mentioned previously, these are clone-specific and typical findings. It is hoped that this will lead to a better understanding of somatic cell cloning. Most research is currently being conducted on placenta at term, but future research to elucidate underlying mechanisms is expected using placenta at earlier stages.

EFFECTS OF DONOR CELLS

Investigation of factors causing low birth rates and a high incidence of abnormalities in cloning is not an easy task, but finding conditions leading to an improvement in these parameters can provide important clues. Using mice with uniform genetic and biological backgrounds can be effective in reaching these goals. We have conducted large-scale studies using mouse cloning technology to examine the effects of donor cell genetic background and cell type on cloning efficiency. Statistical analysis of the data shows that the combination of genetic background and cell type has a significant effect on production efficiency of viable clones.[13] With a combination of newborn Sertoli cells and (C57BL/6 × 129/Sv-ter) F1 genetic background, a constant birth rate of about 10% per embryo transfer was demonstrated. In microinsemination using postmeiotic round spermatids (with the same genome as spermatozoa), the birth rates are 10 to 25%. This suggests that the genotype and donor combined genome may be almost normally reprogrammed by nuclear transfer. Donor cells containing the 129 strain genome usually produce favorable results, and establishment of embryonic stem (ES) cells from the 129 strain is also good, thus suggesting high genomic plasticity.

How does the degree of cell differentiation affect cloning efficiency? Currently, the donor cells with the highest cloning efficiency in mice are embryonic stem cells. The reason is probably that the *Oct-3/4* gene, which is involved in maintaining high pluripotency, is expressed in donor cells before nuclear transfer.[14] *Oct-3/4* gene expression begins during the morula stage, and in ES cell clones, *no* reprogramming is needed for gene activation. These findings in ES clones also suggest that adult tissue-specific stem cells with pluripotency might also

have high cloning efficiency. Therefore, our laboratory conducted a nuclear transfer cloning study using hematopoietic stem cells and terminally differentiated
lymphocytes with the same genotype. Surprisingly, the results showed poor
development of the cloned embryos from hematopoietic stem cells, whereas good
results were obtained with cloned embryos from lymphocytes (unpublished data).
Thus, at least in nuclear transfer cloning, genomic plasticity does not seem to
correlate with the degree of cell differentiation.

FUTURE DIRECTIONS

Eight years have passed since the birth of the first somatic cell cloned sheep from
fetal fibroblasts. However, somatic cell cloning still faces continuing problems
of low efficiency rates and a high incidence of abnormalities. The key to solving
these problems is a better understanding of the mechanisms of germ cell development. Of particular interest is why and how the genome of germ cells at fertilization (or before and after) is normally reprogrammed. Future research, including
epigenetic and biochemical studies of germ cells and embryological studies by
nuclear transfer cloning, is expected to answer these questions.

CONCLUSIONS

Almost ten years have passed since the first animal was produced by somatic cell
nuclear transfer cloning. However, there have still been no decisive technological
breakthroughs, as evidenced by continuing low birth rates and a high incidence
of pathological abnormalities. The difficulty, of course, lies in using the genomic
reprogramming mechanism of germ cells in somatic cell genomes. Phenomeno-
logical explanation of the abnormalities observed in cloned animals derived from
somatic cell nuclear transfer is alone insufficient to resolve these problems. For
this technology to find wider application in industry and medicine, it is essential
that these abnormalities be classified and their clinical manifestations and incidence be objectively evaluated. In this regard, careful investigation of clone-
specific and typical abnormalities (occurring with greater than chance probability)
is particularly important. Solutions to these problems require more than simply
improving the peripheral technology. Therefore, researchers in nuclear transfer
cloning must examine these phenomena at the molecular, gene expression, and
clinicopathological levels; select suitable models; and use objective evaluation
and statistical analysis.

REFERENCES

1. Gurdon, J.B., Byrne, J.A., and Simonsson, S., Nuclear reprogramming and stem
   cell creation. *Proc. Natl. Acad. Sci. USA*, 100, 11819, 2003.
2. Campbell, K.H.S., Nuclear transfer in farm animal species. *Sem. Cell. Dev. Biol.*,
   10, 245, 1999.

3. Tamashiro, K.L.K. et al., Cloned mice have an obese phenotype not transmitted to their offspring. *Nat. Med.*, 8, 262, 2002.

4. Ogonuki, N. et al., Early death of mice cloned from somatic cells. *Nat. Genet.*, 30, 253, 2002.

5. Young, L.E. et al., Epigenetic change in *IGF2R* is associated with fetal overgrowth after sheep embryo culture. *Nat. Genet.*, 27, 153, 2001.

6. Inoue, K. et al., Faithful expression of imprinted genes in cloned mice. *Science*, 295, 297, 2002.

7. Inoue, K. et al., Tissue-specific distribution of donor mitochondrial DNA in cloned mice produced by somatic cell nuclear transfer. *Genesis*, 39, 79, 2004.

8. Tanaka S. et al., Placentomegaly in cloned mouse concepti caused by expansion of the spongiotrophoblast layer. *Biol. Reprod.*, 65, 1813, 2001.

9. Shimozawa, N. et al., Abnormalities in cloned mice are not transmitted to the progeny. *Genesis*, 34, 203, 2002.

10. Schaetzlein, S. et al., Telomere length is reset during early mammalian embryogenesis. *Proc. Natl. Acad. Sci. USA*, 101, 8034, 2004.

11. Miyashita, N. et al., Normal telomere lengths of spermatozoa in somatic cell-cloned bulls. *Theriogenology*, 59, 1557, 2003.

12. Boiani, M. et al., Pluripotency deficit in clones overcome by clone-clone aggregation: Epigenetic complementation? *EMBO J.*, 22, 5304, 2003.

13. Inoue, K. et al., Effects of donor cell type and genotype on the efficiency of mouse somatic cell cloning. *Biol. Reprod.*, 69, 1394, 2003.

14. Boiani, M. et al., Oct4 distribution and level in mouse clones: Consequences for pluripotency. *Genes Dev.*, 16, 1209, 2002.

10 Exchangeable Gene Trapping

Kimi Araki

TABLE OF CONTENTS

GENE TRAPPING—PRINCIPLE, ADVANTAGES, AND DISADVANTAGES

The whole genome sequences for both humans and mice are now near to completion, and thousands of genes were identified within the last several years. However, sequence information by itself is not sufficient for identifying gene functions *in vivo,* and mutational analysis is used as a powerful and efficient approach. Gene trapping in mouse embryonic stem (ES) cells was developed as a powerful means of inducing insertional mutations, identifying responsible genes, and analyzing the function of genes *in vivo.*[1–8]

The principle of gene trapping is simple (Figure 10.1). Gene trap vectors contain a promoterless reporter gene (the *βgeo* gene in Figure 10.1) that is preceded by a splice acceptor. The vectors are introduced into the ES cell genome, and upon integration into an endogenous gene, a fusion transcript between the endogenous gene and the reporter gene is produced. In many cases, the insertion of a gene trap vector results in gene disruption. The cDNA sequence of the trapped (disrupted) gene is easily determined by rapid amplification of cDNA 5-ends (5RACE)[9] and direct sequencing using primers specific for the reporter gene. Furthermore, genomic DNA flanking the integrated trap vector can easily be obtained by the plasmid rescue method.[10] Because the databases for both mouse cDNA and genomic DNA have improved, homology searches quickly reveal trapped genes, their sequences, genomic structures, and also chromosomal localizations. With the genomic information, PCR primers for genotyping of the trapped allele can immediately be produced. This simple identification of trapped genes is a great advantage over mutagenesis using chemicals or X-rays.

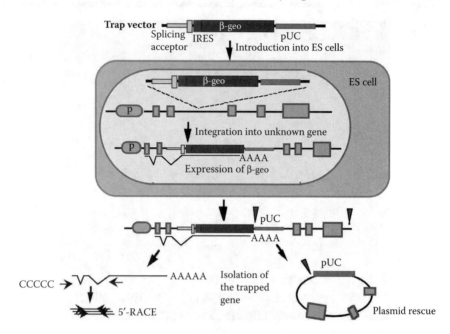

FIGURE 10.1 Gene trapping. The two essential elements of gene trap vectors are a reporter gene to trap endogenous transcription and a selectable marker gene to obtain colonies in which the vector is integrated into the genome. This figure shows a gene trap vector containing the *β-geo* gene, namely a β–galactosidase (reporter)/neomycin-resistance (selectable marker) fusion gene. Upon integration into the genome, the reporter gene is only expressed when it becomes integrated into a gene. The transcript of the trapped gene can be identified by 5-RACE and the genomic locus around the integrated site can be obtained by the plasmid rescue method.

To date, many kinds of gene trap vectors with various reporter genes and selectable markers have been reported. We used the βgalactosidase/neomycin-resistance fusion gene *β-geo*[11,12] because of its high trapping efficiency. When gene trap vectors containing the *β-geo* gene are used, trapped colonies can only appear when the *β-geo* gene is integrated and expressed. Therefore, in almost all the G418-resistant colonies obtained, the integrated trap vector has successfully "trapped" a gene, although the trapped genes are restricted to those genes expressed in ES cells.

The process for producing gene trap mouse lines requires skill compared to that for chemical mutagenesis. Figure 10.2 shows the procedure for gene trapping performed in our laboratory. In order to facilitate the production of chimeric mice, we use TT2 ES cells[13] from which chimeric mice can be produced economically and efficiently through aggregation with morulae from outbred ICR (CD-1) mice. After germ line transmission, the gene trap lines can be analyzed for mutant phenotypes and also for the expression patterns of trapped genes, which can be traced by simple histochemical staining for the reporter gene (usually x-gal staining if the β-galactosidase gene of *E. coli* is used).

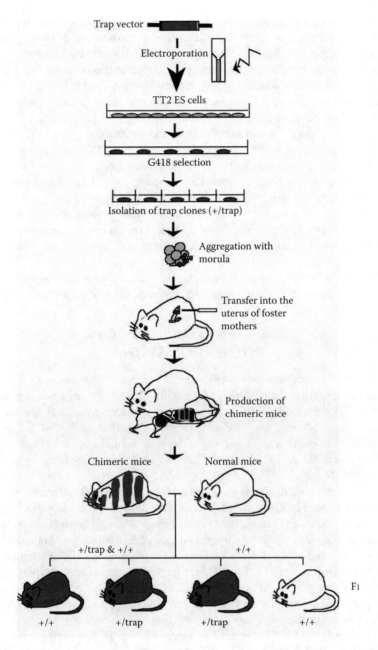

FIGURE 10.2 Procedure for the establishment of gene trap mouse lines in our laboratory. We use electroporation to introduce a gene trap vector into ES cells. After G418 selection, colonies are picked and analyzed for the integration pattern of the gene trap vector. Clones carrying a single copy of the gene trap vector are selected and used for the production of chimeric mice by aggregation with a morula. Gene trap mouse lines are established through mating with normal mice.

Thus, the gene trapping approach has a number of advantages for functional analysis of the genome. However, it has one shortcoming, that is, the relative inability to induce subtle or gain-of-function mutations. Because we cannot remanipulate the integrated gene trap vectors, further functional analyses of the trapped genes require additional gene targeting or the production of transgenic mice. If we can replace the reporter gene with any other DNA fragment using established gene trap clones, various modifications of the trapped alleles become possible. For example, a null trapped allele can be changed into a hypomorphic or dominant-negative allele by inserting a mutated cDNA. A gene trap line showing a very specific expression pattern can be utilized as a tool for the ectopic expression of another gene. In particular, if the *cre* gene of bacteriophage P1[14,15] could be inserted, this would be very useful for the production of various Cre-mice as a tool for conditional gene knock-out.[16] Exchange of the reporter gene for fluorescent protein genes will also be useful for sorting the gene-expressing cells or the production of compound-heterozygote mice by crossing two gene trap lines.

In order to engineer trapped alleles, we have developed an exchangeable gene trap vector, in which the *Cre-Lox* site-specific recombination system is utilized.

CRE-MUTANT LOX SITE-SPECIFIC INTEGRATION SYSTEM

The *Cre-lox*P recombination system was derived from bacteriophage P1[14,15] and became prominent as the most powerful tool for achieving many types of genome engineering in ES cells.[17,18] Cre recombinase catalyzes reciprocal site-specific recombination between two specific 34-bp sites called *lox*P.[14,15] Because the recombination reaction does not require any additional cofactors, this system has been used successfully in yeast,[19] plants,[20] cultured mammalian cells,[19,21,22] and mice.[23–26]

Cre recombinase mediates both intramolecular (excisive) and intermolecular (integrative) recombination. In integrative recombination, a circular DNA carrying a *lox*P site is inserted into a *lox*P site on a chromosome. However, this reaction is quite inefficient, because the integrated DNA has *lox*P sites at both ends and is easily removed again through excisive recombination if the Cre recombinase is still present (Figure 10.3, left). This means that removal of the reporter gene in the trap vector is easy but insertion of a DNA fragment into the position of the reporter gene is difficult. In order to promote integrative recombination, two different kinds of mutant *lox* systems have been developed.

The first is the LE/RE mutant *lox* system (Figure 10.3, middle), which was originally devised by Albert et al.[27] The *lox*P site is composed of an asymmetric 8-bp spacer flanked by 13-bp inverted repeats that the Cre recombinase recognizes and binds. Albert et al. introduced five nucleotide changes into the left 13-bp element (LE mutant *lox*, *lox*71) or the right 13-bp element (RE mutant *lox*, *lox*66). Recombination between an LE mutant *lox*71 and an RE mutant *lox*66

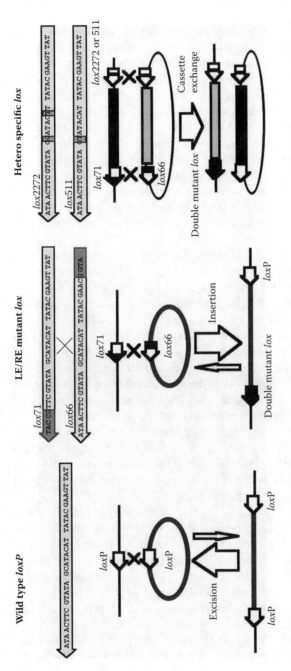

FIGURE 10.3 Mutant *lox* sites. Left: Sequence of wild-type *loxP* and its recombination reaction. Because the integrated product after Cre-mediated recombination itself becomes the substrate for Cre-mediated excisive recombination, the reaction tends toward excision. Middle: LE/RE mutant *lox* system. The pink-colored regions represent the mutations. RE mutants have mutations in the right inverted repeat region, and LE mutants have mutations in the left inverted repeat region. Through single crossover recombination between LE and RE mutant *lox* sites, a wild-type *loxP* site and a double mutant *lox* site carrying mutations in both ends are produced. Because the Cre enzyme has a lower affinity for the double mutant *lox* site, recombination of the product hardly occurs, and the recombination reaction tends toward integration. Right: Heterospecific *lox* system. Heterospecific *lox* sites have mutation(s) in the spacer region in the center. *lox* sites with different spacer regions are never recombined, so the product of the cassette exchange reaction through double crossover is stable.

produces a wild-type *lox*P site and an LE+RE mutant site which is poorly recognized by Cre recombinase, resulting in a reduction in reexcision and a shift to stable integration.

The second is a heterospecific *lox* system, which is called Recombinase Mediated Cassette Exchange (RMCE)[28,29] or the double *lox* strategy[30] (Figure 10.3, right). Heterospecific *lox* sites have mutation(s) in the 8-bp spacer region where strand exchange takes place. Recombination does not occur between two *lox* sites with different spacer regions, whereas *lox* sites with the identical spacer region recombine efficiently.[15] *Lox*511 was reported as early as 1986 by Hoess et al.[15] and contains a single base substitution. *Lox*2272 was identified by Lee and Saito[31] during systemic mutation analysis of the *lox*P spacer region and has two base substitutions. These authors showed that *lox*2272 never recombined with the wild-type *lox*P site, whereas *lox*511 recombined with *lox*P at low frequency in *in vitro* recombination reactions. In the heterospecific *lox* system, a genomic DNA segment flanked by heterospecific *lox* sites is replaced with another floxed cassette located on a transfected plasmid vector by Cre-mediated recombination.[28–30,32,33]

We examined which mutant *lox* system produced the most efficient site-directed integration in ES cells.[34] Figure 10.4 shows the experimental outline used to assess the recombination efficiencies. ES cell lines carrying target *lox* site(s) (middle maps in Figure 10.4) were initially established, and then targeting plasmids (upper maps in Figure 10.4) and the Cre expression vector (pCAGGS-Cre[35]) were co-introduced through electroporation into the established cell lines. Target *lox* site(s) on the ES genome were placed on the 3′ side of the CAG promoter,[36] and the targeting plasmids contained a promoter-less *lacZ* gene and the complete *neo* gene with a promoter and poly(A) signal. In this system, both random and targeted recombinants become G418-resistant, but only the targeted colonies are stained blue with X-gal, because the *LacZ* gene is inserted downstream of the CAG promoter through targeted recombination. The percentage of blue colonies represents the targeted recombination efficiency. In *lox*P–*lox*P or *lox*71–*lox*66 single crossover recombination, the frequency of targeted integration was monitored, and in double crossover recombination using heterospecific *lox* sites, the frequency of cassette exchange replacement was monitored.

The results are shown in Figure 10.4 (right). When wild-type *lox*P sites were used, the frequency of site-specific integration events was quite low (less than 0.5%). With the LE/RE mutant system using *lox*71 and *lox*66, the frequency of site-specific integration reached about 15%, indicating that the LE/RE mutant system is more effective than wild-type *lox*P sites.[37] The cassette exchange recombination via double *lox* crossover yielded 2- to 3-fold higher efficiency than insertion recombination via a single *lox* crossover. In a comparison between *lox*511 and *lox*2272, the *lox*2272-targeting plasmid gave approximately 1.5-fold higher efficiency than the *lox*511-targeting plasmid. Thus, we established a site-specific cassette exchange system in ES cells using a combination of the LE/RE mutant *lox* and *lox*2272.[34]

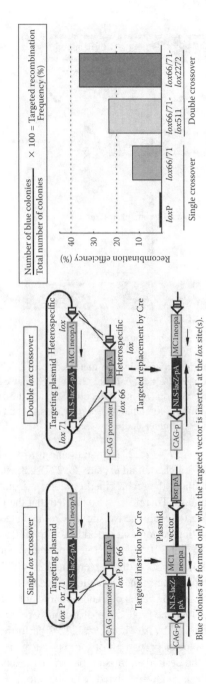

FIGURE 10.4 Evaluation of recombination efficiency. Left: Single crossover recombination experiment. ES cell lines carrying a single copy of *loxP* or *lox71*-CAG promoter-*blasticidin-resistance* gene (*bsr*)-pA were established, and co-electroporated with a targeting plasmid and the Cre expression vector, pCAGGS-Cre. After G418 selection, the colonies were stained with X-gal and the percentage of blue colonies was scored. Middle: Double crossover recombination experiment. ES cell lines carrying a single copy of *lox71*-CAG-*bsr* gene-pA-heterospecific *lox* were established, and co-electroporated with a targeting plasmid and pCAGGS-Cre. After G418 selection, the colonies were stained with X-gal and the percentage of blue colonies was scored. Right: Targeted recombination efficiency. The recombination between *lox66/71* and *lox2272* shows the highest efficiency.

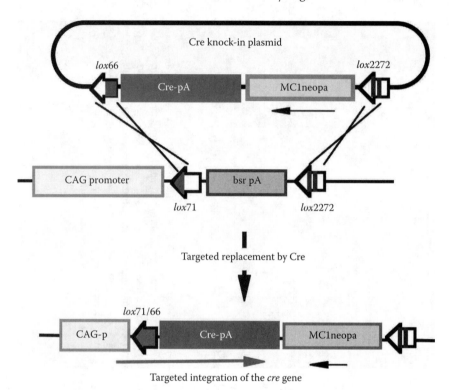

FIGURE 10.5 Targeted integration of the *Cre* gene by Cre-mediated recombination. The Cre-knock-in vector carries *lox*66-Cre-pA-MC1neopA-*lox*2272. The Cre-knock-in vector and Cre-expression vector were co-electroporated into ES cells carrying the target *lox* sites, CAG-*lox*71-*bsr*-pA-*lox*2272. Through recombination between *lox*66/71 and *lox*2272, the *Cre* gene is inserted under the CAG promoter and expressed.

Because the recombination of a *lox*71–*lox*2272 cassette and a *lox*66–*lox*2272 cassette results in the production of a double mutant *lox* site–*lox*2272 cassette, which is never recombined by Cre recombinase, the recombined products are quite stable even in the presence of Cre recombinase. Therefore, we expected that the *Cre* gene could even be integrated into a transcriptionally active site by Cre-mediated recombination. We constructed a *Cre*-targeting vector, and introduced it to an ES cell line carrying *lox*71 and *lox*2272 sites (Figure 10.5). After electroporation and G418 selection, 126 colonies were present, and 36 of these were randomly picked and analyzed for their integration patterns. Four clones were identified as the targeted integrants, and they expressed the integrated *Cre* gene.[34] Thus, our mutant *lox* cassette exchange system can be a useful tool for the production of various Cre-mice. Gene trapping with the reporter cassette flanked by *lox*71 and *lox*2272 is performed as the first step, and after analysis of the expression pattern of the reporter gene, we can choose the gene trap lines showing the desired expression pattern and exchange the reporter cassette for the *Cre* gene through Cre-mediated recombination.

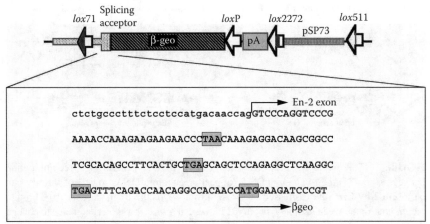

Sequence of the junction of splice acceptor and βgeo

FIGURE 10.6 Exchangeable trap vector, pU-17. A map of pU-17 and the sequence of the junction between the *En-2* splicing acceptor and the initiation codon of the *βgeo* gene are shown. Lowercase and uppercase letters represent intron and exon sequences, respectively. Three stop codons and the start codon of the *βgeo* gene are boxed.

PU-17 EXCHANGEABLE TRAP VECTOR

We constructed a new gene trap vector using a combination of the LE/RE mutant *lox* and heterospecific *lox* sites. Figure 10.6 shows the gene trap vector pU-17, which was designed as a promoter trap vector using the *βgeo* gene and carries three kinds of mutant *lox* sites for replacement. In pU-17, three in-frame stop codons were placed upstream of the ATG of the *βgeo* gene in the splicing acceptor sequence. Theoretically, the integration of the vector should be restricted in the upstream of the start codon of a trapped gene. Thus, this vector was expected to work as a promoter trap, and should be convenient for expressing cDNAs under the control of the trapped genes. We introduced pU-17 into ES cells, isolated over 500 gene trap clones, and identified the trapped genes by 5RACE. Analysis of the insertion positions confirmed that pU-17 integrated in the introns adjacent to the exon containing the start codon of a trapped gene in more than 80% of the trap clones.

Figure 10.7 represents the possible replacement patterns in pU-17 trap clones. Through recombination between *lox*71 and *lox*P, any cDNA or genomic DNA of interest can be inserted and expressed. In this case, we can obtain targeted integrants at very high frequencies (50–90%) by using polyadenylation (pA) signal trap selection, in which the puromycin-resistance (*Puro*) gene in the targeting vector does not have a pA signal, and only fuses to the pA signal on the trap vector upon targeted integration of the *Puro* gene, thereby making the cells drug resistant. Through replacement between *lox*71 and *lox*2272, we can change the expression pattern of the trapped gene by inserting an exogenous promoter sequence, and the *Cre* gene can also be inserted and expressed as mentioned above. Because the

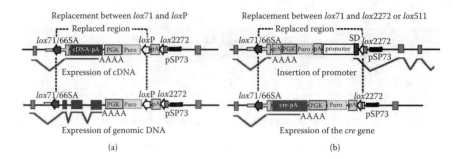

FIGURE 10.7 Possible replacement patterns with pU-17. (a) Through recombination between *lox*66/71 and *lox*P, any cDNA (upper) or genomic DNA (lower) of interest can be inserted by Cre, and expressed under the promoter of the trapped gene. (b) Through recombination between *lox*66/71 and *lox*2272, we can insert a promoter sequence to change the expression pattern of the trapped gene (upper). In addition, the *Cre* gene can be inserted and expressed (lower).

inserted *Cre* gene is transcribed by the endogenous promoter, the expression of the *Cre* gene would be exactly controlled in a spatiotemporal manner.

Figure 10.8 summarizes the exchangeable gene trap system. Our exchangeable gene trap system can overcome the limitations of conventional gene trapping, which can induce only loss-of-function mutations. Through replacement of the reporter gene, we can change original null trap alleles into point mutations, hypomorphic alleles, dominant-negative mutations, and so on. Furthermore, by combining with other site-specific systems such as Flp-*FRT*,[38] exchange into conditional alleles would be possible. Thus, the exchangeable promoter trap system with pU-17 will be an ideal means for large-scale mutagenesis.

REFERENCES

1. Gossler, A. et al., Mouse embryonic stem cells and reporter constructs to detect developmentally regulated genes, *Science* 244 4903, 463, 1989.
2. Gossler, A. and Zachgo, J., Gene and enhancer trap screens in ES cell chimeras. In *Gene Targeting: A Practical Approach*, Joyner, A., ed. Oxford University Press, Oxford, 1993, pp. 181–227.
3. Niwa, H. et al., An efficient gene-trap method using poly A trap vectors and characterization of gene-trap events, *J Biochem (Tokyo)* 113 3, 343, 1993.
4. Evans, M.J., Carlton, M.B.L., and Russ, A.P., Gene trapping and functional genomics, *Trends Genet* 13 9, 370, 1997.
5. Hicks, G.G. et al., Functional genomics in mice by tagged sequence mutagenesis, *Nat Genet* 16 4, 338, 1997.
6. Chowdhury, K. et al., Evidence for the stochastic integration of gene trap vectors into the mouse germline, *Nucleic Acids Res* 25 8, 1531, 1997.
7. Bonaldo, P. et al., Efficient gene trap screening for novel developmental genes using IRES β geo vector and *in vitro* preselection, *Exp Cell Res* 244 1, 125, 1998.

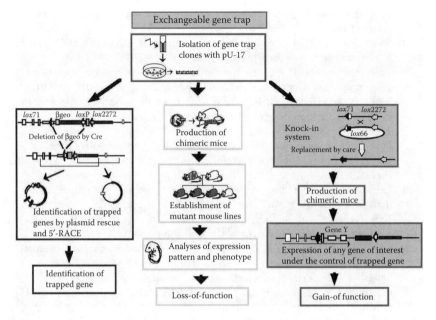

FIGURE 10.8 Scheme for exchangeable gene trapping. We can induce loss-of-function mutations by inserting the trap vector pU-17 and establish mutant mouse lines through chimeric mice production and germ line transmission (middle). Trapped genes can be easily cloned by 5′-RACE, and the integration sites can be identified by the plasmid rescue method (left). Note that the reporter gene can be replaced with any other gene of interest to induce gain-of-function mutations (right).

8. Zambrowicz, B.P. and Friedrich, G.A., Comprehensive mammalian genetics: History and future prospects of gene trapping in the mouse, *Int J Dev Biol* 42 7, 1025, 1998.

9. Townley, D.J. et al., Rapid sequence analysis of gene trap integrations to generate a resource of insertional mutations in mice, *Genome Res* 7 3, 293, 1997.

10. Araki, K. et al., Exchangeable gene trap using the *Cre*/mutated *lox* system, *Cell Mol Biol (Noisy-le-grand)* 45 5, 737, 1999.

11. Friedrich, G. and Soriano, P., Promoter traps in embryonic stem cells: A genetic screen to identify and mutate developmental genes in mice, *Genes Dev* 5 9, 1513, 1991.

12. Voss, A.K., Thomas, T., and Gruss, P., Efficiency assessment of the gene trap approach, *Dev Dyn* 212 2, 171, 1998.

13. Yagi, T. et al., A novel ES cell line, TT2, with high germline-differentiating potency, *Anal Biochem* 214 1, 70, 1993.

14. Hoess, R.H., Ziese, M., and Sternberg, N., P1 site-specific recombination: Nucleotide sequence of the recombining sites, *Proc Natl Acad Sci U S A* 79 11, 3398, 1982.

15. Hoess, R.H., Wierzbicki, A., and Abremski, K., The role of the *loxP* spacer region in P1 site-specific recombination, *Nucleic Acids Res* 14 5, 2287, 1986.

16. Gu, H. et al., Deletion of a DNA polymerase β gene segment in T cells using cell type-specific gene targeting, *Science* 265 5168, 103, 1994.

17. Rossant, J. and Nagy, A., Genome engineering: The new mouse genetics, *Nat Med* 1 6, 592, 1995.

18. Sauer, B., Inducible gene targeting in mice using the *Cre/lox* system, *Methods* 14 4, 381, 1998.
19. Sauer, B. and Henderson, N., Targeted insertion of exogenous DNA into the eukaryotic genome by the Cre recombinase, *New Biol* 2 5, 441, 1990.
20. Dale, E.C. and Ow, D.W., Gene transfer with subsequent removal of the selection gene from the host genome, *Proc Natl Acad Sci U S A* 88 23, 10558, 1991.
21. Sauer, B. and Henderson, N., Site-specific DNA recombination in mammalian cells by the Cre recombinase of bacteriophage P1, *Proc Natl Acad Sci USA* 85 14, 5166, 1988.
22. Fukushige, S. and Sauer, B., Genomic targeting with a positive-selection *lox* integration vector allows highly reproducible gene expression in mammalian cells, *Proc Natl Acad Sci U S A* 89 17, 7905, 1992.
23. Lakso, M. et al., Targeted oncogene activation by site-specific recombination in transgenic mice, *Proc Natl Acad Sci U S A* 89 14, 6232, 1992.
24. Araki, K. et al., Site-specific recombination of a transgene in fertilized eggs by transient expression of Cre recombinase, *Proc Natl Acad Sci U S A* 92 1, 160, 1995.
25. Orban, P.C., Chui, D., and Marth, J.D., Tissue- and site-specific DNA recombination in transgenic mice, *Proc Natl Acad Sci U S A* 89 15, 6861, 1992.
26. Dragatsis, I. and Zeitlin, S., A method for the generation of conditional gene repair mutations in mice, *Nucleic Acids Res* 29 3, E10, 2001.
27. Albert, H. et al., Site-specific integration of DNA into wild-type and mutant *lox* sites placed in the plant genome, *Plant J* 7 4, 649, 1995.
28. Bouhassira, E.E., Westerman, K., and Leboulch, P., Transcriptional behavior of LCR enhancer elements integrated at the same chromosomal locus by recombinase-mediated cassette exchange, *Blood* 90 9, 3332, 1997.
29. Feng, Y.Q. et al., Site-specific chromosomal integration in mammalian cells: Highly efficient CRE recombinase-mediated cassette exchange, *J Mol Biol* 292 4, 779, 1999.
30. Soukharev, S., Miller, J.L., and Sauer, B., Segmental genomic replacement in embryonic stem cells by double *lox* targeting, *Nucleic Acids Res* 27 18, e21, 1999.
31. Lee, G. and Saito, I., Role of nucleotide sequences of loxP spacer region in Cre-mediated recombination, *Gene* 216 1, 55, 1998.
32. Bethke, B. and Sauer, B., Segmental genomic replacement by Cre-mediated recombination: Genotoxic stress activation of the p53 promoter in single-copy transformants, *Nucleic Acids Res* 25 14, 2828, 1997.
33. Kolb, A.F., Selection-marker-free modification of the murine beta-casein gene using a *lox*2272 [correction of *lox*2722] site, *Anal Biochem* 290 2, 260, 2001.
34. Araki, K., Araki, M., and Yamamura, K., Site-directed integration of the *Cre* gene mediated by Cre recombinase using a combination of mutant lox sites, *Nucleic Acids Res* 30 19, e103, 2002.
35. Araki, K. et al., Efficiency of recombination by Cre transient expression in embryonic stem cells: Comparison of various promoters, *J Biochem (Tokyo)* 122 5, 977, 1997.
36. Niwa, H., Yamamura, K., and Miyazaki, J., Efficient selection for high-expression transfectants with a novel eukaryotic vector, *Gene* 108 2, 193, 1991.
37. Araki, K., Araki, M., and Yamamura, K., Targeted integration of DNA using mutant *lox* sites in embryonic stem cells, *Nucleic Acids Res* 25 4, 868, 1997.
38. Fiering, S. et al., An "in-out" strategy using gene targeting and FLP recombinase for the functional dissection of complex DNA regulatory elements: Analysis of the beta-globin locus control region, *Proc Natl Acad Sci U S A* 90 18, 8469, 1993.

11 Genetic Monitoring of Mice

Hideki Katoh

TABLE OF CONTENTS

INTRODUCTION:
GENETIC CONTAMINATIONS REPORTED IN THE PAST

There are many strains that are genetically controlled and well characterized worldwide. The number of strains of mice and rats are 436[1] and 220.[2] They were developed under strict genetic controls such as sister–brother mating. However, there are three primary causes that change gene (DNA) and homozygosity in inbred strains: (1) mutation, (2) residual heterozygosity, and (3) contamination (unexpected outcrossing).

Kahan et al.[3] reported that commercially supplied BALB/c mice showed histocompatibility and isoenzyme differences caused by genetic contamination. The abstract from that paper is as follows.

143

"BALB/c mice obtained commercially were found to differ significantly from the standard phenotype of BALB/c strain mice. Isoenzyme tests and H-2 haplotype analyses indicated that the majority of mice from two of the three sources tested appeared mixed, frequently heterozygous, and did not consistently express either the expected H-2 or glucose phosphate isomerase type."

This fact was introduced as a big problem in the journal *Science*.[4]

We reported that some NZB substrains in Japan were genetically contaminated as shown in Table 11.1. Five of 26 substrains showed different alleles from those of the standard NZB strain at three loci in #503 and #401, and at four loci in #302, #305, and #311. These were NZB substrains derived from the original strain (#302) at different times. These allele differences could be brought about by genetic contamination with other strains. Recently, genetic contamination in 129 substrains was also reported.[5–7]

One of the serious problems for maintaining ordinary inbred strains is the genetic contamination described above. The genetic homogeneity of inbred strains will be lost, and animal experiments with contaminated strains will result in the complications described below.

SPONTANEOUS MUTANT AND GENETICALLY ENGINEERED MOUSE PHENOTYPES VARY WITH GENETIC BACKGROUND

It was reported that phenotypes, especially disease-related ones, change on different genetic backgrounds when incipient and true congenic strains are created. Sundberg et al.[8] reported that the flaky skin phenotypes with seven parameters (acanthosis, hypergranulosis, orthokeratosis, parakeratosis, intracorneal microabscess, dilated dermal capillaries, and dermal inflammation) varied on different genetic backgrounds as shown in Table 11.2. We observed that a second allelic mutation (*fsn^Jic*) of the flaky skin locus showed different phenotypes when this mutant allele was transferred onto BALB/cByJcl, C3H/HeJJcl, C57BL/6JJcl, and DBA/2JJcl by back-crossing (Katoh, unpublished data).

According to TBASE[9] the number of genetically engineered strains of targeted mutant mice and transgenics in 2003 were 1103 and 758, respectively.

With the development of genetically engineered mice, we face different problems from those seen in the ordinary inbred strains. One is the heterogeneity of genetic backgrounds and the other is the change in phenotypes on different genetic backgrounds. The author has surveyed genetically engineered mouse strains listed in the JAX catalogue.[10] About 20% of 246 transgenics and about 30% of 511 targeted mutant stocks were back-crossed to create congenic strains. Therefore, they were noninbred at that time and heterogeneous at many loci differing between donor and recipient strains.

Thus, a considerable number of genetically engineered mice will be in the same genetic situations as genetically contaminated mice. There is no doubt

TABLE 11.1
Typical Genetic Contamination Observed in NZB Substrains in Japan

	Chromosome No. and gene loci*																		
	1	1	1	2	3	4	4	5	6	7	7	8	8	9	9	9	11	17	17
Substrain	Idh1	Pep3	Akp1	Hc	Car2	Mup1	Gpd1	Pgm1	Ldr1	Gpi1	Hbb	Es1	Es2	Thy1	Mod1	Trf	Es3	H2K	H2D
Control	a	c	b	0	a	a	b	b	a	a	d	b	b	b	b	b	c	d	d
#302	a	c	b	*1*	a	*b*	b	b	a	a	*s*	*a*	b	b	b	b	c	d	d
#305	a	c	b	*1*	a	*b*	b	b	a	a	*s*	*a*	b	b	b	b	c	d	d
#311	a	c	b	*1*	a	*b*	b	b	a	a	*s*	*a*	b	b	b	b	c	d	d
#401	a	c	*a*	0	a	*b*	b	b	a	a	d	*a*	b	b	b	b	c	d	d
#503	a	*b*	*a*	0	a	a	b	b	a	a	d	*a*	b	b	b	b	c	d	d

*See Table 11.3. Allele shown in bold differs from that of the control strain.

TABLE 11.2
Flaky Skin (*fsn*) Phenotypes Modified by Genetic Backgrounds[a]

Strain	Acanthosis	Hyper-granulosis	Ortho-keratosis	Para-keratosis	Intra-corneal Micro-abscess	Dilated Dermal Capillaries	Dermal inflam-mation
A/J	S	Mi	S	F	F	Y	Mo-S
BALB/cByJ (N7)	Mo	S	Mo	R	R	Y	Mi-Mo
A/J × BALB/cByJ	Mo-S	Mi	S	F	F	Y	Mo-S
C57BL/6J (N6)	S	Mo	S	F	F	Y	Mo

[a] S: severe; Mo: moderate; Mi: mild; F: focal; R: rare; Y: yes.
Source: Modified data of Sundberg.[8]

that if we desire reasonable and stable results in animal experimentation, we have to develop and maintain inbred and congenic strains with appropriate genetic quality control.

GENETIC MONITORING
(GENETIC QUALITY TESTING) OF LIVE MICE

Genetic monitoring systems were developed to ensure genetic quality of experimental animals at the Central Institute for Experimental Animals (Kawasaki, Japan) designated as the ICLAS Monitoring Center by ICLAS (International Council for Laboratory Animal Science) in 1979.[11] Since then, we have been responsible for developing the genetic monitoring system for live animals as follows.

MARKERS FOR GENETIC MONITORING

There are various types of genetic markers that can be used for genetic testing. They are divided into six groups according to their biological functions, phenotypes, and detecting methods as follows: (1) biochemical markers (*Hbb*, *Gpi1*, etc.), (2) immunological markers (*H2*, *Thy1*, etc.), (3) morphological markers (coat colors, etc.), (4) cytogenetic markers (C bands, etc.), (5) molecular genetic markers (minisatellites, microsatellites, etc.), and (6) pharmacogenetic markers (alcohol preference, etc.). Well-known inbred strains such as C57BL/6J and DBA/2J have been characterized for these markers. Almost all data are available in a mouse genetic data book[12] or online at Mouse Genome Informatics.[13]

When markers for genetic quality testing are selected from these six groups, the following conditions should be considered: (1) Exact (clearly detected using easier techniques), (2) Easy (quickly detected using easier techniques), (3) Efficient (polymorphic among inbred strains), and (4) Economic (detected using cheaper materials and equipment). We call these specific requirements the 4Es by taking the first letter from each category. We concluded that biochemical and immunological markers were among the six types of markers that were satisfactory for meeting the goals of the 4Es. The ICLAS Monitoring Center selected 28 loci (19 biochemical and 9 immunological markers) for mice as shown in Table 11.3.

Recently, microsatellite DNA markers detected using PCR (polymerase chain reaction) were used for gene mapping with linkage analyses. We attempted to use microsatellite markers for genetic quality testing of cell lines derived from experimental animals. When we selected microsatellite markers as markers for genetic quality testing, we paid special attention to the first E (Exact) of the 4Es, because it was demonstrated that not all microsatellite DNA markers show genetic stability of their PCR products (Katoh, unpublished data). After characterizing many microsatellite markers on every chromosome in the mouse, we have finally selected 20 microsatellite markers showing genetic stability (data not shown).

SETS OF MARKERS FOR GENETIC QUALITY TESTING

Sets of markers for genetic quality testing are classified into three profiles according to their purposes.

1. *Genetic Profile:* Many markers are investigated for each strain and the genetic differences from the standard strain are clarified. The genetic profile of an inbred strain consists of genotypes of the markers listed in Table 11.3 and certifies genetic condition (genetic quality) of a given mouse strain. Other biochemical and immunological markers selected on the basis of the requirements for genetic quality testing may be added on demand. There is no limitation with respect to the number of markers, but the markers shown in Table 11.3 will be adequate for the purpose mentioned above.

2. *Monitoring Profile:* A monitoring profile is prepared using a set of markers selected on the basis of genetic quality testing. The ICLAS Monitoring Center selected 19 markers in mice as shown in Table 11.3. The monitoring profile of the strain certifies genetic condition (genotypes) of the strain (Table 11.4).

3. *Critical Subset Profile*: A critical subset profile is prepared using a set of restricted markers selected to identify a limited one among inbred strains maintained or produced in a room or a facility. As shown in Figure 11.1, six typical albino mouse strains can be differentiated using a critical subset consisting of only four biochemical markers. They are enough to discriminate one from another. If genetic contamination occurs between them, it is easy to expose it using a critical subset.

TABLE 11.3
Genetic Markers for Genetic Quality Testing in Mice

Chromosome No.	Gene Symbol and Gene Name (Old Symbol and gene name in parenthesis)		Genetic Profile	Monitoring Profile
		Biochemical markers (19 loci)		
1	*Idh1*	Isocitrate dehydrogenase-1	×	×
1	*Pep3*	Peptidase-3	×	×
1	*Akp1*	Alkaline phosphatase-1	×	×
3	*Car2*	Carbonic anhydrase-2	×	×
4	*Mup1*	Major urinary protein-1	×	×
4	*Gpd1*	Glycerol-3-phosphate dehydrogenase 1 (soluble)	×	×
5	*Pgm1*	Phosphoglucomutase-1	×	×
6	*Ldr1*	Lactate dehydrogenase regulator-1	×	×
7	*Gpi1*	Glucose phosphate isomerase-1	×	×
7	*Hbb*	Hemoglobin beta chain complex	×	×
8	*Es1*	Esterase-1	×	×
8	*Es2*	Esterase-2	×	×
9	*Mod1*	Malic enzyme, supernatant	×	×
9	*Trf*	Transferrin	×	×
11	*Hba*	Hemoglobin alpha chain complex	×	
11	*Es3*	Esterase-3	×	×
14	*EsD (Es10)*	Esterase D/formylglutathione hydrolase (Esterase-10)	×	
14	*Nptx1 (Np1)*	Neuronal pentraxin (Nucleoside phosphorylase-1)	×	
17	*Glo1*	Glyoxalase-1	×	
		Immunological markers (9 loci)		
2	*Hc*	Hemolytic component (synonyms:C5)	×	×
6	*Cd8a (Ly-2)*	Antigen, alpha chain (Lymphocyte antigen 2)	×	
6	*Cd8b1 (Ly-3)*	Antigen, beta chain (Lymphocyte antigen 3)	×	
9	*Thy1*	Thymus cell antigen 1, theta	×	×
12	*IghC*	Immunoglobulin heavy chain constant region	×	
17	*H2K*	Histocompatibility 2, K region	×	×
17	*H2D*	Histocompatibility 2, D region	×	×
17	*C3*	Complement component-3	×	
19	*Cd5 (Ly1)*	CD5 antigen (Lymphocyte antigen 1)	×	

| Six albino mice | Test markers | | | | Strain |
	Idh1 ⇒	Hbb ⇒	Es3 ⇒	Mod1	
	a	d	c	nd	A
	a	d	a	nd	BALB/c
	a	s	c	nd	DDD/1
	b	d	nd	b	AKR
	b	d	nd	a	NZW
	b	s	nd	a	SJL

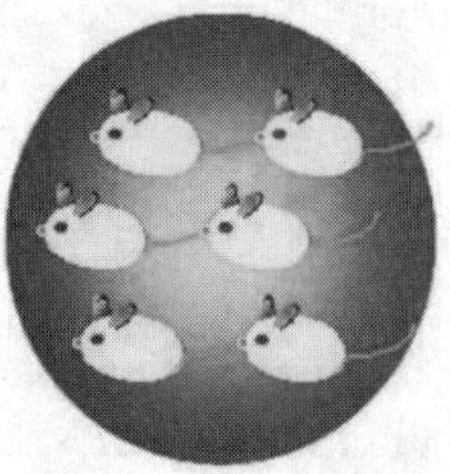

FIGURE 11.1 A critical subset for strain identification among six albino mice; *a–d* and *s* are alleles at tested genes, *Idh1*, *Hbb*, *Es3*, and *Mod1*. nd: not done.

CATEGORIES OF GENETIC QUALITY TESTING

Depending on different purposes, genetic tests are divided into three categories as follows.

1. *Characterization:* This is carried out to study or demonstrate genotypes of inbred strains and also to define genetic profiles of newly bred strains. This test is performed once when strains are newly introduced or established.
2. *Genetic Monitoring I:* This is periodically carried out to confirm the monitoring profile of a given inbred strain in order to demonstrate that genetic contamination or mistakes do not occur in the strain. Testing is performed on a scheduled basis to determine if genetic quality of the strain has been maintained.
3. *Genetic Monitoring II:* This is carried out in order to determine whether genetic contamination or mistakes occurred. Several markers showing genetic polymorphisms among the strains are selected as the test items from monitoring markers (Table 11.3).

SCHEDULE OF GENETIC MONITORING

Test Frequency

We recommend that in the case of strain maintenance and long-term production, three kinds of tests be performed in accordance with the following schedule.

Scheduled testing is recommended in the case of strain maintenance and long-term production, but spot testing can be used in short-term cases or when the colony is small.

Immediately after introduction of new strains, these strains are characterized to confirm genotypes of the strain. These characteristics are monitored every several years after introduction. If the introduced strain is correct, the strain is monitored periodically (once a year) at prescribed schedules (see below).

When individuals with different coat colors appear or the litter sizes sharply increase in inbred strains, there is a possibility of genetic contamination. At this point a critical subset test should be performed immediately.

Monitoring Schedule

In the case of strain maintenance populations, a pair of female and male at $F(n)$ generation surviving the next generation $F(n + 1)$ within the line or six mice from the $F(n + 1)$ generation are selected for testing. In the case of production populations, six animals are selected at random from the population. In the case of closed colonies, 50 mice are selected at random from the populations. See Table 11.5.

GENETIC MONITORING FOR CELL LINES, GAMETES, AND EMBRYOS

Genetic quality of *in vitro* cultured cells derived from experimental animals as well as live animals should be checked, because the genetic quality can be destroyed by the following factors: mixing different cell lines, mishandling, and natural changes occurring on chromosomes and DNA. The first and the second problems can be prevented and identified using genetic techniques. The third factor is inevitable. These errors will be causes of producing genetically changed (contaminated) cell lines and spreading them to other facilities. When genetic monitoring is applied to *in vitro* cultured cell lines, we should pay special attention to the species identification as well as the strain identification. Some of the methods used for this purpose are: (1) karyotyping (chromosome analysis), (2) isozyme analysis using electrophoresis, (3) DNA fingerprinting (minisatellite DNA), and (4) chromosome painting using specific DNA probes. Although these approaches are useful for identification of the animal species, the techniques and procedures are specific and complicated.

Alternatively, genetic resource banks preserving gametes (sperm and ova) and the early stage of embryos (two-cell, eight-cell embryos, etc.) were recently established and are now used worldwide. Repositories have responsibilities to produce and distribute transgenic mice, targeted mutant mice, and strains carrying various mutations. Even in embryo and sperm banks, one should always consider mishandling of straws or cryotubes containing sperm and embryos, because mishandling will lead to mixes between different strains which will result in genetic contamination.

We proposed that cells, gametes, and early-stage embryos should be identified by: species, sex, and strain. PCR (polymerase chain reaction) has been developed

TABLE 11.4
Monitoring Profiles of the Typical Inbred Strains of the Mouse

								Chromosome No. & Gene loci											
	1	1	1	2	3	4	4	5	6	7	7	8	8	9	9	9	11	17	17
Strains	Idh1	Pep3	Akp1	Hc	Car2	Mup1	Gpd1	Pgm1	Ldr1	Gpi1	Hbb	Es1	Es2	Thy1	Mod1	Trf	Es3	H2K	H2D
A	a	b	b	0	b	a	b	a	a	a	d	b	b	b	a	b	c	k	d
AKR	b	b	b	0	a	a	b	a	a	a	d	b	b	a	b	b	c	k	k
BALB/c	a	a	b	1	b	a	b	a	a	a	d	b	b	b	a	b	a	d	d
CBA/J	b	b	a	1	b	a	b	a	a	b	d	b	b	b	b	a	c	k	k
CBA/N	b	b	b	1	a	a	b	b	b	b	d	b	b	b	b	a	c	k	k
C3H/He	a	b	b	1	b	a	b	a	a	b	d	b	b	b	a	b	c	k	k
C57BL/6	a	a	a	1	a	b	a	a	a	b	s	a	a	b	b	b	a	b	b
DBA/1	b	b	a	1	a	a	a	b	a	a	d	b	b	b	a	b	c	q	q
DBA/2	b	b	a	0	b	a	b	b	a	a	d	b	b	b	a	b	c	d	d
KK	a	b	b	0	a	b	a	a	a	b	d	b	a	b	a	b	c	b	b
NZB	a	c	b	0	a	a	b	b	a	a	d	b	b	b	b	b	c	d	d

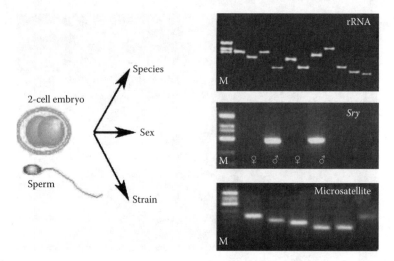

FIGURE 11.2 Three classifications (species, sex, and strain) for identification of embryos and sperm; M: DNA size marker; *rRNA*: ribosomal RNA gene; *Sry*: sex determining region of Y chromosome.

to amplify a part of the desired gene (DNA sequence). Using PCR, microsatellite DNA markers dispersed on the mouse genome, which are detected as simple sequence length polymorphisms (SSLPs), were developed to study linkage of genes, especially disease-related genes in mice.

Species Identification

Naito et al.[14] developed a method discriminating animal species with a ribosomal ribonucleic acid gene (*rRNA*). They designed and synthesized a pair of primers to detect a variable region of DNA coding 28S rRNA. They successfully identified various animal species with PCR products of their rRNA as shown in Figure 11.2. We confirmed that this method could be applied to identify the animal species of the cell lines of rats.[15]

Sex Identification

To identify sexes of cell lines, three pairs of primers for detecting DNA sequences of the Y chromosome of mice and rats, respectively, were synthesized according to the literature. As shown in Figure 11.2, mice and rat females were discriminated from males by not observing Y-specific PCR products in agarose gel.

Strain Identification

We performed a preliminary study to select the most effective microsatellite markers that identify as many mouse strains as possible (see Table 11.6). We

TABLE 11.5
Monitoring Schedule (Example)[a]

Period (year)	1	2	3	4	5
Type of test	C	MI	MII	MI	MII
No. animals	2	2 or 6	2 or 6	2 or 6	2 or 6

[a] C: Characterization; MI: Monitoring I; MII: Monitoring II.

TABLE 11.6
Monitoring Profiles of the Typical Inbred Strains of the Mouse

Strain	Origin	D3Mit54	D5Mit18	D6Mit15	D8Mit50
1. C57BL/6	Jic	1	1	1	4
2. MSM	Ms	1	1	1	5
3. NC/Nga	Jic	1	1	3	3
4. MRL/MpJ-*lpr*	Jic	1	2	1	1
5. BALB/cA	Jic	1	2	3	2
6. IQI	Jic	1	2	3	3
7. C3H/HeN	Crj	1	2	3	4
8. SWR/J	Ms	1	2	4	1
9. NFS/N	Jic	1	2	4	3
10. NOD/Shi-*scid*	Jic	1	2	4	3
11. NZB	Crj	2	1	1	2
12. AKR/J	Jic	2	2	1	2
13. NZW	Crj	2	2	1	2
14. 129/J	Ms	2	2	3	2
15. DBA/2J	Jic	2	2	3	3

selected markers that are detected with the same PCR condition. We selected four microsatellite markers, *D3Mit54*, *D5Mit18*, *D6Mit15*, and *D8Mit50*. If we have four markers with two alleles, they are theoretically enough to discriminate 16 strains. As shown in Table 11.6, 15 typical inbred strains were divided into 13 groups. Therefore, genetic monitoring should be applied at appropriate stages of embryos and gametes to confirm genetic accuracy of the strains.

Genetic monitoring of embryos and gametes are performed using genomic DNA distributed in all tissues and organs. As shown in Figure 11.3, we successfully performed genetic quality testing of embryos using eight-cell stage embryos.[16]

GENOTYPING OF TRANSGENES AND TARGETED GENES

We introduced the fluorescence *in situ* hybridization (FISH) technique for chromosomal mapping of transgenes. Figure 11.4 is an example of chromosomal mapping

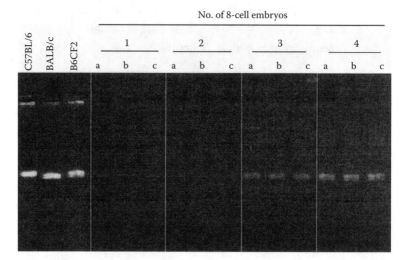

FIGURE 11.3 Genotyping of the MIT marker (*D5Mit18*) performed using eight-cell embryos. A method of preparation of template DNA for PCR was established using the lysis buffer (1 × PCR reaction buffer supplemented with proteinase K at a concentration of 40 µg/ml). Tests were performed three times; a, b, and c are experiment numbers.

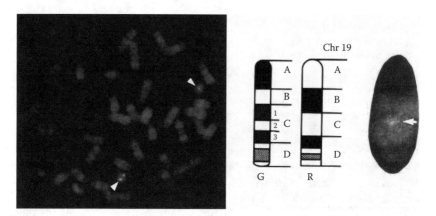

FIGURE 11.4 Chromosomal mapping of human angiotensinogen transgene performed using FISH. Chromosomes were stained with propidium iodide (PI) using the R-banding technique. On the middle of mouse Chr 19, the yellow fluorescence signals shown by a white arrowhead indicate the location of the transgene.

of a transgene by FISH. This figure demonstrates the exact chromosomal localization of the human angiotensinogen transgene on mouse Chromosome 19. We also performed gene mapping by mating experiments in order to confirm the chromosome number and to clarify the map position of transgenes. Three microsatellite DNA markers and the transgene were linked in the order: *D19Mit28*-(11.5cM)-Tg-(6.47cM)-*D19Mit40*-(0.72cM)-*D19Mit13*.[17]

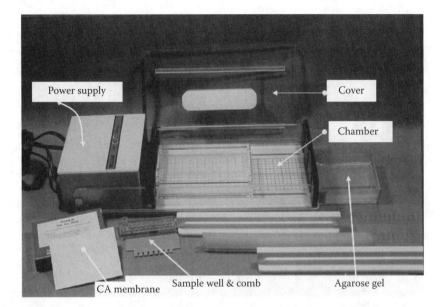

FIGURE 11.5 A genetic monitoring kit developed by the ICLAS Monitoring Center. Proteins and enzymes are detected with cellulose acetate membrane and SSLP markers with agarose gel. CA: cellulose acetate.

DEVELOPMENT OF A GENETIC MONITORING KIT

Genetic monitoring should be performed using standardized techniques. We developed a genetic monitoring kit that is useful for beginners and includes the minimum materials required and an operation manual. Protein and enzyme staining kits are supplied separately on demand. The kit can also be used as an educational tool in order to demonstrate genetic polymorphisms of proteins and enzymes in mice and rats.

CONCLUSION

Genetic monitoring was developed to ensure genetic quality of inbred strains of mice. As described in this chapter, we established a genetic monitoring system including not only genetic techniques but also procedures and concepts for testing genetic quality. Simultaneously, we successfully applied this system to test not only inbred strains of mice but also for closed colonies, cultured cell lines, sperm, and embryos.

Recently, genetic resource repositories were established worldwide. Shipping now occurs not only of live animals but also of frozen embryos and sperm preserved in liquid nitrogen. In many cases, researchers want to know the genetic background of the strain and genotype of an interesting gene immediately upon receiving such genetic resources. Based on genetic monitoring systems described in this chapter, a procedure for assurance of genetic quality will be available at each repository or facility and utilized depending on the resources and specimens available (sperm, embryos, or live animals).

REFERENCES

1. Festing MFW: Inbred strains of mice and their characteristics (http://www.informatics. jax.org/external/festing/search_form.cgi), 1996.
2. Festing MFW: Inbred strains of rat and their characteristics (http://www.informatics. jax.org/external/festing/search_form.cgi), 1998.
3. Kahan B: Histocompatibility and isozyme differences in commercially supplied BALB/c mice. *Science* 217: 379–381, 1982.
4. Fox J: Scientist sues over genetically impure mice. *Science* 221: 625–628, 1983.
5. Simpson EM, Linder CC, Sargent EE, Davisson MT, Mobraaten LE, Sharp JJ: Genetic variation among 129 substrains and its importance for targeted mutagenesis in mice. *Nat. Genet.* 16, 19–27, 1997.
6. Threadgill DW, Yee D, Matin A, Nadeau JH, Magnuson T: Genealogy of the 129 inbred strains: 129/SvJ is a contaminated inbred strain. *Mamm. Genome* 8: 390–393, 1997.
7. Festing MF, Simpson EM, Davisson MT, Mobraaten LE: Revised nomenclature for strain 129 mice. *Mamm. Genome*, 10: 836, 1999.
8. Sundberg JP, Boggess D, Shultz LD, Beamer WG: The flaky skin (*fsn*) mutation, Chromosome ? In: *Handbook of Mouse Mutations with Skin and Hair Abnormalities* (Ed. by Sundberg JP), 253–268, CRC Press, Boca Raton, FL, 1994.
9. TBASE: http://tbase.jax.org/docs/tbase.html.
10. The Jackson Laboratory: *2000 JAX Catalogue*, 2003.
11. Nomura T, Esaki K, Tomita T: *ICLAS Manual for Genetic Monitoring of Inbred Mice.* University of Tokyo Press, Tokyo, 1984.
12. Lyon MF, Rastan S, Brown SDM (eds.): *Genetic Variants and Strains of the Laboratory Mouse.* Third Edition, Oxford University Press, Oxford, 1996.
13. The Jackson Laboratory: Mouse Genome Informatics (http:www.informatics. jax.org/), 2003.
14. Naito E, Dewa K, Yamanouchi H, Kominami R: Ribosomal ribonucleic acid (rRNA) gene typing for species identification. *J. Forensic Sci.* 37: 396–403, 1992.
15. Katoh H, Muguruma K, Watanabe Y, Ishikawa S, Ebukuro M, Nomura T, Nakagawa Y, Tanaka N: Genetic quality testing of cell lines derived from laboratory rats by polymerase chain reaction. *Transplantation Proc.* 29, 1709–1712, 1997.
16. Katoh H, Oda K, Hioki K, Muguruma K: A genetic quality testing system for early stage embryos in the mouse. *Exp. Anim.* 52: 397–400, 2003.
17. Watanabe Y, Ebukuro M, Yagami K, Sugiyama F, Ishida J, Murakami K, Nomura T, Katoh H: Chromosome mapping of human angiotensinogen gene and human renin gene by fluorescence *in situ* hybridization (FISH) in transgenic mice. *Exp. Anim.* 45: 265–269, 1996.

12 Microbiological Monitoring of Laboratory Mice

James R. Fahey

TABLE OF CONTENTS

INTRODUCTION

Performing microbiological surveillance on a deliberately scheduled, routine basis is the cornerstone of a sound animal health program. The data derived from routine microbiological surveillance provides animal health professionals the benchmark information they require to determine whether the housing, husbandry, and care they provide are optimal for the type of animals they maintain. In this regard, genetically engineered mice (GEM) cannot be viewed in the same manner as standard inbred mouse strains in that they present unique health monitoring challenges.

These challenges must be addressed when devising plans for microbiologic surveillance of GEM colonies. For example, many GEM strains and stocks have known deficiencies in innate or acquired immunological function whereas others have subtle physiological or metabolical alterations that affect their interactions with pathogens.[1] Thus, GEM strains and stocks can manifest unforeseen clinical signs of disease as well as disease outcomes when infected with organisms whose pathogenic potential is well known in standard inbred mouse strains.[2,3]

A good example of this is the finding of Rheg et al.[4] who reported that mice of the CBA/CaJ-Tg (TcrβV8.1) and B10.BR-H-2^k –T18^a /SgSnJ-Tg (TcrβV8.1) strains, both of which possess rearranged T-cell receptor β-chains, shed mouse hepatitis virus (MHV) for two years despite the concurrent existence of anti-MHV antibodies in their sera. Thus, despite evidence of an anti viral immune response, the infected GEM strains were unable to clear infectious virus to the extent that they were not contagious to other mice. This is an atypical response to MHV in immunocompetent mice (putatively, the β-chain rearrangement did not confer immunological deficiencies) and suggests that mice of these particular genotypes require stringent health surveillance for viral pathogens. Perhaps frequent serological testing (weekly, monthly) combined with ancillary diagnostic methods such as PCR would enhance the probability of detecting MHV-infected mice of these genotypes. Furthermore, restricted housing such as isolators or high barrier rooms should be considered for mice of these strains. A similar case involving prolonged shedding of MHV has recently been reported for a GEM strain with a targeted null mutation of tumor necrosis factor, although a role for this cytokine in the clearance of MHV from mice has yet to be determined.[5]

The two cases described above are limited to MHV infections of mice, in other words, a viral pathogen–mouse interaction. Nonetheless, they still exemplify the types of host–pathogen interactions that must be considered when devising a microbiological monitoring plan for animal facilities housing GEM strains and stocks. These examples do not take into account interactions between commensal or opportunistic microorganisms or even genetically modified pathogens and their genetically modified murine hosts.

We can get an inkling of these interactions from the many published studies demonstrating *Helicobacter hepaticus* interactions with GEM strains and stocks. Generally speaking, these studies show that some mouse strains are adversely affected by *H. hepaticus* and others are not,[6] yet overall, it is clear that the outcomes of these host–parasite interactions remain largely unpredictable. Given the number of GEM strains and stocks now existing around the world and the numerous microbes known to associate with mice as benign commensals or latent opportunists, the probability of undesirable host–parasite interactions is enormous. Moreover, it can be disturbing to ponder all of the potential negative interactions that can occur. Nonetheless, these interactions must be anticipated when designing microbiological surveillance programs for GEM strains.

MICROBIAL EXCLUSION PLANNING

In laboratory animal medicine, health maintenance of rodent colonies is undertaken as an integrated management program of which microbiological monitoring is one phase. It is not within the scope of this chapter to discuss all facets of this integrated approach, but it must be emphatically stated that risk assessment, in terms of planning to exclude undesirable organisms from your facilities and to contain opportunistic microorganisms, or overt pathogens that are currently present, is an inherent part of this process. Risk assessment should include discussion of the type and design of mouse facilities, caging systems, room entry and exit policies, personnel, materials handling, and so on. It also includes the development of an exclusion list of microbes and a plan for responding when a microbe from your exclusion list is discovered in your facilities.

The criteria for development of a microbial exclusion list varies with each institution; nonetheless, there are several basic considerations common to all institutions that utilize mice for research. First of all, determine which organisms must absolutely be excluded from your facility because, for example, (1) they are zoonotic, (2) are avirulent but have serious research effects, or (3) are known to be virulent and spread rapidly throughout mouse colonies. In facilities with large populations of GEM strains and stocks, recall that commonly known mouse pathogens may induce diseases of unknown severity or outcome, so it is imperative to review the scientific literature for evidence of interactions between specific mouse-associated microorganisms and the GEM strains and stocks within your facility. Additionally, the individuals planning for microbiological surveillance at your facility should be familiar with any known immune deficiencies or other genetic modifications that predispose the specific GEM strains within your facilities to infection.

Another general consideration in developing a microbial exclusion list is to determine whether there are microorganisms in your facility that are acceptable in some of your mouse rooms, but not others because of the specific GEM strains and stocks within those rooms. For example, if mice with *Staphylococcus aureus* are present in some rooms within your facility, determine whether you need to go through the expense and time required to rederive the affected colonies to remove this organism. This determination should be based on the potential consequences of the presence of the organism, your ability to contain the organism within the stipulated animal rooms, and your willingness to act if the organism spreads. In the case of *Staphylococcus aureus*, it behaves as an opportunist in mice primarily inducing disease in young and immunodeficient animals.[3] It may have unpredictable interactions with some GEM strains and stocks. You can designate any microorganism as undesirable in your facility to protect GEM strains and stocks even though the microorganism is not generally known to induce disease in other mouse strains.

White et al.[7] have suggested an extensive list of criteria for risk evaluation on a per microorganism basis. For example, they suggest reviewing prevalence of microorganisms, species specificity, research/disease effects, manipulation of mice and access to them, replaceability/availability (of mice), and several other criteria. This list is a good guideline for laboratory animal health professionals who are developing a microbial exclusion list for GEM facilities. Additional written resources for information about microbial monitoring are the Federation of European Laboratory Animal Science Associations (FELASA) guidelines for health monitoring of rodent and rabbit colonies,[8] *Infectious Diseases of Mice and Rats*,[9] and *Laboratory Animal Medicine*.[3] The Jackson Laboratory Web site (http://www.jax.org) lists the organisms that we test for in our mouse colonies. To find this and additional information on the way in which we categorize and respond to mouse pathogens refer to the "Mouse Strain Information" heading on the Jackson Laboratory Web site and click on the "Animal Health" heading.

ELEMENTS OF ROUTINE
MICROBIOLOGIC MONITORING

Operationally, four principal areas need to be addressed when setting up a micro-biological monitoring program for GEM colonies: sampling frequency, sample size determination, sampling strategy, and methods of sample testing. Clearly, sampling issues are very important in the overall framework of microbiologic monitoring.

Sampling frequency can be done on any basis—for example, monthly, quar-terly, or biannually—but in all cases should be driven by current microbial prevalence or risk of contamination. Generally speaking, selection of specific sampling intervals should not overtax your ability to collect and process samples nor put you on the brink of economic ruin. Sampling frequency should be greatest for the organisms most likely to contaminate a mouse facility because the organisms are already present, or because of current practices that increase the risk of contamination from other sources. For example, if your facility allows unrestricted access to mouse rooms, or the movement of personnel, mice, or supplies from contaminated to noncontaminated areas, the frequency of sampling should be high. Sampling should be the least frequent for uncommon organisms with a low likelihood of being found, such as Polyoma virus or K virus. Further-more, high barrier rooms with strict policies for entrance and exit of personnel and supplies are less likely to be contaminated than a conventionally operated mouse room, and therefore require less frequent sampling for microbial contami-nation. In the event of a disease outbreak, sampling frequency will initially be very high if you utilize a test and cull procedure. Note, however, that in this event, the frequency of sampling will change over time as infected animals are removed from the contaminated colony.

Hypothetically, estimation of an appropriate sample size for monitoring of GEM colonies can be accomplished by utilizing statistical formulae.[10–14] However, the universal existence of multiple constraints such as unknown microbial preva-lence rates within facilities, test sensitivities and specificities less than 100%, limitations on resource availability (small or very valuable colonies), and limited monitoring budgets at most GEM facilities, render the underlying assumptions of these statistical formulae extremely difficult to meet.[12,15] Therefore, some compromise is inevitable when determining the appropriate number of mice to take for each sampling period.

The type of cage used to house GEM further complicates the issue of determining an appropriate sample size for health monitoring. This is because the statistical tests mentioned above are based on the probability of detecting at least one infected animal in a study population in which each sampling unit (individual mice) has an equal chance of becoming infected. Microisolator cages (designated as cages with tight-fitting filtered tops), which are the predominant cage type in many modern mouse facilities, reduce the probability of detecting at least one infected mouse because they essentially eliminate the opportunity for random spread of infectious agents throughout an animal room. Ostensibly, this is an important and desirable function of these cages. Nonetheless, this same characteristic changes the paradigm of random spread of disease. The paradigm that applies to mouse rooms with microisolator cages is that disease spread is more likely within and not between cages.[12] In addition, within microisolator cage-containing facilities, the microisolator cage and not the mouse becomes the sampling unit. The fact that microisolator cages reduce cage-to-cage disease spread means that disease prevalence within a population of microisolators in a given room is likely to be very low. The design of the statistical tests used for sample size determination requires that in the case of low prevalence of disease, a large number of samples should be taken. This makes it economically difficult to adequately test a mouse room that utilizes microisolator cages.

Microisolator cages also add some complexity to the question of sampling strategy. When utilizing probability sampling, the ideal strategy for finding at least one infected animal in a population is random sampling. Theoretically, random sampling is unbiased in that each sampling unit in the population has an equal chance of being selected for testing. As described above, in animal rooms that utilize microisolator cages, the cage and not the mouse is the sampling unit, so it is cages that must be randomly sampled. Because it is not economically feasible to test every animal within each cage picked for sampling, the probability of picking an infected mouse for testing becomes an issue. If you are fortunate enough to pick an infected animal and it tests positive for disease, you cull the entire cage because you make the assumption that all the mice have been exposed to the agent and are likely to be infected. But, in the case that fewer than 100% of the mice in a cage are infected, picking an uninfected mouse for testing gives a false negative result for that cage. If you do not test that cage again for a long time, the infection goes undetected, but the room is considered free of infection because no positive results have been found. Thus, using microisolator cages as sampling units is risky as a sampling strategy unless all of the mice in the microisolator cage are tested at each sampling interval.

What can be done to enhance microbiological monitoring of microisolator caging? There are several options that have the same basic rationale: maximize your monitoring by enhancing the probability that the sampled mice are exposed to contaminated fomites or mice that may be present in an animal room. One option is to utilize sentinel mice that are exposed to either pooled animals from several areas of the animal room, or to pooled dirty bedding from representative areas of the room. In either case you should be able to easily trace back the origin of the pooled mice or bedding. When using pooled bedding, keep in mind that not all mouse pathogens are transmitted from bedding and may require a different form of exposure.[16] Thus, pooled bedding should be reserved for those pathogens known to be transmitted by this fomite. Pooling mice should be done at weaning age (and of same sex) to reduce the chances of fighting. The pooled mice should then be aged for at least four weeks with the sentinels.

An alternative option is to pool weanlings from several preselected areas of a room and test these mice directly after the appropriate exposure period. Unlike sentinels, colony mice are truly representative of the animals housed in the room and can be taken from surplus litters that will not be used for experimentation or breeding. Immunocompetent, age- and sex-matched sentinels can be placed with pooled colony mice if the pooled weanlings are immunodeficient. In each case described above, you are amplifying the probability of finding at least one infected animal by pooling from different microisolator cages that represent different areas of a mouse room. The downside of pooling schemes is that you are diluting the presence of pathogens by pooling potentially contaminated mice or materials. Theoretically, this may diminish the likelihood of detecting infected mice. However, it seems clear that compared to randomly sampling individual cages, exposing pooled colony mice to each other, or exposing sentinel mice to pooled colony mice or pooled bedding increases the probability that an infection will be spread and, thus, be more easily detected. Pooling samples may not be appropriate for every institution and should be reviewed as part of risk assessment.

Testing samples from mice for detection of pathogens falls into five principal categories: serology, molecular biology, parasitology, bacteriology, and histopathology. The application of these methodologies depends on several criteria including the type of organism being sought; the immunocompetence of the host; and the tissue, body fluid, or excretion that is to be tested. For routine monitoring of mice to detect any pathogen that may be present in a given colony or room, the testing is broad-based and encompasses all of the general categories of pathogenic organisms, including bacteria, fungi, viruses, and parasites. Thus, each test category mentioned above would be employed for routine monitoring.

Serology is reserved primarily for detection of viral infections of mice; however, there are also serological tests commercially available for *Mycoplasma* spp. and several other bacteria that infect mice. The predominant serological tests used today are the Enzyme Linked Immunosorbent Assay (ELISA) and Immunofluorescence Assay (IFA). Both test formats have reasonably good diagnostic sensitivity and specificity, but differ in their ease of use as screening tools. The ELISA, because of its multi-well format and capacity for automation, is more useful for screening large numbers of serum samples for evidence of antiviral or antibacterial antibodies. The IFA is frequently used as a confirmatory test for serum samples that have given positive results in ELISA tests. IFA is not useful for large-scale testing because of the requirement for observation of each test slide on a fluorescent microscope by an individual trained to read the pattern of fluorescence.

With regard to molecular diagnostics for mouse pathogens, PCR is by far the most heavily utilized method. The extraordinary diagnostic and analyte sensitivity make it very useful for detecting low genome copy numbers of pathogenic organisms from virtually any sample. One disadvantage for detection of mouse pathogens by PCR is that, for viruses in particular, some are only found in target tissues for a short period of time and therefore can give false negative results if the tissue sample is not taken in a timely manner during active infection. A second disadvantage is that there are PCR inhibitors in some mouse tissues that can give false negative results if not properly controlled for. A third disadvantage to using PCRs is that the methodology is somewhat complicated and therefore easy to confound by technical errors. For this reason, it is also important to run positive controls with each set of samples tested by PCR.

Testing mice for bacteria or parasites employs the same standard techniques that are used for testing humans and domestic animals. Thus, culture of samples on nutrient media and selective media are routinely employed to detect bacterial infections of mice, whereas fecal flotation, and direct microscopy of fecal preparations are utilized for detection of protozoan and metazoan mouse parasites. External parasites of the skin and hair are observed by examination of the mouse under a dissecting microscope. Finally, histopathology is often used as an adjunct method for detection of infections in mice. The identification of tissue lesions that suggest the presence of specific organisms can be very useful for determining whether an animal has been infected. In addition, there are histopathology techniques, such as *in situ* PCR and immunohistochemistry, that can be used to confirm the identity of organisms within tissue lesions.

SUMMARY

Microbiological monitoring of GEM colonies presents a unique challenge. Infectious organisms whose behavior is well described in standard inbred mouse strains may have a totally different interaction with a given GEM strain or stock. Thus, planning for microbiological monitoring of GEM colonies requires well-thought-out risk planning. A microbial exclusion list that outlines the organisms that you will and will not accept in your GEM facility should be drawn up. When undertaking microbiological monitoring of GEM colonies, sampling issues are extremely important. There are statistical formulae that can predict appropriate sample sizes, yet they are difficult to apply under many circumstances. Microisolator cages, although very good at preventing disease spread, also limit optimal sampling strategies. In microisolator facilities, pooling mice from different areas of a room or exposing sentinels to pooled bedding may be the best option for enhancing detection of disease. Finally, there are five standard laboratory test systems that are used for routine health monitoring of mice. All of these systems are applicable to microbiological monitoring of GEM colonies.

REFERENCES

1. Barthold, S.W., "Muromics": Genomics from the perspective of the laboratory mouse, *Comp. Med.*, 52, 206, 2002.
2. Zeiss, C.J., Mutant mouse pathology: An exercise in integration, *Lab Anim.* (NY), 31, 34, 2002.
3. Jacoby, R.O., Fox, J.G., and Davisson, M., Biology and diseases of mice, in *Laboratory Animal Medicine*, second edition, Academic Press, San Diego, 2002, chap. 3.
4. Rehg, J.E., Blackman, M.A., and Toth, L.A., Persistent transmission of mouse hepatitis virus by transgenic mice, *Comp. Med.,* 51, 369, 2001.
5. Pullium, J.K., et al. [Authors are Homberger, Benjamin, Dillehay, and Huerkamp], Confirmed persistent mouse hepatitis virus infection and transmission by mice with a targeted null mutation of tumor necrosis factor to sentinel mice using short-term exposure, *Comp. Med.*, 53, 439, 2003.

6. Foltz, C.J., et al., Spontaneous inflammatory bowel disease in multiple mutant mouse lines: Association with colonization by *Helicobacter hepaticus*, *Helicobacter*, 3, 69, 1998.
7. White, W.J., et al., Current strategies for controlling/eliminating opportunistic microorganisms, *ILAR J.*, 39, 291, 1998.
8. Nicklas, W., et al., Recommendations for health monitoring of rodent and rabbit colonies in breeding and experimental units, *Lab. Anim.* (UK), 36, 20, 2002.
9. National Research Council, *Infectious Diseases of Mice and Rats*, National Academy Press, Washington, DC, 1991.
10. Dubin, S. and Zietz, S., Sample size for animal health surveillance, *Lab Anim.* (NY), 20, 29, 1991.
11. Selwyn, M.R. and Shek, W.R., Sample sizes and frequency of testing for health monitoring in barrier rooms and isolators, *Contemp. Topics,* 33, 56, 1994.
12. Clifford, C.B., Samples, sample selection, and statistics: Living with uncertainty, *Lab Anim.* (NY), 30, 26, 2001.
13. Committee on Long-Term Holding of Laboratory Rodents, Long-term holding of laboratory rodents, ILAR News, L1-L25, 1976.
14. Thrusfield, M., *Veterinary Epidemiology*, second edition, Blackwell Science, Edinburgh, 1995.
15. Weisbroth, S.H., et al., Opportunistic infections in rats and mice, *ILAR J.,* 39, 272, 1998.
16. Cundiff, D.D., et al., Failure of a soiled bedding sentinel system to detect cilia-associated respiratory bacillus infection in rats, *Lab. Anim. Sci.,* 45, 219, 1995.

13 Effect of Intestinal Flora on Phenotype

Seiko Narushima and Kikuji Itoh

TABLE OF CONTENTS

Human and animal intestines harbor 100 trillion bacteria, comprising both beneficial and harmful bacteria. The intestinal bacteria are known to play an important role both in the health and the disease of the host. Genetically engineered animals are now considered to be indispensable models of human disease, because these animals clearly show the function of particular genes. However, it has been reported that no lesions or clinical signs are observed in these transgenic (Tg) or knock-out (KO) mice under germ-free (GF) or barrier-sustained conditions. Much of the data indicate that intestinal flora modify not only host responses as one of the accessory factors, but also gene expression in these mice. Russell and Burch proposed the concept of genotype, phenotype, and dramatype.[1] The genotype is modified by the developmental environment to form the phenotype, and the dramatype is characterized as the phenotype with proximate environmental factors added. According to the definition of dramatype, intestinal flora affect both the phenotype and the dramatype of a host, because intestinal flora are a factor in both the developmental and proximate environment. In this chapter, the importance of the intestinal flora in the phenotype and dramatype of genetically engineered mice is revealed.

Transgenic (Tg) mice carrying the human mutant transthyretin gene, MT-hMet30, are model animals for familial amyloidotic polyneuropathy.[2,3] Although amyloid deposition in organs was frequently observed in the Tg mice under conventional conditions, amyloidosis was not observed after these mice were kept clean by Caesarean section and maintained in barrier facilities. Furthermore, it was also reported that the intestinal flora in conventional mice from different facilities resulted in a different incidence of amyloid deposition in 6.0-hMet30 Tg mice.[4] These results strongly suggest that the control of intestinal flora is imperative for amyloid deposition in this Tg mouse model, and that the intestinal bacteria affect the pathophysiology of Tg mice as a proximate environment.

Although the etiology of inflammatory bowel disease (IBD) remains unclear, it is well known that genetic and environmental factors as well as microbial factors are involved in the development of IBD.[5] Recently, various GEM models for IBD including interleukin-2 KO, interleukin-10 KO, T-cell receptor αKO, and HLA-B27/β2m Tg models were proposed.[6–8] Interestingly, in many of these GEM models, no IBD lesions were observed when they were housed under GF conditions, although they spontaneously develop intestinal inflammation after birth under conventional conditions.[9] In IL-10-deficient mice, association with certain bacteria including *Eneterococcus faecalis*[10] or *Helicobacter hepaticus*[11,12] were able to induce colitis, whereas administration of *Lactobacillus* species protected the mice against the development of colitis.[13] It can be hypothesized that intestinal flora may modify IBD lesions in these animal models.

EFFECT OF INTESTINAL FLORA ON TUMORIGENESIS IN GEM MICE

The effect of microflora on the development of intestinal adenocarcinoma in T-cell receptor chain (TCR) and p53 double-knock-out mice was reported by Kado et al.[14] They produced GF groups of the double-knockout-mice and compared the incidence of hyperplasia, dysplasia, and adenocarcinoma between germfree and conventionalized mice. There was no development of adenocarcinoma of the colon among the germ-free groups. However, in the conventionalized group, adenocarcinomas in the ileocecum and cecum were detected in 70% of the animals tested. This result clearly indicates that intestinal bacteria play a major role in the development of adenocarcinoma of the colon in these animals.

COLON TUMORS IN GNOTOBIOTIC RASH2 TRANSGENIC MICE

RasH2 mice are transgenic mice carrying human prototype C-Ha-*ras* genes and are reported to show high sensitivity to various genotoxic carcinogens.[15] In previous studies, we proposed that the rasH2 transgenic mouse is also a useful model for determining the efficacy of anticancer agents, probiotics, and functional foods. The yogurt diet significantly reduced the number of colorectal tumors induced by 1,2–dimethylhydrazine (DMH) in RasH2 mice. In another experiment using rasH2 mice, we evaluated the effect of apple pectin or a culture condensate of *Bifidobacterium longum* on the development of colorectal tumors induced by DMH. The number of colorectal tumors was significantly lower in rasH2 mice fed an apple pectin-supplemented diet or the culture condensate.[16] Then we produced gnotobiotic (GB) rasH2 mice and compared the incidence of colorectal tumors induced by DMH injection. GF-rasH2 mice were inoculated with various mouse fecal suspensions or mixtures of bacteria isolated from mouse feces (Table 13.1). The combination of these four strains of bacteria showed the highest incidence in liver tumors in C3H/He GB mice.[17]

TABLE 13.1
Organisms Associated in the Transgenic Mice

Group	Organisms	
4 strains	*Escherichia coli* M66	Highest incidence of liver
	Enterococcus faecalis M266TA	tumors in C3H/He mice
	Clostridium paraputrificum VPI 6372	
	C. paraputrificum VPI 1584	
Lact.	*Lactobacillus acidophilus* strain 129	Most common species in mice
	L. murinus strain 91	
	L. fermentum strain 106	
E. Coli + Clost.	*Escherichia coli* E-17	Normalization of cecal size
	Clostridia derived from mouse feces	
AC stock	*Escherichia coli* E-17	Standard flora for SPF mice
	Lactobacillus acidophilus strain 129	
	L. fermentum strain 106	
	Clostridia derived from mouse feces	
	Bacteroides derived from mouse feces	

Three strains of *Lactobacilli* species are the most common species in mice, and are necessary for normalization of the mouse intestine. The combination of *E. coli* and *Clostridia* was able to convert germ-free mice to normal state, especially the cecal size of the mice. The intestinal flora of the AC stock group was composed of *E. coli*, three strains of *Lactobacilli*, *Clostridia*, and also *Bacteroidaceae*. The combination of these bacterial groups led to elimination of *Pseudomonas aeruginosa* from the intestine.[18] We use AC stock as seed flora for SPF mice as the minimum requirement for the normalization of GF mice. The incidence of colorectal tumors was high in both Tg- and non-Tg-GF mice (100%). In Tg-SPF mice and Tg-GB mice associated with basic mouse flora consisting of *Escherichia coli*, *Lactobacilli*, *Bacteroidaceae*, and *Clostridia*, the incidence of colorectal tumors was as high as that in GF mice. The mean tumor score in SPF-Tg mice was significantly higher than that in GF-Tg mice, although the incidences were 100% in both groups. In GF mice, the incidence in both Tg and non-Tg mice were 100%.

On the other hand, the incidence in SPF Tg mice was 100%, but 0% in SPF non-Tg mice. The tumor incidence and the score in Tg- and non-Tg-GB mice varied depending on the bacterial combination present in their intestines (Figures 13.1 and 13. 2). In four-strain Tg mice, the tumor score was higher than that in GF-Tg mice and other ex-germ-free Tg mice. It is considered that the four-strain group stimulated the tumor score in RasH2 mice as was seen in the liver tumors in C3H/He mice. The score of AC stock Tg mice was lower than that of SPF mice. There are two possible explanations for the tumor score being higher in SPF mice than in AC stock; AC stock might require more complex flora to increase the tumor score, or the difference in the developing environment of these two groups. AC stock mice start to interact with the intestinal flora after

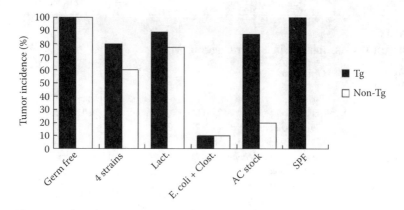

FIGURE 13.1 Incidence of colorectal tumors induced by 1,2–dimethylhydrazine in gnotobiotic transgenic mice.

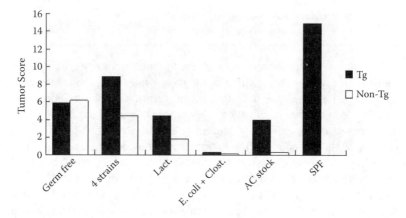

FIGURE 13.2 Tumor score of colorectal tumors induced by 1,2–dimethylhydrazine in gnotobiotic transgenic mice.

birth. Briefly, in Tg mice, intestinal flora stimulated the tumor development, but inhibited it in non-Tg mice. However, the tumor development in Tg-GB mice showed wide variation, depending on their intestinal flora. In all non-Tg-GB mice, intestinal flora inhibited the development of colorectal tumors. These results indicate that presence of human c-Ha-*ras* genes and intestinal bacteria substantially modify colorectal tumorigenesis induced by DMH.[19]

STANDARDIZATION OF INTESTINAL FLORA

Finally, the importance of controlling the intestinal flora of Tg and KO mice for each experiment should be clarified. In most animal facilities, SPF buildings are strictly controlled microbiologically, so that neither pathogenic microbes nor normal intestinal bacteria invade the barrier facility (Figure 13.3). It is reported that

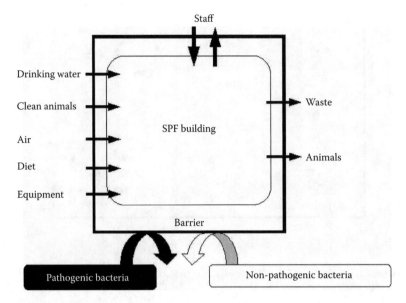

FIGURE 13.3 Barrier system for laboratory animals.

there is a clear difference in the composition of fecal flora between conventional and SPF mice.[20] In SPF mice, higher numbers of aerobes, *Pseudomonas aeruginosa*, as well as spore-forming clostridia were detected, whereas anaerobes were detected in lower numbers. It is suggested that aerobes and clostridia more easily invade these barriers than anaerobes, especially ones that are extremely oxygen sensitive. Also the variation of species in each bacterial group in SPF mice was lower than that in conventional mice. Furthermore, we have experienced samples of abnormal composition of fecal flora in barrier-sustained mouse colonies. High numbers of *Enterobacteriaceae* and *P. aeruginosa*, and abnormally high numbers of *C. perfringens* were observed in the intestines of the barrier-sustained colonies. Surprisingly, no *Lactobacilli* were detected, although *Lactobacillus* is the most common bacterial group in the mouse intestine.

What then is the minimum requirement for normalization of mouse flora? Itoh et al. examined the populations of indigenous bacteria necessary for the normalization of mice by evaluating the *E. coli* populations in the cecum and the size of the cecum as the parameters for normalization (Figure 13.4). The bacteria capable of reducing the cecal size of germ-free mice to normal state were *Clostridia*, and the combination of *Clostridia* and *Lactobacilli* was necessary for the reduction of the numbers of *E. coli* populations in the cecum of the normal state.[21] Most of the clostridia harbored in the mouse intestine are extremely oxygen sensitive and are called fusiform bacteria.[22] Another important requisite factor in normalization of intestinal flora is the colonization resistance against bacteria that produce opportunistic infection. The combination of *Lactobacilli*, *Clostridia*, and *Bacteroidaceae* was required to reach complete colonization resistance against *P. aeruginosa* in GB mice.[18]

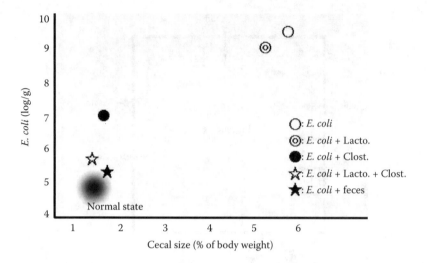

FIGURE 13.4 Control of *E. coli* populations and cecal size by indigenous bacteria in gnotobiotic mice.

In Tg and KO mice as models of human diseases, intestinal flora, as an environmental factor, strongly affect not only host responses, but also gene expressions in these mice, because these types of animals are more sensitive to environmental factors than the same background control animals. If we are indifferent to the intestinal flora of SPF mice, there would be a high possibility of existence of various types of flora compositions in each SPF mouse colony. Our results strongly suggest that one should use standard intestinal flora, consisting of the basic normalization for germ-free mice, to produce SPF colonies of Tg and KO mice.

REFERENCES

1. Russell, W. M. and Burch, R. L., *The Principles of Humane Experimental Technique*, Methuen, London, 1959.
2. Shimada, K., Maeda, S., Murakami, T., Nishiguchi, S., Tashiro, F., Yi, S., Wakasugi, S., Takahashi, K., and Yamamura, K., Transgenic mouse model of familial amyloidotic polyneuropathy, *Mol Biol Med* 6 (4), 333–343, 1989.
3. Yi, S., Takahashi, K., Naito, M., Tashiro, F., Wakasugi, S., Maeda, S., Shimada, K., Yamamura, K., and Araki, S., Systemic amyloidosis in transgenic mice carrying the human mutant transthyretin (Met30) gene. Pathologic similarity to human familial amyloidotic polyneuropathy, type I, *Am J Pathol* 138 (2), 403–412, 1991.
4. Noguchi, H., Ohta, M., Wakasugi, S., Noguchi, K., Nakamura, N., Nakamura, O., Miyakawa, K., Takeya, M., Suzuki, M., Nakagata, N., Urano, T., Ono, T., and Yamamura, K., Effect of the intestinal flora on amyloid deposition in a transgenic mouse model of familial amyloidotic polyneuropathy, *Exp Anim* 51 (4), 309–316, 2002.

5. Podolsky, D. K., Lessons from genetic models of inflammatory bowel disease, *Acta Gastroenterol Belg* 60 (2), 163–165, 1997.
6. Sadlack, B., Merz, H., Schorle, H., Schimpl, A., Feller, A. C., and Horak, I., Ulcerative colitis-like disease in mice with a disrupted interleukin-2 gene, *Cell* 75 (2), 253–261, 1993.
7. Bhan, A. K., Mizoguchi, E., Smith, R. N., and Mizoguchi, A., Colitis in transgenic and knockout animals as models of human inflammatory bowel disease, *Immunol Rev* 169, 195–207, 1999.
8. Dianda, L., Hanby, A. M., Wright, N. A., Sebesteny, A., Hayday, A. C., and Owen, M. J., T cell receptor-alpha beta-deficient mice fail to develop colitis in the absence of a microbial environment, *Am J Pathol* 150 (1), 91–97, 1997.
9. Sellon, R. K., Tonkonogy, S., Schultz, M., Dieleman, L. A., Grenther, W., Balish, E., Rennick, D. M., and Sartor, R. B., Resident enteric bacteria are necessary for development of spontaneous colitis and immune system activation in interleukin-10-deficient mice, *Infect Immun* 66 (11), 5224–5231, 1998.
10. Balish, E. and Warner, T., *Enterococcus faecalis* induces inflammatory bowel disease in interleukin-10 knockout mice, *Am J Pathol* 160 (6), 2253–2257, 2002.
11. Fox, J. G., Gorelick, P. L., Kullberg, M. C., Ge, Z., Dewhirst, F. E., and Ward, J. M., A novel urease-negative *Helicobacter* species associated with colitis and typhlitis in IL-10-deficient mice, *Infect Immun* 67 (4), 1757–1762, 1999.
12. Kullberg, M. C., Ward, J. M., Gorelick, P. L., Caspar, P., Hieny, S., Cheever, A., Jankovic, D., and Sher, A., *Helicobacter hepaticus* triggers colitis in specific-pathogen-free interleukin-10 (IL-10)-deficient mice through an IL-12- and gamma interferon-dependent mechanism, *Infect Immun* 66 (11), 5157–5166, 1998.
13. Madsen, K. L., Doyle, J. S., Jewell, L. D., Tavernini, M. M., and Fedorak, R. N., *Lactobacillus* species prevents colitis in interleukin 10 gene-deficient mice, *Gastroenterology* 116 (5), 1107–1114, 1999.
14. Kado, S., Uchida, K., Funabashi, H., Iwata, S., Nagata, Y., Ando, M., Onoue, M., Matsuoka, Y., Ohwaki, M., and Morotomi, M., Intestinal microflora are necessary for development of spontaneous adenocarcinoma of the large intestine in T-cell receptor beta chain and p53 double-knockout mice, *Cancer Res* 61 (6), 2395–2398, 2001.
15. Yamamoto, S., Urano, K., and Nomura, T., Validation of transgenic mice harboring the human prototype c-Ha-ras gene as a bioassay model for rapid carcinogenicity testing, *Toxicol Lett* 102–103, 473–478, 1998.
16. Ohno, K., Narushima, S., Takeuchi, S., Itoh, K., Mitsuoka, T., Nakayama, H., Itoh, T., Hioki, K., and Nomura, T., Inhibitory effect of apple pectin and culture condensate of *Bifidobacterium longum* on colorectal tumors induced by 1,2-dimethylhydrazine in transgenic mice harboring human prototype c-Ha-ras genes, *Exp Anim* 49 (4), 305–307, 2000.
17. Mizutani, T. and Mitsuoka, T., Effect of intestinal bacteria on incidence of liver tumors in gnotobiotic C3H/He male mice, *J Natl Cancer Inst* 63 (6), 1365–1370, 1979.
18. Itoh, K., Urano, T., and Mitsuoka, T., Colonization resistance against *Pseudomonas aeruginosa* in gnotobiotic mice, *Lab Anim* 20 (3), 197–201, 1986.
19. Narushima, S., Itoh, K., Mitsuoka, T., Nakayama, H., Itoh, T., Hioki, K., and Nomura, T., Effect of mouse intestinal bacteria on incidence of colorectal tumors induced by 1,2-dimethylhydrazine injection in gnotobiotic transgenic mice harboring human prototype c-Ha-ras genes, *Exp Anim* 47 (2), 111–117, 1998.

20. Itoh, K., Mitsuoka, T., Sudo, K., and Suzuki, K., Comparison of fecal flora of mice based upon different strains and different housing conditions, *Z Versuchstierkd* 25 (3), 135–146, 1983.
21. Itoh, K. and Freter, R., Control of Escherichia coli populations by a combination of indigenous *Clostridia* and *Lactobacilli* in gnotobiotic mice and continuous-flow cultures, *Infect Immun* 57 (2), 559–565, 1989.
22. Itoh, K. and Mitsuoka, T., Characterization of clostridia isolated from faeces of limited flora mice and their effect on caecal size when associated with germ-free mice, *Lab Anim* 19 (2), 111–118, 1985.

14 Helicobacter pylori and Stomach Cancer

Masae Tatematsu, Tetsuya Tsukamoto, and Tsutomu Mizoshita

TABLE OF CONTENTS

RELATION BETWEEN *H. PYLORI* INFECTION AND STOMACH CANCER IN MONGOLIAN GERBILS TREATED WITH CHEMICAL CARCINOGENS

Helicobacter pylori (Hp) is the main cause of active chronic gastritis and peptic ulcers in man, and WHO/IARC (World Health Organization/International Agency for Research on Cancer) concluded that Hp is a "definite carcinogen" in 1994.[1] Regarding animal models for Hp infection, Hirayama et al. proved that persistent infection of Hp in Mongolian gerbils[2] enhances glandular stomach carcinogenesis in Mongolian gerbils treated with chemical carcinogens.[3–5] This system was approved as a good model in terms of the similarity to human stomach cancers including well-differentiated adenocarcinoma (Figure 14.1a) and signet-ring cell carcinoma (Figure 14.1b). Nozaki et al.[6] showed that a high-salt diet enhances the effects of Hp infection on gastric carcinogenesis, and these two factors act synergistically to promote development of stomach cancers, the bacteria exerting a much stronger influence. In addition, early acquisition of infection significantly

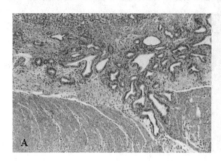 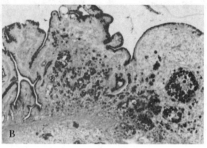

FIGURE 14.1 (See color insert following page 210.) Gastric carcinomas in the glandular stomach of Hp-infected Mongolian gerbils treated with chemical carcinogen N-methyl-N-nitrosourea. (a) H&E staining of a well-differentiated adenocarcinoma. (b) Alcian blue-periodic acid Schiff staining of a signet-ring cell carcinoma.

increases gastric chemical carcinogenesis in Mongolian gerbils, as compared to the case with late infection, suggesting the importance of childhood infection in human beings.[7]

In the gerbil model, the effect of the Hp eradication therapy was also analyzed in detail. Eradication diminishes enhancing effects of Hp infection on glandular stomach carcinogenesis in Mongolian gerbils,[8] especially in early stages of inflammation.[9]

Taking into account these results, we consider that Hp is a strong promoter of gastric carcinogenesis rather than an initiator, and thus eradication of Hp may be one of the most effective methods for prevention for neoplastic development in the stomach.

PHENOTYPIC CLASSIFICATION OF INTESTINAL METAPLASIA AND REGULATION OF GASTRIC AND INTESTINAL EPITHELIAL PHENOTYPIC EXPRESSION BY SPECIFIC TRANSCRIPTION FACTORS

Intestinal metaplasia (IM) was extensively studied as a possible premalignant condition in the human stomach. However, many questions remain regarding its pathogenesis as well as its relationship to gastric cancers. IM also occurs in the glandular stomach of Mongolian gerbils with Hp infection. Regarding classification of IM, several proposals were made by pathologists,[10–12] but although they are generally accepted, they are based solely upon intestinal properties and do not take into account the gastric properties that are still preserved in association.[13]

We proposed a new classification of IM on the basis of cellular differentiation status using both gastric and intestinal epithelial phenotypic markers.[13–15] Thereby, IM can be divided into two major types: a gastric-and-intestinal-mixed phenotype (GI-IM) and a solely intestinal phenotype (I-IM). A mixture of gastric and intestinal phenotypes occurs at the cellular level as well as at the glandular level, involving cells of the gastric phenotype together with those of intestinal phenotype as mentioned above for GI-IM. Inada et al. demonstrated that a gradual metaplastic shift from pyloric mucosa, through GI-IM, to I-IM occurs in the human stomach with chronic gastritis.[14] We also showed an analogous shift from GI-IM to I-IM in the rat stomach upon treatment with X-rays.[16] Therefore, IM subtypes may be considered not as independent entities, but rather a sequence of pathological states with a gradual change from stomach to intestinal character.[13,15] Regarding the phenotypic expression of endocrine cells, the phenotypes are in line with their epithelial counterparts in normal gastric mucosa and IM. A shift from gastric to intestinal phenotypic expression was also confirmed in endocrine cells as alternations of pyloric to intestinal metaplastic glandular ducts.

It is very important to clarify what genes regulate the gastric and intestinal epithelial phenotypic expression in the shift from gastric to intestinal character. Caudal-type homeobox (*Cdx*) 1 and *Cdx2* are mammalian members of the caudal-related homeobox gene family, which are believed to be important for the maintenance of intestinal epithelial cells.[17–22] Regarding gastric phenotypic expression, *Sox2* harboring an *Sry*-like high-mobility group box is a gastric-specific transcription factor.[15,23] Tsukamoto et al. showed that *Sox2* decreases gradually from normal pyloric, through GI-IM, to I-IM glandular duct, and *Cdx1* and *Cdx2* appear concomitantly in IM. Downregulation of *Sox2*, in addition to ectopic expression of *Cdx* genes, was suggested as an important factor for the development of IM.[15]

REVERSIBILITY OF TUMORLIKE LESIONS INDUCED BY HP IN MONGOLIAN GERBILS

Heterotopic proliferative glands (HPGs) frequently develop with Hp infection in the gerbil animal model (Figure 14.2a,b). They can be induced solely by Hp infection in the glandular stomach and are obviously reduced by eradication (Figure 14.2c); their characteristics include: (1) organized polarity of component cells; (2) formation of large cystic dilatations containing mucin, often with calcification, surrounded by collagen fibers; (3) shedding of epithelial cells and necrosis at the tips of lesions; (4) relation to high-grade inflammation with infiltration of inflammatory cells; and (5) organized polarity of the proliferating zone.[24] These features are quite different from those of well-differentiated adenocarcinomas, and HPGs appear related to severe gastritis, rather than being malignant in character.[24] HPGs consist of all components of epithelial cells including mucous and chromogranin A-positive endocrine cells, indicating their origin as stem cells (Figure 14.2d). In contrast, adenocarcinomas consist of only mucous components, suggesting their origin as progenitor cells (Figure 14.3). However, they also showed a shift from gastric to gastric-and-intestinal mixed or solely intestinal phenotypes during the overall course of Hp infection,[24] very similar to the IM case.

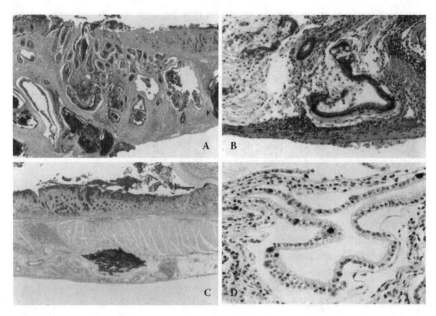

FIGURE 14.2 (See color insert following page 210.) The heterotopic proliferative glands (HPGs) in the glandular stomach of Hp-infected MGs. (a) HPGs with IM penetrating stomach mucosa (AB-PAS). (b) Higher magnification of HPG glands lined with intestinal metaplastic epithelium surrounded with collagen fibers and inflammatory cells (H&E). (c) Mucin remnant left after eradication (AB-PAS). (d) The chromogranin A-positive cells were observed among mucous epithelial cells in the HPG gland (chromogranin A immunohistochemistry).

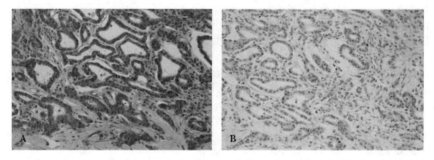

FIGURE 14.3 (See color insert following page 210.) The expression of chromogranin A is not detected in the cancerous lesions of the glandular stomach of MGs. (a) H&E staining. (b) Chromogranin A immunohistochemistry.

IS IM A PREMALIGNANT LESION IN GASTRIC CANCER?

It has been suggested that differentiated gastric carcinomas arise from mucosa with IM, but that undifferentiated gastric cancers originate from mucosa without IM, in view of morphological similarities between each cancer and the surrounding mucosa.[25–28] However, previous studies on phenotypic expression of individual intestinal metaplastic or stomach cancer cells pointed to several contradictions to this hypothesis.[13,16,29–38] In particular, Kawachi et al. showed that the phenotypes of microcarcinomas (defined as lesions less than 3.0 mm in diameter) and their surrounding mucosa are unrelated.[30] We also showed that *Cdx2* expression and the phenotypes of microadenocarcinomas are independent of the characteristics of the backgound epithelium.[38] Therefore, it remains unclear whether IM is indeed a premalignant lesion for gastric cancer.

GASTRIC AND INTESTINAL PHENOTYPIC EXPRESSION OF GASTRIC CANCER CELLS

It is widely believed that the phenotypic expression of tumor cells is the same as that of the tissue of origin. Histologically, human gastric cancers present as two major groups, the "intestinal" and "diffuse" types of Lauren,[25] which, respectively, nearly correspond to the "differentiated" and "undifferentiated" types of Nakamura et al.[26] and Sugano et al.[39] Although these classifications are widely used, they are not appropriate for studies of the histogenesis of gastric carcinomas and phenotypic expression at the cellular level, because they confuse intestinal phenotypic cancer cells with "diffuse" structure and the gastric phenotype with the "intestinal" type of Lauren.[13] Using gastric and intestinal epithelial cell markers, it is possible to analyze the phenotypic expression of each gastric cancer cell.[33,35,36,38,40–44] Stomach cancer cells can thereby now be clearly classified into gastric phenotype malignant cells, resembling pyloric gland cells and surface mucous cells, and intestinal phenotype malignant cells, such as goblet and intestinal absorptive cells.

We previously showed that gastric cancers at early stages, independent of the histological type, consist mainly of gastric phenotypic cancer cells, and a shift from gastric to intestinal phenotypic expression is clearly observed with progression in experimental animal models.[31,32,45,46] We and others have also reported a similar phenomenon in human gastric cancers.[33,35,36,40,41,47] The shift from gastric to intestinal phenotypic expression in differentiated cancers occurs at an earlier stage than in undifferentiated lesions.[36,38,40]

Regarding the expression of *Cdx* genes, we provided evidence that *Cdx1* and *Cdx2* are indispensable for intestinal phenotypic expression in gastric cancer cells.[42] In addition, we also showed that *Cdx2* is a useful prognostic marker for advanced gastric cancers.[43] It should further be noted that it may be expressed in very early stages of gastric carcinogenesis in association with a shift from gastric to intestinal phenotypic expression.[38] Similarly, *Sox2* may play an important role in the gastric phenotypic expression of gastric cancer cells.[44,48]

Taking into account the available results in combinations, we can conclude that a shift from gastric to intestinal phenotypic expression occurs with progression in gastric cancers as well as in IM and HPG. In addition, gastric and intestinal specific transcription factors such as *Sox2* and *Cdx1*/*Cdx2*, respectively, are strongly associated with this shift, presumably in all cases.

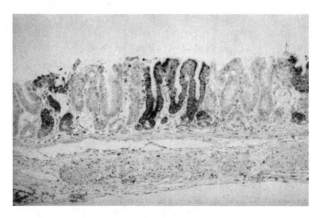

FIGURE 14.4 Analysis of cellular origin in normal pyloric mucosa in a C3H↔BALB/c chimera mouse with an antibody against C3H strain-specific antigen (CSA). Each gland is composed entirely of CSA positive or negative cells.

STEM CELLS OF GASTRIC AND INTESTINAL MUCOSA AND CLONAL GROWTH OF GASTRIC CANCER

As argued above, gastric cancer cells show heterogeneity in terms of gastric and intestinal epithelial phenotypic markers. To answer the question of whether this variation arises because of genesis from different types of cells or from a multipotential cell, we employed histological markers for analysis of mosaicism in chimeric mice. Antibodies strictly recognizing C3H strain-specific antigens (CSAs) established by Kusakabe et al. enable immunohistochemical discrimination of C3H cells in histological sections of chimera mouse tissues.[49–52] In normal gastric and intestinal mucosa of C3H/HeN↔BALB/c chimera mice, each glandular duct is composed entirely of CSA positive or negative cells, and no mixed glandular ducts are found, indicating that each individual glandular duct is derived from a single progenitor cell[50,51] (Figure 14.4). In addition, gastric cancers also develop from single cells, based on data from clonality analysis in C3H/HeN↔BALB/c chimeric mice. Therefore, we consider that all of the different types of epithelial cells of stomach and intestinal metaplasia are derived from a single progenitor cell in each glandular duct.

Our hypothesis for pathways of intestinal metaplasia and intestinalization of gastric carcinomas based on pathological findings with human material and experimental results with rodent models is summarized in Figure 14.5. Intestinalization progresses from gastric to intestinal phenotypic expression in noncancerous and cancerous tissue independently.

CONCLUSION

We consider that Hp is a strong promoter of gastric carcinogenesis rather than an initiator. We also believe that all of the different types of stomach epithelial

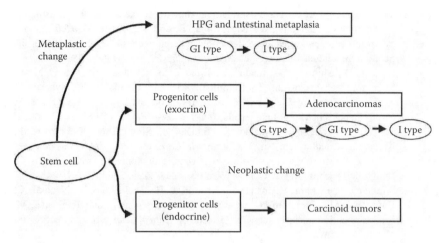

FIGURE 14.5 Schematic illustration of the hypothesis for pathways of HPGs, IMs, and intestinalization of gastric carcinoma based on the pathological findings. HPGs and IMs occur in consequence of the alternation of the stem cell level, and the origin of gastric cancers is a progenitor cell differentiating mucous cells. Intestinalization progresses from G through GI, to I type in noncancerous and cancerous tissue independently.

cells are derived from a single progenitor cell in each glandular duct. Similarly, cancers develop from single cells, based on data from clonality analysis in the C3H/HeN⇔BALB/c chimeric mice. A shift from gastric to intestinal epithelial phenotypic expression occurs with progression in gastric cancers, independent of the histological type, as well as in IM and HPG, this being associated with altered expression of the specific transcription factors, *Sox2* and *Cdx1/Cdx2*, which we hypothesize to play important roles in regulation of the phenotype.

REFERENCES

1. IARC, L. Schistosomes, Liver flukes and Helicobacter pylori, *IARC Monographs on the Evaluation of Carcinogenic Risks to Humans*. Vol. 61: pp. 1–241, 1994.
2. Hirayama, F., Takagi, S., Yokoyama, Y., Iwao, E., and Ikeda, Y. Establishment of gastric *Helicobacter pylori* infection in Mongolian gerbils. *J. Gastroenterol.*, 31 Suppl 9: 24–28, 1996.
3. Tatematsu, M., Yamamoto, M., Shimizu, N., Yoshikawa, A., Fukami, H., Kaminishi, M., Oohara, T., Sugiyama, A., and Ikeno, T. Induction of glandular stomach cancers in *Helicobacter pylori*-sensitive Mongolian gerbils treated with N-methyl-N-nitrosourea and N-methyl-N′-nitro-N-nitrosoguanidine in drinking water. *Jpn. J. Cancer Res.*, 89: 97–104, 1998.
4. Sugiyama, A., Maruta, F., Ikeno, T., Ishida, K., Kawasaki, S., Katsuyama, T., Shimizu, N., and Tatematsu, M. *Helicobacter pylori* infection enhances N-methyl-N-nitrosourea-induced stomach carcinogenesis in the Mongolian gerbil. *Cancer Res.*, 58: 2067–2069, 1998.

5. Shimizu, N., Inada, K., Nakanishi, H., Tsukamoto, T., Ikehara, Y., Kaminishi, M., Kuramoto, S., Sugiyama, A., Katsuyama, T., and Tatematsu, M. *Helicobacter pylori* infection enhances glandular stomach carcinogenesis in Mongolian gerbils treated with chemical carcinogens. *Carcinogenesis*, 20: 669–676, 1999a.

6. Nozaki, K., Shimizu, N., Inada, K., Tsukamoto, T., Inoue, M., Kumagai, T., Sugiyama, A., Mizoshita, T., Kaminishi, M., and Tatematsu, M. Synergistic promoting effects of *Helicobacter pylori* infection and high-salt diet on gastric carcinogenesis in Mongolian gerbils. *Jpn. J. Cancer Res.*, 93: 1083–1089, 2002a.

7. Cao, X., Tsukamoto, T., Nozaki, K., Tanaka, H., Shimizu, N., Kaminishi, M., Kumagai, T., and Tatematsu, M. Earlier *Helicobacter pylori* infection increases the risk for the N-methyl-N-nitrosourea-induced stomach carcinogenesis in Mongolian gerbils. *Jpn. J. Cancer Res.*, 93: 1293–1298, 2002.

8 Shimizu, N., Ikehara, Y., Inada, K., Nakanishi, H., Tsukamoto, T., Nozaki, K., Kaminishi, M., Kuramoto, S., Sugiyama, A., Katsuyama, T., and Tatematsu, M. Eradication diminishes enhancing effects of *Helicobacter pylori* infection on glandular stomach carcinogenesis in Mongolian gerbils. *Cancer Res.*, 60: 1512–1514, 2000.

9. Nozaki, K., Shimizu, N., Ikehara, Y., Inoue, M., Tsukamoto, T., Inada, K., Tanaka, H., Kumagai, T., Kaminishi, M., and Tatematsu, M. Effect of early eradication on *Helicobacter pylori*-related gastric carcinogenesis in Mongolian gerbils. *Cancer Sci.*, 94: 235–239, 2003.

10. Kawachi, T., Kogure, K., Tanaka, N., Tokunaga, A., and Sugimura, T. Studies of intestinal metaplasia in the gastric mucosa by detection of disaccharidases with "Tes-Tape". *J. Natl. Cancer Inst.*, 53: 19–30, 1974.

11. Jass, J. R. and Filipe, M. I. A variant of intestinal metaplasia associated with gastric carcinoma: A histochemical study. *Histopathology*, 3: 191–199, 1979.

12. Filipe, M. I., Munoz, N., Matko, I., Kato, I., Pompe-Kirn, V., Jutersek, A., Teuchmann, S., Benz, M., and Prijon, T. Intestinal metaplasia types and the risk of gastric cancer: A cohort study in Slovenia. *Int. J. Cancer*, 57: 324–329, 1994.

13. Tatematsu, M., Tsukamoto, T., and Inada, K. Stem cells and gastric cancer—Role of gastric and intestinal mixed intestinal metaplasia. *Cancer Sci.*, 94: 135–141, 2003.

14. Inada, K., Nakanishi, H., Fujimitsu, Y., Shimizu, N., Ichinose, M., Miki, K., Nakamura, S., and Tatematsu, M. Gastric and intestinal mixed and solely intestinal types of intestinal metaplasia in the human stomach. *Pathol. Int.*, 47: 831–841, 1997.

15. Tsukamoto, T., Inada, K., Tanaka, H., Mizoshita, T., Mihara, M., Ushijima, T., Yamamura, Y., Nakamura, S., and Tatematsu, M. Down regulation of a gastric transcription factor, Sox2, and ectopic expression of intestinal homeobox genes, *Cdx1* and *Cdx2*: Inverse correlation during progression from gastric/intestinal-mixed to complete intestinal metaplasia. *J. Cancer Res. Clin. Oncol.*, 130: 135–145, 2004.

16. Yuasa, H., Inada, K., Watanabe, H., and Tatematsu, M. A phenotypic shift from gastric-intestinal to solely intestinal cell types in intestinal metaplasia in rat stomach following treatment with X-rays. *J. Toxicol. Pathol.*, 15: 85–93, 2002.

17. Silberg, D. G., Sullivan, J., Kang, E., Swain, G. P., Moffett, J., Sund, N. J., Sackett, S. D., and Kaestner, K. H. Cdx2 ectopic expression induces gastric intestinal metaplasia in transgenic mice. *Gastroenterology*, 122: 689–696, 2002.

18. Silberg, D. G., Swain, G. P., Suh, E. R., and Traber, P. G. *Cdx1* and *Cdx2* expression during intestinal development. *Gastroenterology*, 119: 961–971, 2000.

19. Freund, J. N., Domon-Dell, C., Kedinger, M., and Duluc, I. The *Cdx-1* and *Cdx-2* homeobox genes in the intestine. *Biochem. Cell Biol.*, 76: 957–969, 1998.

20. Soubeyran, P., Andre, F., Lissitzky, J. C., Mallo, G. V., Moucadel, V., Roccabianca, M., Rechreche, H., Marvaldi, J., Dikic, I., Dagorn, J. C., and Iovanna, J. L. Cdx1 promotes differentiation in a rat intestinal epithelial cell line. *Gastroenterology*, 117: 1326–1338, 1999.

21. Mallo, G. V., Rechreche, H., Frigerio, J. M., Rocha, D., Zweibaum, A., Lacasa, M., Jordan, B. R., Dusetti, N. J., Dagorn, J. C., and Iovanna, J. L. Molecular cloning, sequencing and expression of the mRNA encoding human *Cdx1* and *Cdx2* homeobox. Down-regulation of *Cdx1* and *Cdx2* mRNA expression during colorectal carcinogenesis. *Int. J. Cancer*, 74: 35–44, 1997.

22. Mizoshita, T., Inada, K., Tsukamoto, T., Kodera, Y., Yamamura, Y., Hirai, T., Kato, T., Joh, T., Itoh, M., and Tatematsu, M. Expression of *Cdx1* and *Cdx2* mRNAs and relevance of this expression to differentiation in human gastrointestinal mucosa—With special emphasis on participation in intestinal metaplasia of the human stomach. *Gastric Cancer*, 4: 185–191, 2001.

23. Ishii, Y., Rex, M., Scotting, P. J., and Yasugi, S. Region-specific expression of chicken Sox2 in the developing gut and lung epithelium: Regulation by epithelial-mesenchymal interactions. *Dev. Dyn.*, 213: 464–475, 1998.

24. Nozaki, K., Shimizu, N., Tsukamoto, T., Inada, K., Cao, X., Ikehara, Y., Kaminishi, M., Sugiyama, A., and Tatematsu, M. Reversibility of heterotopic proliferative glands in glandular stomach of *Helicobacter pylori*-infected Mongolian gerbils on eradication. *Jpn. J. Cancer Res.*, 93: 374–381, 2002b.

25. Lauren, P. The two histological main types of gastric carcinoma: diffuse and so-called intestinal-type carcinoma: An attempt at a histo-clinical classification. *Acta Pathol. Microbiol. Scand.*, 64: 31–49, 1965.

26. Nakamura, K., Sugano, H., and Takagi, K. Carcinoma of the stomach in incipient phase: Its histogenesis and histological appearances. *Gann*, 59: 251–258, 1968.

27. Correa, P. A human model of gastric carcinogenesis. *Cancer Res.*, 48: 3554–3560, 1988.

28. Correa, P. *Helicobacter pylori* and gastric carcinogenesis. *Am. J. Surg. Pathol.*, 19 Suppl. 1: S37–43, 1995.

29. Egashira, Y., Shimoda, T., and Ikegami, M. Mucin histochemical analysis of minute gastric differentiated adenocarcinoma. *Pathol. Int.*, 49: 55–61, 1999.

30. Kawachi, H., Takizawa, T., Eishi, Y., Shimizu, S., Kumagai, J., Funata, N., and Koike, M. Absence of either gastric or intestinal phenotype in microscopic differentiated gastric carcinomas. *J. Pathol.*, 199: 436–446, 2003.

31. Tatematsu, M., Furihata, C., Katsuyama, T., Hasegawa, R., Nakanowatari, J., Saito, D., Takahashi, M., Matsushima, T., and Ito, N. Independent induction of intestinal metaplasia and gastric cancer in rats treated with N-methyl-N′-nitro-N-nitrosoguanidine. *Cancer Res.*, 43: 1335–1341, 1983.

32. Tatematsu, M., Katsuyama, T., Furihata, C., Tsuda, H., and Ito, N. Stable intestinal phenotypic expression of gastric and small intestinal tumor cells induced by N-methyl-N′-nitro-N-nitrosoguanidine or methylnitrosourea in rats. *Gann*, 75: 957–965, 1984.

33. Tatematsu, M., Furihata, C., Katsuyama, T., Miki, K., Honda, H., Konishi, Y., and Ito, N. Gastric and intestinal phenotypic expressions of human signet ring cell carcinomas revealed by their biochemistry, mucin histochemistry, and ultrastructure. *Cancer Res.*, 46: 4866–4872, 1986a.

34. Hattori, T. Development of adenocarcinomas in the stomach. *Cancer*, 57: 1528–1534, 1986a.

35. Tatematsu, M., Ichinose, M., Miki, K., Hasegawa, R., Kato, T., and Ito, N. Gastric and intestinal phenotypic expression of human stomach cancers as revealed by pepsinogen immunohistochemistry and mucin histochemistry. *Acta Pathol. Jpn.*, 40: 494–504, 1990a.

36. Tatematsu, M., Hasegawa, R., Ogawa, K., Kato, T., Ichinose, M., Miki, K., and Ito, N. Histogenesis of human stomach cancers based on assessment of differentiation. *J. Clin. Gastroenterol.*, 14 Suppl. 1: S1–7., 1992b.

37. Matsukuma, A., Mori, M., and Enjoji, M. Sulphomucin-secreting intestinal metaplasia in the human gastric mucosa. An association with intestinal-type gastric carcinoma. *Cancer*, 66: 689–694, 1990.

38. Mizoshita, T., Tsukamoto, T., Inada, K., Ogasawara, N., Hirata, A., Kato, S., Joh, T., Itoh, M., Yamamura, Y., and Tatematsu, M. Immunohistochemically detectable Cdx2 is present in intestinal phenotypic elements in early gastric cancers of both differentiated and undifferentiated types, with no correlation to non-neoplastic surrounding mucosa. *Pathol. Int.*, 54: 392–400, 2004b.

39. Sugano, H., Nakamura, K., and Kato, Y. Pathological studies of human gastric cancer. *Acta Pathol. Jpn.*, 32 Suppl. 2: 329–347, 1982.

40. Yoshikawa, A., Inada Ki, K., Yamachika, T., Shimizu, N., Kaminishi, M., and Tatematsu, M. Phenotypic shift in human differentiated gastric cancers from gastric to intestinal epithelial cell type during disease progression. *Gastric Cancer*, 1: 134–141, 1998.

41. Yamachika, T., Inada, K., Fujimitsu, Y., Nakamura, S., Yamamura, Y., Kitou, T., Itzkowitz, S. H., Werther, J. L., Miki, K., and Tatematsu, M. Intestinalization of gastric signet ring cell carcinomas with progression. *Virchows Arch.*, 431: 103–110, 1997.

42. Mizoshita, T., Inada, K., Tsukamoto, T., Nozaki, K., Joh, T., Itoh, M., Yamamura, Y., Ushijima, T., Nakamura, S., and Tatematsu, M. Expression of the intestine-specific transcription factors, Cdx1 and Cdx2, correlates shift to an intestinal phenotype in gastric cancer cells. *J. Cancer Res. Clin. Oncol.*, 130: 29–36, 2004a.

43. Mizoshita, T., Tsukamoto, T., Nakanishi, H., Inada, K., Ogasawara, N., Joh, T., Itoh, M., Yamamura, Y., and Tatematsu, M. Expression of Cdx2 and the phenotype of advanced gastric cancers: Relationship with prognosis. *J. Cancer Res. Clin. Oncol.*, 129: 727–734, 2003.

44. Tsukamoto, T., Mizoshita, T., Mihara, M., Tanaka, H., Takenaka, Y., Yamamura, Y., Nakamura, S., Ushijima, T., and Tatematsu, M. Sox2 expression in human stomach adenocarcinomas with gastric and gastric-and-intestinal-mixed phenotypes. *Histopathology*, in press.

45. Tatematsu, M., Katsuyama, T., Fukushima, S., Takahashi, M., Shirai, T., Ito, N., and Nasu, T. Mucin histochemistry by paradoxical concanavalin A staining in experimental gastric cancers induced in Wistar rats by N-methyl-N-nitro-N-nitrosoguanidine or 4-nitroquinoline 1-oxide. *J. Natl. Cancer Inst.*, 64: 835–843, 1980b.

46. Yuasa, H., Hirano, K., Kodama, H., Nakanishi, H., Imai, T., Tsuda, H., Imaida, K., and Tatematsu, M. Immunohistochemical demonstration of intestinal-type alkaline phosphatase in stomach tumors induced by N-methyl-N′-nitro-N-nitrosoguanidine in rats. *Jpn. J. Cancer Res.*, 85: 897–903, 1994.

47. Bamba, M., Sugihara, H., Kushima, R., Okada, K., Tsukashita, S., Horinouchi, M., and Hattori, T. Time-dependent expression of intestinal phenotype in signet ring cell carcinomas of the human stomach. *Virchows Arch.*, 438: 49–56, 2001.

48. Li, X. L., Eishi, Y., Bai, Y. Q., Sakai, H., Akiyama, Y., Tani, M., Takizawa, T., Koike, M., and Yuasa, Y. Expression of the SRY-related HMG box protein SOX2 in human gastric carcinoma. *Int. J. Oncol.*, 24: 257–263, 2004.

49. Kusakabe, M., Yokoyama, M., Sakakura, T., Nomura, T., Hosick, H. L., and Nishizuka, Y. A novel methodology for analysis of cell distribution in chimeric mouse organs using a strain specific antibody. *J. Cell Biol.*, 107: 257–265, 1988.

50. Tatematsu, M., Fukami, H., Yamamoto, M., Nakanishi, H., Masui, T., Kusakabe, N., and Sakakura, T. Clonal analysis of glandular stomach carcinogenesis in C3H/HeN<->BALB/c chimeric mice treated with N-methyl-N-nitrosourea. *Cancer Lett.*, 83: 37–42, 1994.

51. Tatematsu, M., Masui, T., Fukami, H., Yamamoto, M., Nakanishi, H., Inada, K., Kusakabe, M., and Sakakura, T. Primary monoclonal and secondary polyclonal growth of colon neoplastic lesions in C3H/HeN<-->BALB/c chimeric mice treated with 1,2–dimethylhydrazine immunohistochemical detection of C3H strain-specific antigen and simple sequence length polymorphism analysis of DNA. *Int. J. Cancer*, 66: 234–238, 1996.

52. Tsukamoto, T., Inada, K., Fukami, H., Yamamoto, M., Tanaka, H., Kusakabe, M., Bishop, C. E., and Tatematsu, M. Mouse strain susceptibility to diethylnitrosamine induced hepatocarcinogenesis is cell autonomous whereas sex-susceptibility is due to the micro-environment: Analysis with C3H <--> BALB/c sexually chimeric mice. *Jpn. J. Cancer Res.*, 91: 665–673, 2000.

15 Professional Use of Mutant Laboratory Mice in Research

John P. Sundberg and Carol J. Bult

TABLE OF CONTENTS

Inbred laboratory mice have been and continue to be the most widely used mammalian species for hair research, investigation, and discovery. Breakthroughs in the field come from identifying genes and their functions from spontaneous or genetically engineered mutant mice. Genetically engineered mice can also be used to prove the purported function of homologous genes in other species, including *Homo sapiens.* Therapeutic testing can be done as mouse models analogous to specific human diseases are developed and validated. For the purpose of this chapter, we focused on diseases affecting the hair follicle; however, any organ system can be substituted using the examples provided here. Many models for human hair diseases are already available[1] and others for more common hair problems, such as androgenetic alopecia, are under development.[2,3] This chapter describes factors to consider when using mice as experimental systems and general approaches for their use to address issues in hair biology. We also provide references to numerous literature and Web-based resources that can be consulted for information on genetics, pathology, and phenotype descriptions of mutant mice as well as international vendors for obtaining mice for use in research.

USES OF LABORATORY MICE IN SKIN AND HAIR RESEARCH

CHARACTERIZING MUTANT MICE

It is beyond the scope of this chapter to provide detailed protocols for establishing colonies and defining the biological phenotypes of inbred or genetically engineered mice. Information on such procedures is published and detailed elsewhere.[4–7] Protocols specific for the evaluation of skin and hair have been detailed by several independent groups.[6–9] Although specific methodologies vary among different laboratories the general approaches are very similar.

An important factor to consider when selecting or characterizing mouse models is to use a well-defined genetic background (e.g., inbred strains) whenever possible to minimize the variability of segregating (mixed genetic) backgrounds. Knowledge of background genes and their associated phenotypes is necessary to differentiate the phenotype for a new mutant from that of the background associated with each contributing strain. Some information regarding the genetic background of a particular mouse can be gleaned from its name. For example, a mouse designated as B6;129 means there is an undetermined genetic component from a C57BL/6 (substrain usually not known) mixed with one of the many 129 substrains.[10] Both strains contribute background genes and therefore diverse phenotypic effects.[6,11] In some cases the mouse strain name will also include a gene symbol and allele designation. For example, 129S-$Rb1^{tm1Tyj/+}$ indicates a 129 strain that is heterozygous for a targeted mutation allele in the retinoblastoma 1 (*Rb1*) gene. The "Tyj" in the allele name is the lab code for the research group of Dr. Tyler Jacks, who generated the allele. The Institute for Laboratory Animal Research (ILAR; http://www4.nas.edu/cls/afr.net) maintains a database of such lab codes. Mouse strain naming conventions are determined by The International Committee on Standardized Genetic Nomenclature for Mice. The official guidelines for mouse naming conventions can be found at the following URL: http://www.informatics.jax.org/mgihome/nomen/.

TABLE 15.1
Examples of Coat Color Variations in C57BL/6J Mice Due to Specific Mutations in Pigment Genes[a]

Chromosome	Gene Symbol	Gene Name/Coat Color
2	A	Agouti
2	A^i	Intermediate agouti
2	A^{sy}	Sienna yellow
2	A^{vy}	Viable yellow
2	A^{w-J}	White-bellied agouti Jackson
2	A^y	Yellow
2	a^{m-J}	Mottled agouti Jackson
2	a^{mt-33J}	Black and tan 33 Jackson
2	a^{td}	Tanoid
7	Tyr^{c-h}	Himalayan
7	Tyr^{c-J}	Albino Jackson
7	Tyr^{c-2J}	Albino 2 Jackson
7	$Tyrp1^{b-cJ}$	Cordovan Jackson
7	$Tyrp1^{b-J}$	Brown Jackson

[a] A complete list of alleles for the non-agouti (*a*), the tyrosinase (*Tyr*), and tyrosinase-related protein (*Tyrp1*) genes can be found by searching the Mouse Genome Informatics database (http://www.informatics.jax.org).

Inbred strains often have mutant genes fixed in the colony and these are designated in the name when the correct and full name is used—a rare occurrence in most of the published literature. For example, white mice are often white because they have the recessive albino (*a*) mutant gene. The C57BL/6J strains are thought to be black. However, many pigment mutations have occurred over the years, so this strain can have various colors (Table 15.1). Another example of strains commonly used in hair biology research are C3H/HeJ and FVB/N and their various substrains. Both of these strains carry the retinal degeneration 1 (*rd1*) gene, making them blind.[6,12] Retinal degeneration is therefore "normal" for these strains and not part of a new mutant phenotype. There are numerous retinal degeneration genes now known.[12,13] Using the full name and looking up the gene symbols and their phenotypes will minimize errors in interpreting the phenotype of new mutant mice.

Thousands of inbred strains of mice are available, and each strain has its own characteristic spontaneous background diseases that can complicate interpretation of a newly identified putative mutant mouse. Some of the more common strains have been studied and reported on in great detail including the C57BL/6,[6,14–26] BALB/c,[27–31] B6CBAF1 (hybrid stock),[16,27,32,33] FVB/N,[6,34,35] 129,[6,36,37] and C3H/HeJ.[22,24,30,31,38,39] Both the C57BL/6 and C3H/He strains and substrains are commonly used in hair biology research. Both are naturally prone to certain skin and hair diseases as well as other diseases. Descriptions of the characteristics of many inbred strains have been compiled by Michael Festing and are posted on the Web by the Mouse Genome Informatics group at The Jackson Laboratory (http://www.informatics.jax.org/external/festing/mouse/INTRO.shtml). The newly established Mouse Phenome Database (MPD; http://www.jax.org/phenome) contains baseline measurement data for a number of phenotypes for the most commonly used inbred strains of mice. The Mouse Tumor Biology Database (http://tumor.informatics.jax.org) includes information specifically on hyperplastic and neoplastic lesions in different strains of mice and eventually will be expanded to include general mouse pathology.[40,41]

The use of normal wild-type (+/+) controls that are age and gender matched for comparison will minimize misinterpretation of strain background diseases. When studying a recessive mutant phenotype, normal littermate mice are often used (+/+ or +/mutant gene, usually referred to as +/? for the genotype) because it is impossible to genotype clinically normal mice.

SPONTANEOUS MUTANTS

If, for example, you have identified a new, bald, spontaneous mutant mouse, one of the first steps in its characterization is to evaluate the clinical features carefully: onset of baldness (neonatal, after the first hair wave, late in life), extent (focal, diffuse), and distribution (dorsal, ventral, tail only, etc.). Histopathology can be extremely useful in the assessment of new mutants because examination of well-prepared slides[13] may reveal classic features of a mutant mouse already characterized in great detail. This was the case in our laboratory for remutations of lanceolate hair (*lah, lahᴶ*)[42,43] and asebia (*Scd1ᵃᵇᴶ, Scd1ᵃᵇ²ᴶ, Scd1ᵃᵇˣʸᵏ*).[44]

To verify that the genetic lesion associated with a newly identified mutant mouse is an allele of a previous mutation requires mating mice to generate offspring. Allelism can be initially confirmed by crossing the mutant strain with an unrelated inbred strain. This mating is followed by screening DNA from mutant F2 hybrids between two unrelated inbred strains for linkage in the anticipated region of the chromosome carrying the known mutation or amplifying DNA fragments by PCR using primer pairs known to be specific for the gene in question.[5,7] This approach limits the number of mice required for mapping the mutation considerably. An alternative mapping approach is to do a genome-wide screen of DNA collected from mutant F2 hybrids generated by crossing the mutant-containing strain with an unrelated inbred strain as described above in the case where no obvious similarity is evident between the new mutant and existing ones.[42,43]

Once the mutant gene locus is identified, current gene maps should be consulted to determine what previously mapped genes are close to the mutant gene locus. The Mouse Genome Informatics Web site (http://www.informatics.jax.org), described later in this chapter and elsewhere in this book, is the primary resource for genetic and physical maps for the laboratory mouse. If a mutation is found near the new mutant locus with a similar phenotype, the known mutant mouse should be crossed with the new mutant, as discussed above, to determine if they are allelic. In the process, candidate genes will become obvious from the map. An advantage of insertional mutagenesis, created by transgenesis, is that the inserted gene can be used as a molecular tool to clone out the flanking sequences for identification of novel genes.[45]

GENETICALLY ENGINEERED MICE

Several technologies have emerged in recent years that allow researchers to introduce specific genes from one organism to another to produce transgenic animals and to silence specific genes using targeted mutagenesis. These genetic engineering methods have become standard methodologies for producing animal models that allow researchers to focus on the molecular and phenotypic effects of specific genes.

Transgenic (Tg) mice are often "overexpressors" of a gene product or may exhibit a phenotype due to insertional mutagenesis. For example, an allele of hairless was created in this latter manner.[45] The process of characterizing these types of mutant mice are similar to the methods described above for spontaneous mutants.

Targeted mutagenesis (Tm, so-called "knock-outs") results in partial or complete loss of function of specific genes. It is important to take advantage of existing and readily available Web-based resources to check for existing mutant mice that have a similar phenotype to the targeted mutation mouse. Many of the spontaneous mouse mutations are functional nulls and have been studied for decades, thereby providing a wealth of biological information. For example, the angora mutant mouse was found to be due to a mutation in the fibroblast growth factor-5 ($Fgf5^{go}$) gene in this manner[46] as was the transforming growth factor alpha ($Tgfa^{wa1}$) for the waved 1 mutant mouse[47,48] and its receptor, epidermal growth factor receptor (Egf^{wa2}) for the waved 2 mutant mouse.[49]

IDENTIFYING MUTANT MICE RESULTING FROM MUTATIONS IN GENES OF INTEREST

As specific genes and proteins are directly associated with the development and function of hair follicles and epidermis, the search goes on for phenotypic deviant mice with mutations in these genes. The genetic engineering techniques described above make it relatively straightforward to create a mouse that carries a mutation in a specific gene. Examining the current gene maps using the Mouse Genome Informatics database provides clues to the existing mutant mice available that might have mutations in the gene of interest. An important consideration is to remember that genetic maps are prepared using data generated by a variety of approaches. Because genetic and physical map distances do not always have a one-to-one correspondence, it is important to look 10 to 20 cM on either side of the gene of interest for relevant genes. Angora ($Fgf5^{go}$) and, more recently, flaky tail (*ft*) were correlated with fibroblast growth factor 5 (*Fgf5*)[46] and filaggrin (*Flg*)[50] using this approach.

Correlation with the Human Disease

Several animal models have now been confirmed at the molecular level for both humans and mice, for example, hairless, *Hr*, with papular atrichia in humans[51, 52] and macaques[54] and the nude, *Foxn1^{nu}*, mouse with a human nude mutation.[53] Coordinated efforts to validate animal models of human disease are required between the molecular geneticists who clone and sequence the genes, veterinary pathologists who know the pathology of mice and many other species, and medical pathologists and dermatologists who are experts in the etiology of specific human diseases. It is essential to initially compare clinical and histologic features between the two species. This is best done by creating a differential diagnosis, a list of most likely human diseases. A thorough review of the literature will generate a list of diagnostic features for the disease of interest. The mice can then be carefully screened for these features. If there is a relatively high correlation between phenotypes in both species, then molecular approaches can be used to define the mutant gene in one species from the known sequence in the other species.

An important caveat to this approach is that the human disease may not be very carefully defined. Often several very similar diseases are lumped together due to overlapping phenotypes. In mice on inbred backgrounds maintained in controlled environments, the phenotype of single gene mutations is usually quite reproducible. Several phenotypic mimics may arise over time. These are mutant mice that arose independently on different backgrounds with similar or identical phenotypes but these mice are not allelic. Mapping places their loci on different chromosomes. These mice provide very powerful tools to work out the multilevel biochemical cascade involved in a particular disease. An example is the tabby (ectodysplasin-A, *Eda^{Ta}*) mouse as a model for anhydrotic ectodermal dysplasia. Two other phenotypic mutants, downless (ectodysplasin-A receptor, *Edar^{dl}*) and crinkled (ectodysplasin-A receptor-associated death domain, *Edaradd^{cr}*), are available and their mutant gene loci map to distinct chromosomes. Tabby has now been shown to be the ligand and downless the receptor.[55-62] All three mutant mouse phenotypes result in loss of the pilosebaceous unit in the tail and specialized hair follicle types at the base of the ears.[4]

Examples provided below illustrate common approaches to working with mutant mice to investigate basic biological processes. One can take advantage of other types of mutant mice that have normal pelage to do manipulations to study hair, hair follicles, or interactions of the pilosebaceous unit with hormones, the immune system, and so on.

Xenographs

A large number of mutant mice now exist with immunodeficiencies.[63] Some, such as nude (*Foxn1^{nu}*), severe combined immunodeficiency (*Prkdc^{scid}*), recombinase activating gene 1 (*Rag1*), and others will accept skin grafts from histoincompatible mice as well as unrelated species as phylogenetically diverse as sea turtles.[64]

Normal and diseased human skin grafts can be implanted and studied or manipulated as xenografts on these types of mice. This is a classic approach to many hair biology and therapeutic studies.[7]

As mutant genes are transferred onto different inbred strains to create congenic strains, it is easy to make double mutant mice on the same inbred background so as to not have to worry about background modifying genes due to mixing of the inbred backgrounds. A panel of double mutant mice that have severe combined immunodeficiency and a variety of hormone deficiency mutant genes is now available.[2] It is now possible to graft human or mouse skin onto these mice to see the effects of hormone absence or excess on hair development.

The alternative to using inbred mice is to purposefully outbreed the colony. This creates great diversity in the genetic background and, therefore, may more accurately resemble the genetic condition associated with human disease.[65] Unfortunately, this strategy increases phenotypic variability between individuals and necessitates the use of large numbers of mice for mapping studies. In addition, phenotypic variations between congenic strains carrying the same mutant gene cannot be studied independently, nor can the modifier gene responsible be mapped and cloned.

Allografts

We have routinely grafted mutant mouse skin onto immunodeficient (*fsn*, *ichq*)[66,67] or histocompatible normal mice[68] to follow development and changes in the mutant phenotype. Neonatal lethal mutations can have their skin phenotype followed progressively after the host dies, as was done with the mouse model for type II harlequin ichthyosis (*ichq*).[67] A surprise was encountered when grafting skin from C3H/HeJ mice with alopecia areata onto normally haired C3H/HeJ recipients. We anticipated that this would yield a useful method to have affected areas (the grafts) adjacent to normal unaffected sites (the recipient mouse skin). This should have provided a useful model for compound testing to expand the relatively few spontaneously affected mice available due to its complex polygenic nature.[69] Instead, by ten weeks after the full thickness skin graft surgery we found the mice developed alopecia areata all over their bodies.[68] This has yielded an inducible model that we can now use to define the pathogenesis of alopecia areata.[70, 71]

Exogenous Compounds

Normal and mutant mice can be treated with exogenous compounds to induce hair growth (cyclosporin A and fibroblast growth factor 7 induce hair growth in nude mice)[72–75] or to develop models that resemble human diseases in some ways. An androgen inducible model was reported in the B6CBAF1 hybrid mouse.[76] This model is not readily available and it turns out that this is not a strain- or hybrid-specific phenomenon. Rather it is a feature of several rodent species[77] and the response varies between inbred strains of mice.[3]

Many other ways have been used to manipulate mice to develop models to test hypotheses, and many of these are described throughout this book.

INTERNET RESOURCES FOR GENETIC AND PHENOTYPIC INFORMATION ABOUT MICE

Reviews are continually being written and updated as the value of mutant mice as research tools is discovered by an increasing number of investigators. These may be broad overviews[1,4,6] or they may focus on a particular type of disease, such as blistering or psoriasiform dermatoses, or even a group of proteins, such as the keratins.[78–80] A simple way to survey the literature on hair biology is to access Medline via the National Center for Biotechnology Information's PubMed system (http://www.ncbi.nlm.nih) followed by bibliographical scans of articles to find book chapters or articles that were published before Medline was established. For example, Chase published a series of papers in the 1950s on hair cycle in the mouse that are often missed when only electronic searches are employed. Yet these studies provide key information on this process.[81–85] There are also many specialized databases and related resources available on the Internet that researchers can access to find information related to mouse genetics and animal resources for skin and hair disease research. The purpose of this section is to give an overview of several of these resources and how they can be used to support the use of mice in the study of skin and hair biology. We have emphasized those resources that focus on genetics, mapping data, phenotype descriptions, pathology, and animal distribution.

Mouse Genome Informatics Web Site

The Mouse Genome Informatics (MGI) Web site (http://www.informatics.jax.org; Figure 15.1) provides access to a comprehensive integrated database on the genetics, genomics, and biology of the laboratory mouse. Three research groups contribute directly to the ongoing development of MGI and to the curation of the data it contains: the Mouse Genome Database (MGD) group,[86] the Gene Expression Database (GXD) group,[87] and the Mouse Genome Sequence database group (MGS).[88] The integrated MGI database, which is updated daily, includes information collected from over 65,000 published references as well as data submitted to the database directly by individual researchers and collaborators. The MGI database provides access to such information as descriptions of known mouse genes using standardized vocabularies, descriptions of phenotypes associated with particular genes, links to nucleotide and protein sequences, mapping data, information on expression patterns of genes, homologies of mouse genes with genes from other species, and links to other relevant databases. Table 15.2 provides some general statistics concerning the database content for MGI. Researchers typically query the MGI database using Web-based query forms. However, a number of specialized data summary reports are also available for viewing or downloading on the MGI FTP (File Transfer Protocol) site.

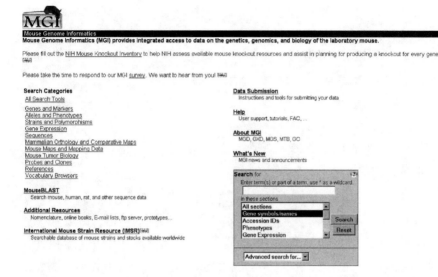

FIGURE 15.1 The Mouse Genome Informatics (MGI; http://www.informatics.jax.org) Web site home page and main menu. The MGI Home Page provides information about the MGI program as well as links to other useful genome-related Web sites. The MGI Main Menu provides links to the various search options and resources maintained by the MGI group. The Quick Gene Search box on the main menu allows researchers to search for genes according to their name or symbol without having to use one of the more complex query forms.

TABLE 15.2
General Statistics of the Mouse Genome Informatics Database Content[a]

Number of references	65,873
Number of genetic markers	43,167
Number of genes	24,971
Number of genetic markers with map positions	25,703
Number of genes with map positions	8,348
Number of genes w/DNA sequence information	20,848
Number of genes w/protein sequence information	12,531
Number of mouse–human homologies	5,711

[a] As of June 18, 2001. Current database statistics can be found at the MGI FTP site (ftp://ftp.informatics.jax.org/pub/informatics/reports/MGD_Stats.sql.rpt). Because the MGI database is updated daily, the statistics change on a daily basis.

SEARCHING MGI BY GENE NAME OR SYMBOL

From the main menu of the MGI Web site, users can perform a Quick Gene Search by typing in the symbol or name of a gene in which they are interested (Figure 15.1). For example, typing in the gene symbol for the alopecia locus, *ap,*

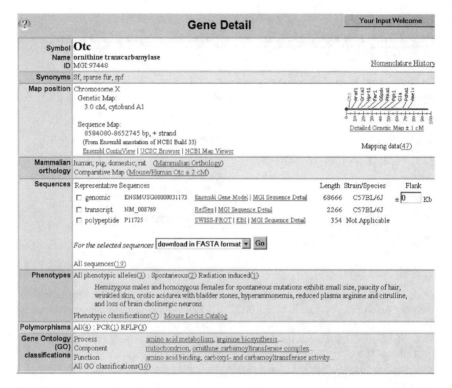

FIGURE 15.2 Gene Detail Page showing some of the information available in MGI on the *Otc* gene.

into the Quick Search text box returns a Gene Detail Page that, in turn, provides links to alleles, references, and phenotypes associated with the alopecia periodica gene.[89] In the case of the *ap* gene, the search reveals that the mutation arose spontaneously in an inbred stock of mice at the National Institute of Genetics in Japan, was described in 1952 by K. Tutikawa, and that the trait has yet to be mapped to a specific chromosome or associated with a specific gene sequence. For genes that are better characterized than the *ap* gene, links are provided to nucleotide and protein sequences, mapping data, phenotype descriptions, allele descriptions, developmental gene expression information, clones that contain the gene, mammalian homology data, and published references about the gene. For example, a Quick Gene Search for the *Otc* (ornithine transcarbamylase) gene reveals a rich set of genetic, genomic, and phenotype information for this gene (Figure 15.2).

An important aspect about searching electronic databases by gene names and symbols is that these labels can change over time. Many genes are known by several different names in the scientific literature and on occasion the same name or symbol is used to refer to different genes. The genes in the MGI database are named according to standards developed by the International Committee on

Mouse Nomenclature (http://www.informatics.jax.org/mgihome/nomen). The MGI curatorial staff surveys the scientific literature for each gene in the database and attempts to assemble a complete list of synonymous names and symbols that have been used to refer to a particular gene. Thus, if a researcher only knows a gene by an older nonstandard name he or she can still use that name or symbol to search the database and will get back appropriate results from the database and also discover what the current official name for that gene is. An example of how gene names and symbols can change over time is illustrated by the genetic nomenclature used to refer to the nude mouse.[90,91] When the genetic locus for the nude mouse phenotype was first reported in 1962, the locus was assigned the gene symbol, *nu*.[90] In 1994 Nehls and colleagues reported that the *nu* locus encoded a winged-helix protein and unofficially named the locus *whn*.[91] The locus was officially named NHF-3 forkhead homolog 1 (*Hfh11*[nu]) in 1995 by the International Committee on Standardized Genetic Nomenclature for Mice when it was proved that *nu* was an allele of the *Hfh11* gene. In 2000, the mouse genetic nomenclature committee approved a renaming of the entire forkhead box family and the gene was renamed once more to its current official name, Forkhead box N1 (*Foxn1*[nu]).[92]

The Web interface to the MGI database also supports sophisticated queries about genes that can be tailored to the interests of individual investigators by allowing them to apply constraints to their searches (Figure 15.3). For example, using the Genes, Markers, and Phenotype query form, searches can be limited to genes found on specific chromosomes or chromosomal regions (or those genes not found on a specific chromosome or chromosomal region) or to those genes associated with specific published references. Being able to access genetic information by chromosome or chromosomal region is particularly helpful to those involved in either positional cloning or quantitative trait loci (QTL) mapping. Researchers also can search using partial gene names or symbols in the event that the exact syntax of the official name or symbol is not known.

A recent addition to the query capacity of MGI is the ability to search the database using controlled vocabularies for molecular function, biological process, and cellular component associated with a gene product. Thus it is now possible to search the database for all genes that have been annotated as being transcription factors or whose gene product is found in mitochondria. The classification terms for these three areas are being developed as part of a consortium of model organism databases called the Gene Ontology Consortium (http://www.geneontology.org).[93] The Gene Ontology classfication terms are automatically provided on the gene detail pages in MGI (Figure 15.2).

Another commonly used query form in MGI allows researchers to generate customized linkage maps of genes and other markers for a particular chromosome. The map shown in Figure 15.4 was generated using MGI's comparative linkage map building tool. The map shows all of the genetic markers for the mouse along a user-defined region of the X chromosome. Along the right-hand side of the graphic, information on known human homologues to the mouse genes and the

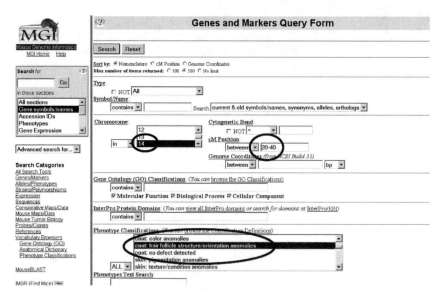

FIGURE 15.3 Example of one of the Web-based query forms used to retrieve information from the MGI database. As many or as few of the fields on the query forms can be filled out so that database users can ask broad queries (e.g., "Show me a list of all genes on chromosome 14") or very specific queries (e.g., "Show me a list of all genes on chromosome 14 that have been associated with alopecia and have been mapped between 20 and 40 cM").

cytogenetic location of those genes are indicated. The MGI database also supports the graphical representation of cytogenetic and physical maps.

SEARCHING MGI BY PHENOTYPE KEYWORDS

It is from the Genes, Markers, and Phenotypes query form (Figure 15.2) that researchers can also search the MGI database by phenotype terms. For example, if the word *skin* is entered as a phenotype term and gene is the type of genetic marker to search for, one would find over 250 entries in the database. It is important to explore different combinations of phenotype keywords to be certain that all of the relevant data in the database are found. If, for example, we had searched by the terms *skin* and *hair* simultaneously, the query would have returned only 61 entries in the database. In general, enter more keywords to narrow a search and fewer keywords to broaden the search. The entries returned in response to phenotype keyword queries are hyperlinked directly to descriptions of the phenotype. For example, one of the genes returned in response to a phenotype query using the terms *skin and hair* is the hair patches (*Hpt*) gene. A researcher wishing to know more about the phenotype associated with *Hpt* would mouse-click on the gene symbol to access the Mouse Locus Catalog (MLC) entry for that gene (Figure 15.5). The records in MLC provide a description of the phenotype and links to specific published references. If there is a

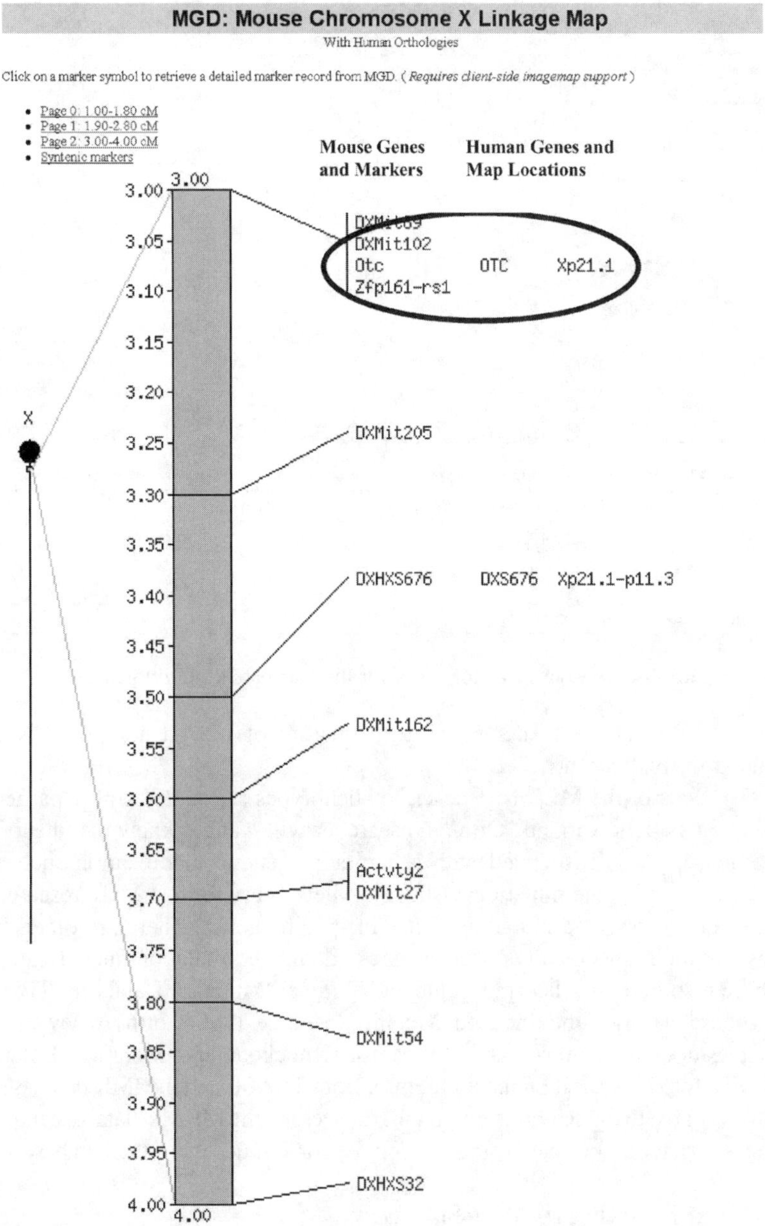

FIGURE 15.4 Example of a linkage map for the region of the X chromosome containing the *Otc* (ornithine transcarboxylase) gene in the mouse. Known human homologues and their cytogenetic locations are indicated to the right of the mouse genes and markers. Images such as these can be reproduced for publications to illustrate the location of new genes or mutant loci. If used, these maps should be cited according to the instructions at the MGI Web site (http://www.informatics.jax.org/mgihome/citation.shtml).

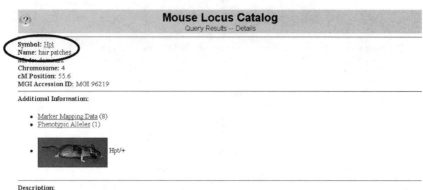

FIGURE 15.5 Mouse Locus Catalog entry for the hair patches (*Hpt*) gene.

picture of the mouse to demonstrate the nature of the phenotype, it is also included in this record.

The focus of the MLC is to describe phenotypes associated with a particular gene and its allelic variants. Often a researcher will want to know the phenotype of a mouse that is associated with a particular genetic modification such as a transgene or a targeted mutation. Using the Allele query form in MGI, researchers can search for targeted mutation alleles in specific genes. There are other databases that focus specifically on transgenic and targeted mutation mice. Examples of these resources are the Transgenic and Targeted Mutation Database (TBASE; http://tbase.jax.org), the Induced Mutant Resource (IMR; http://www.jax.org/resources/documents/imr/), and the Mouse Knockout and Mutation Database (MKMD; http://research.bmn.com/mkmd). Both TBASE and the IMR database are freely available to the scientific community. Access to the MKMD database requires users to register and to pay a subscription fee for full access to the database.

SEARCHING MGI BY SEQUENCE SIMILARITY

In addition to searching the MGI database by gene names or phenotypic keywords, researchers can also search the database using nucleotide or protein sequences via the MouseBLAST sequence similarity server that can be accessed from the MGI Web site main menu (Figure 15.1). If a query sequence matches a gene or protein sequence that is associated with a gene in MGI, the user will be presented with both the information on how well the sequence matches as well as a link directly to the MGI Gene Detail Page (see Figure 15.3 for an example).

MGI USER SUPPORT

The MGI group has full-time user support personnel to assist the research community in using the MGI database and related resources. A hypertext link to the e-mail system for contacting MGI user support is located at the bottom of each query and results page as well as from MGI's home page and main menu. To connect researchers who use mice in their research programs, the MGI group also maintains an electronic bulletin board system called mgi-list (http://www.informatics.jax.org/mgihome/lists/lists.shtml). A similar electronic community communication system is maintained for researchers who use the rat as a model organism (rat-list; http://www.informatics.jax.org/mgihome/lists/lists.shtml). There are currently over 1900 subscribers to mgi-list. The correspondence activity of these bulletin boards is archived and can be searched by keywords. A search of the mgi-list e-mail archive for the terms *skin* and *hair* revealed over 200 messages relating to this topic in mice (Figure 15.6). It is not necessary to be a subscriber to the list service in order to search the archives and read the e-mail correspondence, but you must register in order to post a reply or a question to the list. Researchers can register to join the MGI Email List Service by entering the appropriate information at the Web site or by sending an e-mail message containing the word "help" in the subject or body of the message to lists@lists.informatics.jax.org.

OBTAINING MICE FOR USE IN RESEARCH

The increasing popularity of the laboratory mouse as an animal model for understanding human disease and the increasing accessibility of genetic engineering technologies for targeted manipulation of the mouse genome has led to the need to establish numerous repositories of mice to meet the demands of the research community. The Jackson Laboratory was the prototype archive developed in 1992 in a direct response to the need for improved access by researchers to mouse resources without onerous intellectual property constraints.[94] The Jackson Laboratory's JAX Research Systems (JRS) distributes both standard inbred and genetically engineered mice and provides other services to the scientific community. Although JRS is the largest and most diversified resource for mouse stocks, other facilities have now been set up worldwide both to provide backup and to split the expense of archiving and distributing laboratory mice. In addition to nonprofit or government-run/financed facilities, commercial vendors produce large colonies of inbred and mutant laboratory mice and sometimes other species. A listing of some of the animal distribution resources is provided below. Also listed are institutions and companies that provide services for the generation of transgenic and targeted mutation mice.

Genes and Markers
Query Results -- Summary

Query Summary
Phenotypes Text Search contains *skin and hair*
Sort: by *Nomenclature*
Display Limit: equals *500*

78 matching items displayed

Chr	cM	Genome Coordinates (strand) NCBI Build 33	Symbol, Name
2	89.0	155076880-155114360 (+)	a, nonagouti
8	31.0		acd, adrenocortical dysplasia
12	3.0		Adam17, a disintegrin and metalloproteinase domain 17
Unknown			ap, alopecia periodica
X	44.0	97431680-97529107 (+)	Atp7a, ATPase, Cu++ transporting, alpha polypeptide
Unknown			bldy, baldy
11	58.0		Bsk, bare skin
5		108169449-108203565 (+)	Chek2, CHK2 checkpoint homolog (S. pombe)
19	45.0	43409994-43444103 (-)	Chuk, conserved helix-loop-helix ubiquitous kinase
1			Col5a2, procollagen, type V, alpha 2
13	30.0	62703118-62710210 (-)	Ctsl, cathepsin L
9	2.0		cw, curly whiskers
4	38.8		dep, depilated
11	58.0		Dfl, defolliculated
18	7.04	20916445-20946032 (+)	Dsg3, desmoglein 3
X	2.0	6474314-6482491 (-)	Ebp, phenylalkylamine Ca2+ antagonist (emopamil) binding protein
X	37.0	91379360-91800735 (+)	Eda, ectodysplasin-A
10	29.0	58350589-58425509 (-)	Edar, ectodysplasin-A receptor
13	6.0	12465936-12466262 (-)	Edaradd, EDAR (ectodysplasin-A receptor)-associated death domain
11	9.0	16647773-16809479 (+)	Egfr, epidermal growth factor receptor
4	syntenic		Er, repeated epilation
7	4.0	10905110-10918769 (+)	Ercc2, excision repair cross-complementing rodent repair deficiency, complementation group 2

FIGURE 15.6 Results of searching the MGI-list electronic bulletin board for the keywords: skin and hair.

NONPROFIT INSTITUTIONS

The Jackson Laboratory (TJL)
http://www.jax.org

The Jackson Laboratory maintains over 2500 different varieties of mice, 97% of which are solely distributed by JAX Research Systems. In addition, it maintains a DNA resource of genomic DNA for a number of inbred lines that are sold for PCR and mapping. The Jackson Laboratory recently established a mouse facility in Davis, California in collaboration with the University of California to facilitate the distribution of mice to researchers in western North America.

JAX Mice
http://jaxmice.jax.org/

This is the primary source of mice distributed via the Jackson Laboratory.

Induced Mutant Resource (IMR)
http://www.jax.org/resources/documents/imr/

The IMR focuses on maintaining stocks of mice with induced mutations.

Mouse Mutant Resource (MMR)
http://www.jax.org/resources/documents/mmr/mmrhome.html
The MMR maintains stocks of mice with spontaneous mutations.

TJL DNA Resource
http://www.jax.org/resources/documents/dnares
The TJL DNA Resource distributes genomic DNA isolated from different
  strains and mutant stocks of mice.

European Mouse Mutant Archive (EMMA)
http://www.emma.rm.cnr.it
The European Mouse Mutant Archive is a repository of live and cryopreserved mouse lines, including transgenics, targeted mutations, spontaneous mutants, congenics, and inbred strains. The main site of the
  repository is in Monterontondo, Italy. However, there are a number of
  nodes throughout Europe including those at MRC-Harwell (United
  Kingdom; http://www.mmgu.har.mrc.ac.uk), CNRS-CDTA (France;
  http://cdta.cnrs-orleans.fr), Karolinska Institute (Sweden; http://www.
  ki.se/kfc/meg/Murine/Murine.html), and the Gulbenkian Institute of
  Science (Portugal; www.igc.gulbenkian.pt).

National Institutes of Health Animal Genetic Resource (NIHAGR)
http://www.nih.gov/od/ors/dirs/vrp/s&slst.htm
This Web site provides information regarding the strains of mice and other
  laboratory animals that researchers can obtain from NIH.

Shared Information of Genetic Resources (SHIGEN)
http://www.shigen.nig.ac.jp
The SHIGEN Web site is a central listing of genetic resources available
  in Japan for several model organisms, including mammalian species.

Mouse Genetics Research Facility
http://lsd.ornl.ov/htpages/mgd/mouse_fac.htmlx
Oak Ridge National Laboratory, Oak Ridge, TN
ORNL maintains several hundred mouse stocks including mutants and
  inbred lines. The ORNL Mutant Mouse Database lists all of the strains
  that are propagated or maintained as frozen embryos by the group. A
  request form can be printed out and then mailed or faxed to request
  specific mice.

NICHD Transgenic Mouse Development Facility
http://main.uab.edu/
The NICHD Transgenic Mouse Development Facility is located at the
  University of Alabama at Birmingham. This service will produce transgenic mice on B6SJL hybrid, C57BL/6, or FVB inbred backgrounds.

University of Michigan Transgenic Animal Model Core
http://www.med.umich.edu/tamc/

The University of Michigan Transgenic Animal Model Core is a service for producing transgenic animals. The Web site provides access to a broad range of laboratory protocols associated with transgenic animal production. They also maintain a comprehensive list of Web sites of academic and commercial transgenic facilities around the world.

International Mouse Strain Resource (IMSR)
http://www.jax.org/pub-cgi/imsrlist
http://imsr.har.mrc.ac.uk

The IMSR does not distribute mice. Rather, the Web site provides a single point of entry for users to search the stock lists of multiple animal distribution facilities at once. Currently researchers can search for strains of mice available as live or frozen embryos at The Jackson Laboratory and the Harwell Mammalian Genetics Unit. The current plans for the IMSR are to provide access to the inventories of additional mouse repositories.

COMMERCIAL VENDORS

Charles River Laboratories
http://www.criver.com/

Charles River Laboratories sells genetically defined mice as well as other laboratory animals and provides transgenic animal production services.

Taconic Farms
http://www.taconic.com

Taconic sells genetically defined mice as well as other laboratory animals and provides transgenic animal production services.

Harlan Sprague Dawley
http://www.harlan.com

Harlan sells genetically defined mice as well as many other laboratory animals.

B&K Universal
http://www.bku.com

B&K sells genetically defined mice as well as other laboratory animals and provides transgenic animal production services.

Lexicon Genetics, Inc.
http://www.lexgen.com

Lexicon offers a number of genomics-related services including the large-scale production of transgenic mice.

Incyte Genomics, Inc.

http://www.incyte.com

Incyte offers a number of genomics-related services including the large-scale production of transgenic mice.

DNX Transgenics Sciences

http://www.dnxtrans.com

DNX provides production services for producing transgenic mice and rats and related services.

OTHER WEB-BASED RESOURCES RELEVANT TO SKIN AND HAIR RESEARCH

The following are resources that may provide useful information to researchers in the skin and hair community. These resources are not necessarily geared to basic research with animal models.

Keratin.com

http://www.keratin.com

This Web site is a compendium of information and links to other sites with a focus on hair biology and alopecia. It also provides support for several interactive discussion boards on topics related to hair biology. The focus is on research, not clinical data and information.

Pathbase

http://pathbase.net

This is a group effort of the European Mouse Pathology Consortium and invited pathologists from other continents to provide information on the variety of lesions found in genetically engineered mutant mice, spontaneous mutant mice, and inbred strains. A new subdirectory, skinbase, now focuses on describing and illustrating the phenotypes of spontaneous and genetically engineered mutant mice with skin and hair abnormalities.

PathIT

http://pathit.com/icrt/mouse/index.htm

An atlas of spontaneous tumors in the CD-1 mouse.

Mouse Tumor Biology (MTB) Database

http://tumor.informatics.jax.org

The Mouse Tumor Biology Database is accessible from the MGI Web site's main menu. The purpose of the database is to provide information on the cancer characteristics of genetically defined mice (inbred, transgenics, and targeted mutation). The resource will expand over time to include more general pathobiology of the laboratory mouse and is described in more detail elsewhere in this book.

The Internet Dermatology Society

http://www.telemedicine.org

This Web site is devoted to providing the research community with access to information and resources regarding skin disease. The information on the site is predominantly human and is geared toward the medical/clinical community.

Introduction to Diseases of the Skin

http://matrix.ucdavis.edu/tumors.html

This Web resource is an electronic text for introducing students to the diseases of the skin. The focus of the site is human disease.

Transgenic List

http://www.med.ic.ac.uk/db/dbbm/tglist.htm

The transgenic list is an e-mail list service provided by the Imperial College School of Medicine in London (United Kingdom). The list is open to subscription by any interested party and is not moderated. Only registered users are permitted to read and post messages to the list.

The Whole Mouse Catalog

http://www.rodentia.com/wmc/

A very useful compendium of information and resources related to laboratory mice.

ACKNOWLEDGMENTS

This work was supported in part by grants from the National Institutes of Health (RR173, AR43801, CA34196, CSXD22A, HG01559, HG03300), the National Alopecia Areata Foundation, the North American Hair Research Society, and The Council for Nail Disorders. The MouseBLAST service hosted at the MGI Web site is supported in part by DOE-FG02-99ER62850.

REFERENCES

1. Sundberg JP, King Jr LE. Mouse models for the study of human hair loss. *Dermatol Clin* 1996; 14:619–632.
2. Sundberg JP, Beamer WG, Uno H et al. Androgenic alopecia: *In vivo* models. *Exp Mol Pathol* 1999; 67:118–130.
3. Sundberg JP, King Jr LE, Bascom C. Animal models for male pattern (androgenic) alopecia. *Eur J Dermatol* 2001; 11:321–325.
4. Sundberg JP. *Handbook of Mouse Mutations with Skin and Hair Abnormalities: Animal Models and Biomedical Tools.* Boca Raton, FL: CRC Press, 1994.
5. Sundberg JP, Boggess D. *Systematic Approach to the Evaluation of Mouse Mutations.* Boca Raton, FL: CRC Press, 2000.

6. Ward JM, Mahler JF, Maronpot RR et al. *Pathology of Genetically Engineered Mice*. Ames: Iowa State University Press, 2000.
7. Sundberg JP, Montagutelli X, Boggess D. Systematic approach to evaluation of mouse mutations with cutaneous appendage defects. In: Chuong M. ed. *Cutaneous Appendages*. Austin, TX: R.G. Landes, 1998: 421–435.
8. Paus R, Müller-Röver S, Van der Veen C et al. A comprehensive guide for the recognition and classification of distinct stages of hair follicle morphogenesis. *J Invest Dermatol* 1999; 113:523–532.
9. Müller-Röver S, Handjiski B, Van der Veen C et al. A comprehensive guide for the accurate classification of murine hair in distinct hair cycle stages. *J Invest Dermatol* 2001; 117:3–15.
10. Relyea MJ, Miller J, Boggess D et al. Necropsy methods for laboratory mice: Biological characterization of a new mutation. In: Sundberg JP, Boggess D. eds. *Systematic Approach to Evaluation of Mouse Mutations*. Boca Raton, FL: CRC Press, 2000:57–90.
11. Simpson EM, Linder CC, Sargent EE et al. Genetic variation among 129 substrains and its importance for targeted mutagenesis in mice. *Nat Gen* 1997; 16:19–27
12. Bryton C, Justice M, Montgomery CA. Evaluating mutant mice: Anatomic pathology. *Vet Pathol* 2001; 38:1–19.
13. Smith RS, John SWM, Nishina PM et al. *Systematic Evaluation of the Mouse Eye: Anatomy, Pathology, and Biomethods*. Boca Raton, FL: CRC Press, 2002.
14. Blackwell BN, Bucci TJ, Hart RW et al. Longevity, body weight, and neoplasia in *ad libitum*-fed and diet-restricted C57BL/6J mice fed NIH-31 open formula diet. *Toxicol Pathol* 1995; 23:570–582.
15. Frith CH, Highman B, Burger G et al. Spontaneous lesions in virgin and retired breeder BALB/c and C57BL/6J mice. *Lab Anim Sci* 1983; 33:273–286.
16. Porter DD, Porter HG, Cox NA. Immune complex glomerulonephritis in one-year old C57BL/6 mice induced by endogenous murine leukemia virus and erythrocyte antigens. *J Immunol* 1973; 111:1626–1633.
17. Volk MJ, Pugh TD, Kim MJ et al. Dietary restriction from middle age attenuates age-associated lymphoma development and interleukin 6 dysregulation in C57BL/6 mice. *Cancer Res* 1994; 54:3054–3061.
18. Sheldon WG, Bucci TJ, Blackwell B et al. Effect of *ad libitum* feeding and 40% feed restriction on body weight, longevity, and neoplasms in B6C3F1, C57BL/6, and B6D2F1 mice. In: Mohr U, Dungworth DL, Capen CC et al. eds. *Pathology of the Aging Mouse*. Vol. 2. Washington, DC: ILSI; 21–26.
19. Andrews AG, Dysko RC, Spilman SC et al. Immune complex vasculitis with secondary ulcerative dermatitis in aged C57BL/6NNia mice. *Vet Pathol* 1994; 31:293–300.
20. Dimai HP, Linkhart TA, Linkhart SG et al. Alkaline phosphatase levels and osteoprogenitor cell numbers suggest bone formation may contribute to peak bone density differences between two inbred strains of mice. *Bone* 1998; 22:211–216.
21. Holzgraefe M, Spoerri O. Pathological growth of skull-radiological fine structure in experimental animals. *Neurosurg Rev* 1980; 3:257–259.
22. Sheng MH, Baylink DJ, Beamer WG et al. Histomorphometric studies show that bone formation and bone mineral apposition rates are greater in C3H/HeJ (high-density) than C57BL/6J (low-density) mice during growth. *Bone* 1999; 25:421–429.

23. Sundberg JP, Brown KS, McMahon WM. Chronic ulcerative dermatitis in black mice. In: Sundberg JP. ed. *Handbook of Mouse Mutations with Skin and Hair Abnormalities*. Boca Raton, FL: CRC Press, 1994: 485–492.

24. Sundberg JP, Sundberg BA, King LE. Cutaneous changes in commonly used inbred mouse strains and mutant stocks. In: Mohr U, Dungworth DL, Capen CC et al. eds. *Pathobiology of the Aging Mouse*. Vol. 2. Washington, DC: ILSI, 1996:325–337.

25. Van der Heijden A, van Dijk JE, Lemmens AG et al. Spleen pigmentation in young C57BL mice is caused by accumulation of melanin. *Lab Anim* 1995; 29:459–463.

26. Sundberg JP. Splenic melanosis in black mice. *JAX Notes* 1989; 439:4.

27. Frith CH, Ward JM. *Color Atlas of Neoplastic and Non-Neoplastic Lesions in Aging Mice*. Amsterdam: Elsevier, 1988.

28. Frith CH, Highman B, Burger G et al. Spontaneous lesions in virgin and retired breeder BALB/c and C57BL/6 mice. *Lab Anim Sci* 1983; 33:273–286.

29. Sundberg JP, Brown KS, Bedigian R et al. Ulcerative blepharitis and periorbital abscesses in BALB/cJ and BALB/cByJ Mice. *JAX Notes* 1990; 443:3–4.

30. Brunnert SR. Morphologic response of myocardium to freeze-thaw injury in mouse strains with dystrophic cardiac calcification. *Lab Anim Sci* 1997; 47:11–18.

31. Eaton GJ, Custer RP, Johnson FN et al. Dystrophic cardiac calcinosis in mice: Genetic, hormonal, and dietary influences. *Am J Pathol* 1978; 90:173–186.

32. Maronpot RR, Boorman GA, Gaul BW. *Pathology of the Mouse: Reference and Atlas*. Vienna, IL: Cache River, 1999.

33. Smith RS, Roderick TH, Sundberg JP. Microphthalmia and associated abnormalities in inbred black mice. *Lab Anim Sci* 1994; 44:551–560.

34. Goelz MF, Mahler J, Harry J et al. Neuropathologic findings associated with seizures in FVB mice. *Lab Anim Sci* 1998; 48:34–37.

35. Hennings H, Glick AB, Lowry DT et al. FVB/N mice: An inbred strain sensitive to the chemical induction of squamous cell carcinomas in the skin. *Carcinogenesis* 1993; 14: 2353–2358.

36. Stevens LC. A new inbred subline of mice (129/terSV) with a high incidence of spontaneous testicular teratomas. *J Natl Cancer Inst* 1973; 50:235–242.

37. Zheng QY, Johnson KR, Erway LC. Assessment of hearing in 80 inbred strains of mice by ABR threshold analysis. *Hear Res* 1999; 130:94–107.

38. Hüsler MR, Beamer WG, Boggess D et al. Neoplastic and hyperplastic lesions in aging C3H/HeJ mice. *J Exp Anim Sci* 1998; 38:165–180.

39. Nagy ZM, Misanin JR. Visual perception in the retinal degenerate C3H mouse. *J Comp Physiol Psychol* 1970; 72:306–310.

40. Bult CJ, Krupke DM, Sundberg JP et al. Mouse tumor biology database (MTB): Enhancements and updates. *Nuc Acids Res* 2000, 28:112–114.

41. Bult CJ, Krupke DM, Naf D et al. Web-based access to mouse models of human cancers: The mouse tumor biology (MTB) database. *Nuc Acids Res* 2001; 29:95–97.

42. Montagutelli X, Hogan ME, Aubin G et al. Lanceolate hair (*lah*): A recessive mouse mutation with alopecia and abnormal hair. *J Invest Dermatol* 1996; 107:20–25.

43. Sundberg J, Boggess D, Bascom C et al. Lanceolate hair-J (*lah^J*): A mouse model for human hair disorders. *Exp Dermatol* 2000; 9:206–218.

44. Sundberg JP, Boggess D, Sundberg BA et al. Asebia-2J (*Scd1^{ab2J}*): A new allele and a model for scarring alopecia. *Am J Pathol* 2000; 156:2067–2075.

45. Jones, JM, Elder JT, Simin K et al. Insertional mutation of the hairless locus on mouse chromsome 14. *Mamm Genome* 1993; 4:639–43.
46. Hebert JM, Rosenquist T, Gotz J et al. *Fgf5* as a regulator of the hair growth cycle: Evidence from targeted and spontaneous mutations. *Cell* 1994; 78:1017–1025.
47. Mann GB, Fowler KJ, Gabriel A et al. Mice with a null mutation of the TGF alpha gene have abnormal skin architecture, wavy hair, and curly whiskers and often develop corneal inflammation. *Cell* 1993; 73:249–261.
48. Luetteke NC, Qiu TH, Peiffer RL et al. TGF alpha deficiency results in hair follicle and eye abnormalities in targeted and waved-1 mice. *Cell* 1993; 73:263–78.
49. Luetteke NC, Phillips HK, Qiu TH et al. The mouse waved-2 phenotype results from a point mutation in the EGF receptor tyrosine kinase. *Genes Dev* 1994; 8:399–413.
50. Pressland RB, Boggess D, Lewis SP et al. Loss of normal profilaggrin and filaggrin in flaky tail (*ft/ft*) mice: An animal model for the filaggrin-deficient skin disease ichthyosis vulgaris. *J Invest Dematol* 2000; 115:1072–81.
51. Ahmad W, Panteleyev AA, Christiano AM. The molecular basis of congenital atrichia in humans and mice: Mutations in the hairless gene. *J Invest Dermatol Symp Proc* 1999; 4:240–243.
52. Ahmad W, Haque Faiyaz UHM, Brancolini V et al. Alopecia universalis associated with a mutation in the human hairless gene. *Science* 1998; 279:720–724.
53. Frank J, Pignata C, Panteleyev AA et al. Exposing the human nude phenotype. *Nature* 1999; 398:473–474.
54. Ahmad W, Faiyaz-µl-Haque M, Brancolini V et al. Alopecia universalis associated with a mutation in the human hairless gene. *Science* 1998; 279:720–4.
55. Srivastava AK, Pispa J, Hartung AJ et al. The Tabby phenotype is caused by a mutation in a mouse homologue of the *EDA* gene that reveals novel mouse and human exons and encodes a protein (ectodysplasin-A) with collagenous domains. *Proc Natl Acad Sci* 1997; 94:13069–74.
56. Mikkola M, Pispa J, Pekkanen M et al. Ectodysplasin, a protein required for epithelial morphogenesis, is a novel TNF homologue and promotes cell–matrix adhesion. *Mech Dev* 1999; 88:133–146.
57. Headon DJ, Overbeek PA. Involvement of a novel Tnf receptor homologue in hair follicle induction. *Nat Genet* 1999; 22:315–316.
58. Bayes M, Hartung AJ, Pispa J et al. The anydrotic ectodermal dysplasia gene (*EDA*) undergoes alternative splicing and encodes ectodysplasin-A with deletion mutations in collagenous repeats. *Hum Mol Gene* 1998; 7:1661–1669.
59. Isaacs K, Brown G, Moore GP. Interactions between epidermal growth factor and the Tabby mutation in skin. *Exp Dermatol* 1998; 7:273–280.
60. Monreal AW, Ferguson BM, Headon DJ et al. Mutations in the human homologue of mouse *dl* cause autosomal recessive and dominant hypohidrotic ectodermal dysplasia. *Nat Genet* 1999; 22:315–316.
61. Monreal AW, Zonana J, Ferguson B. Identification of a new splice form of the *EDA1* gene permits detection of nearly all X-linked hypohidrotic ectodermal dysplasia mutations. *Am J Hum Genet* 1998; 63:1253–1255.
62. Ferguson BM, Brockdorf N, Formstone E et al. Cloning of Tabby, the murine homolog of the human *EDA* gene: Evidence for a membrane-associated protein with a short collagenous domain. *Hum Mol Genet* 1997; 6:1589–1594.
63. Shultz LD. Single gene mouse models of immunodeficiency and autoimmune disease. In: Herzenberg LA, Weir DM, Blackwell C. eds. *Weir's Handbook of Experimental Immunology.* Oxford: Blackwell Science, 1996:50.1–150.4.

64. Herbst LH, Sundberg JP, Shultz LD et al. Tumorgenicity of green turtle fibropap-illoma-derived fibroblast lines in immunodeficient mice. *Lab Anim Sci* 1998; 48:162–167.

65. Nomura T. Practical development of genetically engineered animals as human disease models. *Lab Anim Sci* 1997; 47:113–117.

66. Sundberg JP, Dunstan RW, Roop DR et al. Full-thickness skin grafts from flaky skin mice to nude mice: Maintenance of the psoriasiform phenotype. *J Invest Dermatol* 1994; 102:781–788.

67. Sundberg JP, Boggess D, Hogan ME et al. Harlequin ichthyosis (*ichq*). A juvenile lethal mouse mutation with ichthyosiform dermatitis. *Am J Pathol* 1997; 151:293–310.

68. McElwee KJ, Boggess D, King LE et al. Experimental induction of alopecia areata-like hair loss in C3H/HeJ mice using full-thickness skin grafts. *J Invest Dermatol* 1998; 111:797–803.

69. Sundberg JP, Cordy WR, King Jr LE. Alopecia areata in aging C3H/HeJ mice. *J Invest Dermatol* 1994; 102:847–856.

70. McElwee KJ, Silva K, Beamer WG et al. Melanocyte and gonad activity as potential severity modifying factors in C3H/HeJ mouse alopecia areata. *Exp Dermatol* in 2001; 10:420–429.

71. Carroll JM, King LE, McElwee KJ et al. Gene expression profiling of alopecia areata in human disease and a mouse model reveals the importance of T-cell activation via co-stimulation. *J Invest Dermatol* 2002; 119:392–402.

72. Danilenko DM, Ring BD Yanagihara D et al. Keratinocyte growth factor is an important endogenous mediator of hair follicle growth, development, and differentiation. Normalization of the *nu/nu* follicular differentiation defect and amelioration of chemotherapy-induced alopecia. *Am J Pathol* 1995; 147:145–54.

73. Sawada M, Terada N, Taniguchi H, et al. Cyclosporin A stimulates hair growth in nude mice. *Lab Invest* 1987; 56:684–686.

74. Buhl AE, Waldon DJ, Miller BF, et al. Differences in activity of minoxidil and cyclosporin A on hair growth in nude and normal mice. Comparisons of *in vivo* and *in vitro* studies. *Lab Invest* 1990; 62:104–107.

75. Watanabe S, Mochizuki A, Wagatsuma K et al. Hair growth on nude mice due to cyclosporin A. *J Dermatol* 1991; 18:714–719.

76. Parthasarathy S, Malloy V, Matias J et al. The *in vitro* transformation of testosterone and the ability of its 5 alpha-reduced, 16 beta-hydroxylated metabolites to induce hair loss in the androchronogenetic alopecia (AGA) mouse. *J Invest Dermatol* 1992; 98:583.

77. Mezick JA, Genimenico GJ, Liebel FT et al. Androgen-induced delay of hair growth in the golden Syrian hamster. *Br J Dermatol* 1999; 140:1100–1104.

78. Yamanishi K. Gene-knockout mice with abnormal epidermal and hair follicle development. *J Dermatol Sci* 1998; 18:75–89.

79. Ishida-Yamamoto A, Tanaka H, Nakane H et al. Inherited disorders of epidermal keratinization. *J Dermatol Sci* 1998; 18:139–154.

80. Schon MP. Animal models of psoriasis—What can we learn from them? *J Invest Dermatol* 1999; 112:405–410.

81. Chase HB, Rauch H, Smith VW. Critical stages of hair development and pigmentation in the mouse. *Physiol Zool* 1951; 24:1–10

82. Chase HB, Montagna W, Malone JD. Changes in the skin in relation to the hair growth cycle. *Anat Rec* 1953; 116:75–82

83. Chase HB. Growth of the hair. *Physiol Rev* 1954; 34:113–126.
84. Chase HB. The physiology and histochemistry of hair growth. *J Soc Cosmetic Chem* 1955; 6:9–14.
85. Chase HB, Eaton GJ. The growth of hair follicles in waves. *Ann NY Acad Sci* 1959; 83:365–368.
86. Blake, JA, Eppig JT, Richardson JE et al. The Mouse Genome Database (MGD): Expanding genetic and genomic resources for the laboratory mouse. *Nuc Acids Res* 200; 28:108–111.
87. Ringwald, M, Eppig JT, Richardson JE. GXD: Integrated access to gene expression for the laboratory mouse. *Trends Genet* 2000; 16:188–190.
88. Bult CJ, Richardson JE, Blake JA et al. Mouse genome informatics in a new age of biological inquiry. *Proceedings of the IEEE International Symposium on Bio-Informatics and Biomedical Engineering* 2000; 29–32.
89. Tutikawa K. Studies on an apparently new mutant, alopecia periodica, found in the mouse. *Ann Rep Natl Inst Genet Jpn* 1952; 3:9–10.
90. Isaacson JH, Cattanach BM. *Hfh11*. *Mouse News Lett* 1962; 27:31.
91. Nehls M, Pfeifer D, Schorpp M et al. New member of the winged-helix protein family disrupted in mouse and rat nude mutations. *Nature* 1994; 372:103–7.
92. Kaestner KH, Knoeher W, Martinez DE. Unified nomenclature for the winged helix/forkhead transcription factors. *Genes Devel* 2000; 14:142–146.
93. Ashburner M, Ball CA, Blake JA et al. Gene ontology: Tool for the unification of biology. *Nat Genet* 2000; 25:25–9.
94. Sharp JJ, Linder CC, Mobraaten LE. Genetically engineered mice. Husbandry and resources. *Methods Mol Biol* 2001; 158:381–96.

16 Phenotyping Postpartum Mutant Laboratory Mice and Determining Their Value for Human Diseases

John P. Sundberg and Tsutomu Ichiki

TABLE OF CONTENTS

INTRODUCTION

Over the centuries, mice were traditionally considered to be pests by farmers who devised all sorts of methods to trap and poison these mammals because of their devastating effects on crops. Concurrently, people maintained mice as pets and, in particular, mutant mice as curiosities. Mice were also used in research but their value improved as inbred strains were developed in 1917.[1] Early work with mammary cancer determined that a transmissible agent in the milk was the cause in mice.[2] This was one of the early landmark works in cancer biology identifying a retrovirus, the mouse mammary tumor virus, as the etiologic agent. Collections of mutant mice were used as tools to study simple Mendelian genetics, and these collections eventually became repositories and distribution centers. With the advent of genetic engineering, initially as transgenic animals, then targeted mutations (so-called "knock-outs"), to the current gene-switch approaches, many of these spontaneous mutant mice archived in repositories served as the prototypes for phenotyping new, genetically engineered animals.[3,4]

BACKGROUND INFORMATION AND RESOURCES

Access to information on existing inbred strains and mutant stocks is critical to identifying the characteristics of new mutants. Targeted mutations often turn out to be allelic mutations of spontaneous mutant mice that have been evaluated in detail sometimes 50 years ago or earlier.[3] The Jackson Laboratory was established in 1929 and has grown to be the resource with the largest variety of inbred, congenic, and recombinant inbred strains as well as mutant laboratory mice in the world. Methods developed to maintain and distribute these mice have set the standards for nodes established throughout the world for archiving and distributing mice.[1,5] The World Wide Web provides many sites that detail this type of information. General information on mice can be found through The Jackson Laboratory Home Page (http://www.jax.org). Information on mouse genetics (genetic maps, sequences, references, etc.), mutant laboratory mice, phenotypes, and much more can be obtained from the Mouse Genome Informatics Web site (http://www.informatics.jax.org/). New sites focus on cancer genetics and pathology (Mouse Tumor Biology Database, http://tumor.informatics.jax.org/FMPro?-db=TumorInstance&-format=mtdp.html&-view;[6–8] general mouse pathology, http://www.pathbase.net;[9] and phenotypic differences between inbred strains (Mouse Phenome Project, http://aretha.jax.org/pub-cgi/phenome/mpdcgi?rtn=docs/home). These are being expanded to provide strain-specific information on spontaneous background lesions common in some strains but not others, supported by photomicrographs, descriptions, and references. Information on availability of mice for research can also be found through these databases and links (http://jaxmice.jax.org/html/jaxnotes/JAXNoteIndex.shtml; Jax Research Services, JRS). Chapters on informatics in this book describe how to find and use these resources.

RESEARCH APPROACHES TO MUTANT MOUSE CHARACTERIZATION

There are three general approaches to research in which characterization of the phenotype is critical. These are (1) gene-driven, (2) phenotype-driven, and (3) genetic diversity approaches. The current and most popular approach is gene-driven research. Genetic engineering provides a means for controlled over- or under expression of specific genes in defined anatomic sites. Targeted mutagenesis is a method to partially or completely inactivate only one gene. The newer technology that uses various forms of activation of genes such as the *Cre-lox*, *Tet*, or various hormone activation methods, can turn expression of specific genes in defined anatomic locations on or off to cause an effect. This is most useful for embryonic or neonatal lethal phenotypes associated with targeted mutagenesis[10,11] in that it provides a way to obtain live animals that develop anatomic site-specific lesions at specific ages. In all cases, the gene being manipulated is known and therefore a phenotype is predicted and sought. This can lead to over interpretation of the phenotype in an attempt to make the phenotype fit the hypothesis. There are numerous examples of phenotypes resulting from these types of studies that were totally unanticipated by the investigators. Most predicted an *in utero* or neonatal lethal effect, but investigators obtained more of a cosmetic effect in basically normal mice.[12–14]

Phenotype-driven characterization is the traditional approach in mouse biology. Spontaneous mutations were collected for over a century and studied based on obvious clinical features, such as baldness.[15] Methods to induce mutations were used for a long time such as chemical- or radiation-induced mutagenesis (these are described in detail elsewhere in this book). Chemical mutagenesis is popular again because targeted mutations have been systematically created for most known genes. The relatively random point mutations caused by ethylnitrosourea (ENU) or ethylenemethane sulfonate (EMS) produce mutations in potentially novel genes. In all these situations, characterization of the mutant mouse phenotype proceeds while the mutated gene is mapped and defined. Because it is impossible to predict the phenotype, each has to be evaluated in a relatively blind manner.

Genetic diversity begins to more accurately predict the common and serious diseases of humans and domestic animals. Most diseases are due to a combination of both genetic (mutations or polymorphisms in one or more genes) and non-genetic (diet, environment, infectious agents, etc.) events. These become exponentially more difficult to dissect and study as each gene or environmental event is added. By controlling the environment and microbiological status of colonies of mice it is possible to analyze multiple gene effects on a single gene mutation that has a major effect. For example, inactivation of the epidermal growth factor receptor (*Egfr*) gene causes many severe clinical abnormalities including neonatal death. The severity and specificity of lesions varies on different congenic strain backgrounds, indicating that multiple genes regulate its function.[16] These can be more specifically defined when a single disease such as inflammatory bowel disease, bone density, or even autoimmune diseases such as alopecia areata are studied on distinct inbred or congenic backgrounds, especially when multiple studies compare these differences between the same strains, such as C3H/HeJ and C57BL/6J, as was done for these diseases.[17–19] Such studies reveal that large numbers of genes can work together or independently to cause an animal to be susceptible or resistant to a particular disease.

DIFFERENTIATING MUTANT PHENOTYPES FROM INFECTIOUS DISEASES

Any clinical, anatomical, serological, histological, or other feature that differs from normal (for mice this is normal between inbred strains) in an individual mouse, the mouse is considered to be a "phenotypic deviant". The phenotype of a strain or mutant consists of the collective features of that group, and this usually means pathological changes (see Chapter 17 on strain-specific background diseases).

When a phenotypic deviant mouse is provided by an investigator for analysis, one must first determine if this could be due to an infectious process. Most infectious agents behave in a similar manner in most species, so routine diagnostic methods should be used to verify that no infection exists. Descriptions are available on these approaches in laboratory animals in general and mice in particular.[20–22] If an infectious etiology is defined, most closed colonies will be eradicated, the mouse room decontaminated, and new, microbiologically defined (specific pathogen-free) mice reintroduced. If there is no evidence of an infectious agent, then a colony can be created to determine the genetic basis of the mutation and to generate enough mice to do a detailed characterization of the phenotype.

COLONY ESTABLISHMENT

To establish a colony of mutant mice one needs access to the parents or abnormal offspring. Parents that have had one or more abnormal offspring are called "tested breeders", indicating that if the phenotype has a genetic basis, the normal parents have one copy of the mutant gene. These parents can be crossed to generate many offspring. If mutant mice die or cannot breed, the siblings will be either wild-type (+/+) or heterozygous (mutant/+). Intercrossing the siblings will determine which are heterozygous and they can be used to maintain the colony. If the mutant mice do not breed well, then ovaries can be removed from mutant female mice and grafted into histocompatible mice, usually those of the same inbred strain, whose ovaries have been removed, or into immunodeficient mice to bypass histocompatibility issues. Sperm can also be collected from mutant male mice and used to inseminate multiple females.[23,24]

If the phenotype of the mutant gene on a particular inbred background is very difficult to maintain, then creation of a congenic strain may be warranted. Intercrossing the mutant mice on one inbred background with a wild-type mouse on a different inbred background produces F1 mice. These are intercrossed to generate F2 mice, 25% of which will exhibit the phenotype if it is recessive. The mutant mice are crossed with those of the new strain to produce the N1 generation. The process is continued until at least N10 generations are obtained. Using this approach, it is possible to move the mutant gene onto several inbred backgrounds. The one that is most stable and easiest to maintain may be the one on which to focus. This is also used for very specific studies when moving the gene onto a particular background is important. For example, the C3H/HeJ strain often gets alopecia areata.[25] Full thickness grafts from affected mice to unaffected histo-compatible mice of the same strain provided a means to study onset and progression of the disease inasmuch as recipients developed alopecia areata within ten weeks of grafting.[26] Creating congenic strains on the C3H/HeJ background for a multitude of mutations that affect the immune system or apoptosis provided novel tools to study the pathogenesis of alopecia areata.[27]

DETERMINING THE GENETIC NATURE OF THE MUTANT PHENOTYPE

Simple Mendelian genetics implies a recessive, semidominant, or dominant pattern of inheritance. Most that occur spontaneously are recessive, meaning two copies are required to be present to see any sort of effect. If two heterozygous, clinically normal mice are crossed, then 25% of their offspring will be affected. If a wild-type and a heterozygous mouse are crossed, then none of the F1 generation will be affected. For dominant mutations, if a parent is affected then one or more of the F1 generation will be affected. If a gene dose effect is noted, mild disease with one copy and severe disease with two copies, this is considered to be a semidominant gene.

PHENOTYPIC CHARACTERIZATION, GENERAL ISSUES

Systematic evaluation of the phenotype of a mutant mouse is done using traditional diagnostic methods as if working up any routine case submitted to a diagnostic laboratory. Sophisticated physiological studies can be done but these are usually secondary to initial screens. Necropsy methods for the mouse and other laboratory mammals are described in detail elsewhere.[28,29] Careful dissection and collection of representative organs are imperative. Gross photographs of obvious lesions taken with normal age- and gender-matched controls to emphasize the differences are useful. Fixation of tissues varies depending upon the preference of the pathologist reading the final slides. Commonly used fixatives include neutral buffered 10% formalin, Fekete's acid-alcohol-formalin solution, Bouin's solution, and 4% paraformaldehyde.[28,29] The alcohol-based fixatives maintain epitopes better than aqueous-based fixes, making immunohistochemistry more reliable.[30] Usually 20 times the volume of fixative is used and it is changed at the end of the necropsy to remove blood, feces, and other organic material. After overnight fixation, tissues are either transferred to 70% ethanol (acid-alcohol-formalin) or washed in running tap water prior to transferring to ethanol (Bouin's solution) and stored. Alternatively, tissues are trimmed in a very specific manner that varies with the tissue, study, and preference of the pathologist.[28,29] Tissues are then processed routinely, paraffin embedded, sectioned at 5 to 6 μm, stained with hematoxylin and eosin, and examined.

Data can be stored in a relational database and the study downloaded into a spreadsheet for analysis.[31,32] This approach can coordinate medical records, glass slides, paraffin blocks, and additional diagnostic analyses or specialized tests.

From the spreadsheets, graphs can be generated to help determine how the phenotype changes with age.

WHICH MICE AND HOW MANY TO ANALYZE?

How many mice to analyze depends upon whether an inbred strain or outbred stock are being studied as well as how old the mice are when they die of the disease. If the mutant mice are embryonic or neonatal lethals, then few mice will be examined. Transgenic mice on outbred stocks, such as CD-1 mice, have great variability in phenotype between age- and gender-matched animals and many more mice need to be evaluated. It may be impossible to determine the common phenotype and progression of the disease in some of these mice because of marked individual variability. Inbred mice carrying single gene mutations tend to be very monomorphic at defined ages. There may or may not be a sexual dichotomy. Therefore, for initial screening studies using inbred mice with single gene mutations it is possible to limit the numbers to two for each sex at several different time points. Equal numbers of age- and gender-matched controls should also be included. Colony management and genetic drift occur in all situations over time, so histological features of tissues in an inbred strain may change as well. Two males and two females of both mutant and control types ($N = 8$) per time point are a minimum. Major changes in the life of an animal can be used as starting points both to look for changes in phenotypes associated with these changes and with increasing age. In the mouse, these changes are birth (day 0), weaning (three weeks of age), sexual maturity (puberty, six weeks of age), and sexual quiescence (retired breeders, six to eight months of age and sometimes older depending upon strain characteristics).[33] Older mice can be used for the geriatric period (two years of age or greater); however, expense usually precludes using these mice.

The advent of large-scale mutagenesis and genetic engineering programs provides another approach. Phenotyping by behavior, clinical examination, and other nonterminal means are being used as first-stage screening levels. Adult mice at defined ages are necropsied for detailed evaluation. To supplement these types of studies, we selected four months of age, again looking at two males and two females, to do extensive necropsies to look for subtle histologic lesions.

Allelism testing, mating mutant mice with similar phenotypes to determine if they are noncomplementary (mutations in the same gene), should be done if a phenotype is clear, lesions closely resemble those found in well-defined mutant mice, and the mutant mice of interest are available for breeding. Crossing two recessive homozygous mutant mice and obtaining mutant offspring is an example of noncomplementation; a normal gene does not complement the mutant gene to produce a normal offspring. If successful, this will identify the locus and possibly even the gene involved in the new mutant with which you are working.[34,35] If this does not work, then the mutant gene will have to be mapped[36] unless it is already known due to the gene targeting approach.[3]

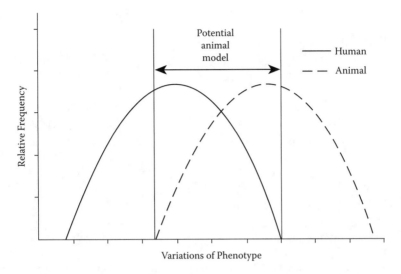

FIGURE 16.1 Comparing clinical disease phenotypes between mutant mice and human patients results in both similarities and differences. Most mutant mice and human diseases have some degree of homology as depicted by this theoretical bell-shaped curve.

COMPARISON WITH HUMAN AND DOMESTIC ANIMAL DISEASES

Comparison of phenotypes between humans and mutant mice is occasionally very straightforward but usually it is a difficult process. One way is to list all the criteria discovered for the mutant mouse under investigation. Candidate diseases in other species can be identified and similar lists created using the literature. Common features between the two will determine the degree of homology.[35,37] Ultimately, identifying the gene mutated in both species is critical to the accurate comparison between species. The hairless genes in mice and humans, when mutated, result in generalized baldness. The correlation with a specific human disease can only be done once the gene is known and the phenotypes compared between humans and mice with mutations only in that gene.[38–40] This approach was expanded to other species as well with the hairless gene.[41]

The results of these types of comparisons can be illustrated using the conceptual graphics in Figure 16.1. If the mouse or any other species with a particular disease is compared with the human disease, there will be some similarities and some differences. Most diseases in mammals fall into this category and are represented by the top of the bell-shaped curve. A few mouse diseases will be exactly identical to specific human diseases and a few will be essentially unique to mice or humans, these representing the bottom of the curve. When such comparisons were made nearly a decade ago, mutant mice could be placed at all points on such a curve[42,43] (Figure 16.2). Using knowledge gained to date, most

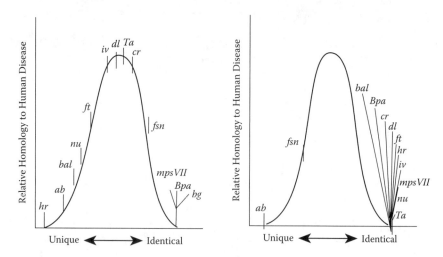

FIGURE 16.2 By fitting specific mutant mouse phenotypes to specific human diseases, in 1989 it was clear that some were unique whereas others were virtually identical to human diseases (left graph). This recently changed (right graph) to nearly identical matching for the same mouse diseases with those in humans because the affected genes were identified so that specific diseases could be directly compared. Outliers are flaky skin (*fsn*), the mutated gene of which has not yet been identified, and asebia. Asebia was determined to be due to a mutation in stearyl coenzyme A desaturase 1. This gene is found in humans but it is expressed in a different part of the body, so mutant phenotypes will not be similar.

of these have now been carefully defined as exact or very close replicates for specific human diseases. The key was to define the mutated gene in both species and then limit comparisons to only patients and mutant mice that have the same mutated gene. Many of the outliers in the patient comparisons were due to other genetic- or nongenetic-based diseases that share some features but not all and were lumped together due to general similarities. Therefore, as we get better at defining human and mouse diseases, the comparisons will become more accurate. In so doing, the mouse continues to prove to be an extremely valuable model for studying all aspects of human disease.

SUMMARY

Systematic evaluation of mutant mice, disregarding preconceptions of experimental design, can be used to identify many more features than focused studies. The better mutant mice are defined, the more appropriately they can be used as animal models for human diseases. Determining the genetic causes at the molecular level for these diseases with direct comparison to human patients with mutations in the same gene have changed many mouse models from being interesting curiosities to identical or nearly identical models for many human genetic-based diseases.

REFERENCES

1. Holstein, J., *The First Fifty Years at the Jackson Laboratory*, The Jackson Laboratory, Bar Harbor, ME, 1979.
2. Staff of the Roscoe B. Jackson Memorial Laboratory, The existence of non-chromosomal influence in the incidence of mammary tumors in mice, *Science* 78, 465–466, 1933.
3. Sundberg, J. P. and Bult, C., Professional use of mutant mice in hair research, in *Handbook of Genetically Engineered Mice*, Sundberg, J. P. and Itchiki, T. ed., CRC Press, Boca Raton 15, 185–209.
4. Sundberg, J. P., The laboratory mouse: a biomedical tool that changed the way we think about medicine. Proc FELASA 8:54–59, 2004.
5. Davisson, M. T. and Sharp, J. J., Repositories of mouse mutations and inbred, congenic, and recombinant inbred strains, in *Systematic Characterization of Mouse Mutations*, Sundberg, J. P. and Boggess, D. eds. CRC Press, Boca Raton, FL, 2000.
6. Bult, C. J., Krupke, D. M., Sundberg, J. P., and Eppig, J. T., Mouse tumor biology database (MTBD): Enhancements and updates, *Nucleic Acids Res* 28, 112–114, 2000.
7. Bult, C. J., Krupke, D. M., Naf, D., Sundberg, J. P., and Eppig, J. T., Web-based access to mouse models of human cancers: The Mouse Tumor Biology (MTB) Database, *Nucleic Acids Res* 29, 95–97, 2001.
8. Naf, D., Krupke, D. M., Sundberg, J. P., Eppig, J. T., and Bult, C. J., The mouse tumor biology database: A public resource for cancer genetics and pathology of the mouse, *Cancer Res* 62, 1235–1240, 2002.
9. Schofield, P. N., Bard, J. B. L., Booth, C., Boniver, J., Covelli, V., Delvenne, P., Ellender, M., Engstrom, W., Goessner, W., Gruenberger, M., Hoefler, M., Hopewell, J., Mancuso, M.-T., Mothersill, C., Potten, C., Rozell, B., Soriola, H., Sundberg, J. P., and Ward, J. M., Pathbase: A database of mutant mouse pathology, *Nucleic Acids Res* 32, D512–515, 2004.
10. Nakamura, M., Tobin, D. J., Richards-Smith, B., Sundberg, J. P., and Paus, R., Mutant laboratory mice with abnormalities in pigmentation: Annotated tables, *J Dermatol Sci* 28, 1–33, 2002.
11. Nakamura, M., Sundberg, J. P., and Paus, R., Mutant laboratory mice with abnormalities in hair follicle morphogenesis, cycling, and/or structure: Annotated tables, *Exp Dermatol* 10, 369–390, 2002.
12. Mann, G. B., Fowler, K. J., Gabriel, A., Nice, E. C., Williams, R. L., and Dunn, A. R., Mice with a null mutation of the tgfa gene have abnormal skin architecture, wavy hair, and curly whiskers and often develop corneal inflammation, *Cell* 73, 249–261, 1993.
13. Luetteke, N. C., Qiu, T. H., Peiffer, R. L., Oliver, P., Smithies, O., and Lee, D. C., tgfa deficiency results in hair follicle and eye abnormalities in targeted and waved–1 mice, *Cell* 73, 263–278, 1993.
14. Hebert, J. M., Rosenquist, T., Gotz, J., and Martin, G. R., FGF5 as a regulator of the hair growth cycle: Evidence from targeted and spontaneous mutations, *Cell* 78, 1017–1025, 1994.
15. Gaskoin, J. S., On a peculiar variety of *Mus musculus*, *Proc Zool Soc London* 24, 38–40, 1856.

16. Threadgill, D. W., Dlugosz, A. A., Hansen, L. A., Tennenbaum, T., Lichti, U., Yee, D., LaMantia, C., Mourton, T., Herrup, K., Harris, R. C., Barnard, J. A., Yuspa, S. H., Coffey, R. J., and Magnusun, T., Targeted disruption of mouse EGF receptor: Effect of genetic background on mutant phenotype, *Science* 269, 230–233, 1995.

17. Farmer, M. A., Sundberg, J. P., Bristol, I. J., Churchill, G. A., Li, R., Elson, C. O., and Leiter, E. H., A major quantitative trait locus on chromosome 3 controls colitis severity in IL-10-deficient mice, *Proc Natl Acad Sci USA* 98 (24), 13820–13825, 2001.

18. Beamer, W. G., Shultz, K. L., Donahue, L. R., Churchill, G. A., Sen, S., Wergedal, J. R., Baylink, D. J., and Rosen, C. J., Quantitative trait loci for femoral and lumbar vertebral bone mineral density in C57BL6J and C3H/HeJ inbred strains of mice, *J Bone Mineral Res* 16, 1195–1206, 2001.

19. Beckwith, J., Cong, Y., Sundberg, J. P., Elson, C. O., and Leiter, E. H., *Cdcs1*, a major colitis susceptibility gene in mice, regulates innate and adaptive immune response to enteric bacterial antigens, Gastroenterology (in press).

20. Percy, D. and Barthold, S., *Pathology of Laboratory Rabbits and Rodents*, Iowa State University Press, Ames, IA, 1993.

21. Lindsey, J. R., Boorman, G. A., Collins, M. J., Hsu, C.-K., VanHoosier, G. L., and Wagner, J. E., *Infectious Diseases of Mice and Rats*, National Academy Press, Washington, DC, 1991.

22. Lindsey, J. R., Boorman, G. A., Collins, M. J., Hsu, C.-K., VanHoosier, G. L., and Wagner, J. E., *Companion Guide to Infectious Diseases of Mice and Rats*, National Academy Press, Washington, DC, 1991.

23. Boggess, D., Cunliffe-Beamer, T. L., and Sundberg, J. P., Colony establishment, in *Systematic Approach to Evaluation of Mouse Mutations*, Sundberg, J. P. and Boggess, D. eds. CRC Press, Boca Raton, FL, 2000, pp. 1–13.

24. Boggess, D., Silva, K. A., Landel, C., Mobraaten, L., and Sundberg, J. P., Approaches to Handling, breeding, strain preservation, genotyping, and drug administration for mouse models of cancer, in *Mouse Models of Human Cancer*, Holland, E. C. ed. John Wiley, Hoboken, NJ, pp. 1–14, 2004.

25. Sundberg, J. P., Cordy, W. R., and King, L. E., Alopecia areata in aging C3H/HeJ mice, *J Invest Dermatol* 102, 847–856, 1994.

26. McElwee, K. J., Boggess, D., King, L. E., and Sundberg, J. P., Experimental induction of alopecia areata-like hair loss in C3H/HeJ mice using full-thickness skin grafts, *J Invest Dermatol* 111, 797–803, 1998.

27. Freyschmidt-Paul, P., McElwee, K. J., Botchkarev, V., Kissling, S., Wenzel, E., Sundberg, J. P., Happle, R., and Hoffmann, R., Fas-deficient C3.MRL-*Tnfrsf6lpr* mice and Fas ligand-deficient C3H/HeJ-*Tnfsf6gld* mice are relatively resistant to the induction of alopecia areata by grafting of alopecia areata-affected skin from C3H/HeJ mice, *J Invest Dermatol Sym Proc* 8:104–108, 2003.

28. Relyea, M. J., Miller, J., Boggess, D., and Sundberg, J. P., Necropsy methods for laboratory mice: Biological characterization of a new mutation, in *Systematic Approach to Evaluation of Mouse Mutations*, Sundberg, J. P. and Boggess, D. eds. CRC Press, Boca Raton, FL, 2000, pp. 57–90.

29. Seymour, R., Ichiki, T., Mikaelian, I., Boggess, D., Silva, K. A., and Sundberg, J. P., Necropsy methods, in *Handbook of Experimental Animals: The Laboratory Mouse*, Hedrich, H. J. ed. Academic Press, London, pp. 495–516, 2004.

30. Mikaelian, I., Nanney, L. B., Parman, K. S., Kusewitt, D., Ward, J. M., Naf, D., Krupke, D. M., Eppig, J. T., Bult, C. J., Seymour, R., Ichiki, T., and Sundberg, J. P., Antibodies that label paraffin-embedded mouse tissues: A collaborative endeavor, *Toxicol Pathol* 32, 1–11, 2004.

31. Sundberg, B. A. and Sundberg, J. P., A database system for small diagnostic laboratories, *Lab Anim* 19, 55–58, 1990.

32. Sundberg, B. A. and Sundberg, J. P., Medical record keeping for project analysis, in *Systematic Approach to Evaluation of Mouse Mutations*, Sundberg, J. P. and Boggess, D. eds. CRC Press, Boca Raton, FL, 2000, pp. 47–55.

33. Green, M. C. and Witham, B. A., *Handbook of Genetically Standardized Jax Mice*, The Jackson Laboratory, Bar Harbor, ME, 1997.

34. Sundberg, J. P., Boggess, D., Sundberg, B. A., Eilersten, K., Parimoo, S., Filippi, M., and Stenn, K., Asebia-2J (Scd^{ab2J}): A new allele and a model for scarring alopecia, *Am J Pathol* 156 (6), 2067–2075, 2000.

35. Sundberg, J., Boggess, D., Bascom, C., Limberg, B. J., Shultz, L. D., Sundberg, B. A., King, L. E., and Montagutelli, X., Lanceolate hair-J (lah^J): A mouse model for human hair disorders, *Exp Dermatol* 9, 206–218, 2000.

36. Montagutelli, X., Determining the genetic basis of a new trait, in *Systematic Approach to Evaluation of Mouse Mutations*, Sundberg, J. P. and Boggess, D. eds. CRC Press, Boca Raton, FL, 2000, pp. 15–33.

37. Sundberg, J. P., Boggess, D., Shultz, L. D., and Beamer, W. G., The flaky skin (*fsn*) mutation, chromosome ?, in *Handbook of Mouse Mutations with Skin and Hair Abnormalities. Animal Models and Biomedical Tools*, Sundberg, J. P. ed. CRC Press, Boca Raton, FL, 1994, pp. 253–268.

38. Ahmad, W., Faiyez ul Haque, Brancolini, V., Tsou, H. C., ul Haque, S., Lam, H., Aita, V. M., Owen, J., deBlaquiere, M., Frank, J., Cserhalmi-Friedman, P. B., Leask, A., McGrath, J. A., Peacocke, M., Ahmad, M., Ott, J., and Christiano, A. M., Alopecia universalis associated with a mutation in the human hairless gene, *Science*, 1998 Jan. 30; 279, (5351) 720–724, 1998.

39. Panteleyev, A. A., Paus, R., Ahmad, W., Sundberg, J. P., and Christiano, A. M., Molecular and functional aspects of the hairless (*hr*) gene in laboratory rodents and humans, *Exp Dermatol* 7, 249–267, 1998.

40. Sundberg, J. P., Price, V. H., and King, L. E., The "hairless" gene in mouse and man, *Arch Dermatol* 135, 718–720, 1999.

41. Ahmad, W., Ratteree, M. S., Panteleyev, A. A., Aita, V. M., Sundberg, J. P., and Christiano, A. M., Atrichia with papular lesions resulting from mutations in the rhesus macaque (*Macaca mulatta*) hairless gene, *Lab Anim* 36, 61–67, 2002.

42. Sundberg, J. P., Mouse mutations: Animal models and biomedical tools, *Lab Anim* 20, 40–49, 1991.

43. Sundberg, J. P., Conceptual evaluation of animal models as tools for the study of diseases in other species, *Lab Anim* 21, 48–51, 1993.

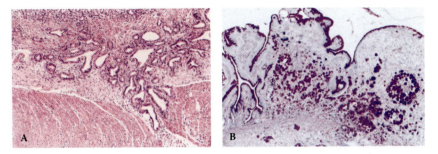

FIGURE 14.1 Gastric carcinomas in the glandular stomach of Hp-infected Mongolian gerbils treated with chemical carcinogen N-methyl-N-nitrosourea. (a) H&E staining of a well-differentiated adenocarcinoma. (b) Alcian blue-periodic acid Schiff staining of a signet-ring cell carcinoma.

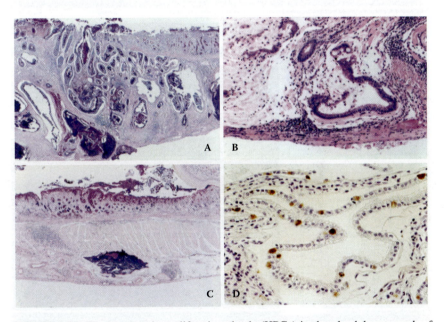

FIGURE 14.2 The heterotopic proliferative glands (HPGs) in the glandular stomach of Hp-infected MGs. (a) HPGs with IM penetrating stomach mucosa (AB-PAS). (b) Higher magnification of HPG glands lined with intestinal metaplastic epithelium surrounded with collagen fibers and inflammatory cells (H&E). (c) Mucin remnant left after eradication (AB-PAS). (d) The chromogranin A positive cells were observed among mucous epithelial cells in the HPG gland (chromogranin A immunohistochemistry).

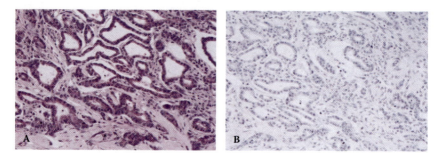

FIGURE 14.3 The expression of chromogranin A is not detected in the cancerous lesions of the glandular stomach of MGs. (a) H&E staining. (b) Chromogranin A immunohisto-chemistry.

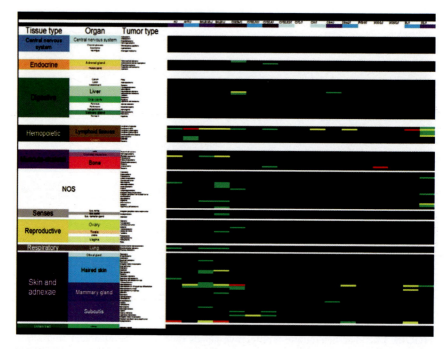

FIGURE 17.1 Examples of the relative frequency of neoplastic disease types in major strains of inbred mice maintained at The Jackson Laboratory between 1987 and 2000. This table is a subdirectory of The Jackson Laboratory Colonies Disease Array Profile. Red indicates common lesions, yellow intermediate, and green rare types. More information can be found on the Mouse Phenome and Pathbase databases as described above and in other chapters in this book.

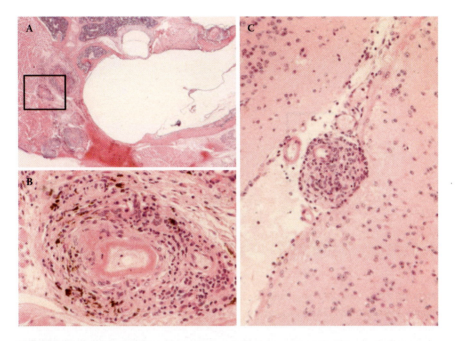

FIGURE 17.3 Polyarteritis nodosa is a relatively common background disease in C57BL/6J mice. The neurological signs of head tilt are due to the arteritis of small- and medium-sized arteries going to the skull. In this case numerous affected arteries were found adjacent to the middle and outer ear (A). Higher magnification of boxed area in A reveals fibrinoid necrosis and surrounding inflammation (B). Cerebral arteries are also affected (C). These lesions were diagnosed in a 270-day-old female C57BL/6J mouse.

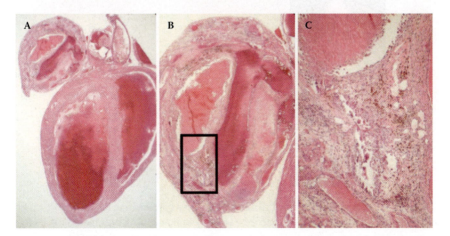

FIGURE 17.4 Mural thrombosis is common in the hearts of BALB strains and substrains. This case is an example of a chronic mineralized left atrial mural thrombus in a 378-day-old female BALB/cJ mouse. Successive increases in magnification reveal the details of this lesion.

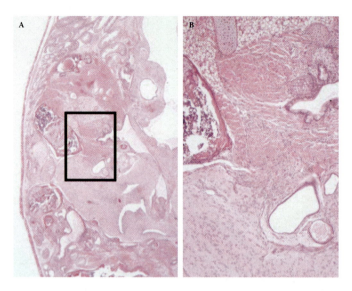

FIGURE 17.6 Testicular teratomas are common unilateral masses found in most 129 strains and substrains. This is a case from a 28-day-old male 129-*Ifngr*^{tm1Agt}/J/*Ifngr*^{tm1Agt}/J mouse. The testicular cancer is a strain background characteristic and has nothing to do with the targeted mutation. Note the normal seminiferous tubules in the upper left area (A) and the majority of the testicle is compressed by a mixture of bone, skeletal muscle, tubules lined by a variety of epithelial cell types, and a large amount of neuropile. The higher magnification of the boxed area in A illustrates this mixture of cell types that are well differentiated and not invasive (B).

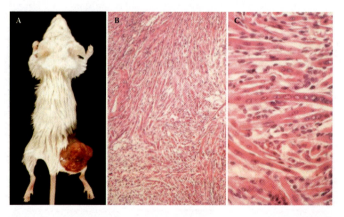

FIGURE 17.7 Well-differentiated rhabdomyosarcomas are rare but found almost exclusively in BALB/cJ mice. This emaciated 118-day-old female BALB/cJ mouse has a large mass on its lower right leg. Removal of the skin revealed the irregular mottled mass (a). Histologically, these lesions consist of well-differentiated skeletal muscle (b,c) that is positive with Bodian, phosphotungstic acid hematoxylin, and iron hematoxylin histologic stains for cross striations and striated muscle actin isoforms by immunohistochemistry confirming that it is of skeletal muscle origin.

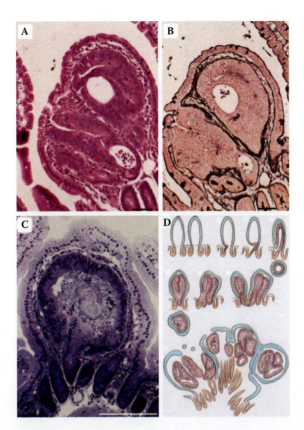

FIGURE 18.1 Adenomatous polyps in the small intestine of the *Apc^{Δ716}* heterozygous mouse. (a) An early adenoma section stained with H&E.[6] (b) A silver staining for the basement membrane of a section adjacent to that in A.[6] (c) Toluidin blue staining of a thin section of a nascent polyp embedded in glycolmethacrylate resin.[10] (d) Schematic processes of polyp formation in the *Apc^{Δ716}* small intestine.[6] Scale bars: 100 μm.

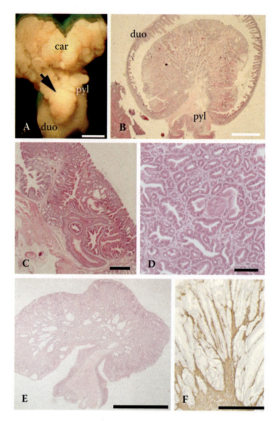

FIGURE 18.2 Hamartomatous polyps in *Smad4*, *Cdx2*, and *Lkb1* heterozygotes, respectively. (A,B) A gastric polyp in a *Smad4+/−* mouse.[43] Dissection micrograph and low-power histology section (H&E) show a hamartoma developed in the pyloric region, protruding into the duodenum (arrow): car, gastric cardia; pyl, pylorus; duo, duodenum. (C,D) Histopathology of a hamartoma of the proximal colon in a *Cdx2+/−* mouse (H&E).[44] (E,F) Histopathology of a gastric hamartoma in an *Lkb1+/−* mouse.[48] Although E shows an H&E staining, the section in F was immunostained for α-smooth muscle actin. Note its arborizing pattern in the polyp stroma. Bars: A, 5 mm; B,E, 2 mm; C,D, 100 μm; and F, 500 μm.

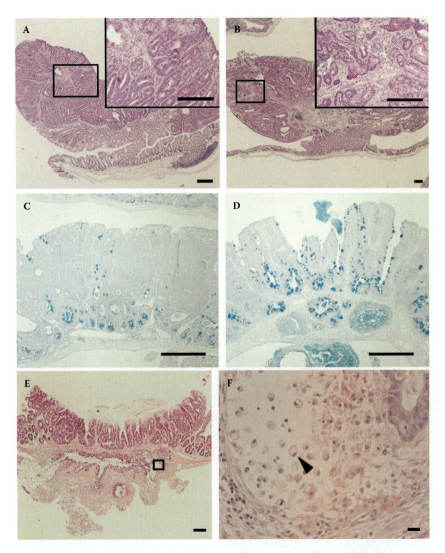

FIGURE 18.3 Adenocarcinomas in the intestines of *cis-Apc*$^{\Delta 716}$*Smad4* heterozygous mice.[54] (A,B) Representative colonic polyps in the *Apc*$^{+/\Delta 716}$ mouse (A; control) and *cis-Apc*$^{+/\Delta 716}$*Smad4*$^{+/-}$ mouse B at 14 weeks of age (H&E). Note from the scale bars that the polyp size is larger in B. (C,D) Small intestinal polyps in the *Apc*$^{+/\Delta 716}$ mouse (C; control) and *cis-Apc*$^{+/\Delta 716}$*Smad4*$^{+/-}$ mouse D at 20 weeks of age (Alucian blue). Note the invasive adenocarcinoma in the smooth muscle layer. (E,F) A more advanced adenocarcinoma in a 20-week-old *cis-Apc*$^{+/\Delta 716}$*Smad4*$^{+/-}$ mouse. Arrowhead in F shows one of the signet-ring cells in the boxed area in E. Bars: A–D, 250 μm; E, 125 mm; and F, 6.3 μm.

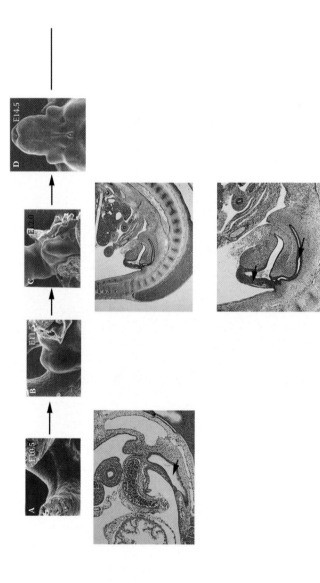

FIGURE 20.2A Mouse genital tubercle (GT) development during embryogenesis; illustration of its development by SEM and sagittal sections of embryos for early hormone independent stages. A slight sign of genital tubercle outgrowth is observed adjacent to the cloaca (A,E) around E10.5 during mouse development. Later, GT outgrowth becomes more prominent and the cloaca is divided into urogenital sinus and hindgut by urorectal septum at E12.5 (F,G).

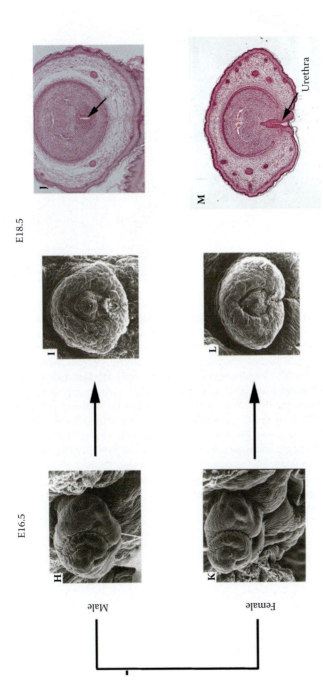

FIGURE 20.2B A characteristic point of the developmental processes for external genitalia formation lies in the hormone-dependent phase of development after E16.5 in mouse development (lower part, H–M). Several corresponding sagittal sections are shown for (E–G). pp: prepuce.

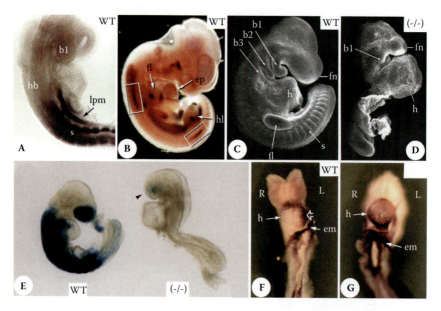

FIGURE 21.2 The RALDH2 enzyme is required for retinoic acid synthesis during early embryogenesis. (A,B) Spatially restricted expression patterns of the *Raldh2* gene in E8.5 and E10.E mouse embryos, respectively. Boxed areas indicate *Raldh2*-expressing motor neuron populations in the brachial and lumbar spinal cord. Whole-mount *in situ* hybridizations with digoxigenin-labeled riboprobes. (C,D) Scanning electron micrographs of E9.5 wild-type and *Raldh2*–/– knock-out embryos, respectively. (E) Deficient activity of a RA-responsive *lacZ* reporter transgene in an E8.5 *Raldh2*–/– embryo (right) compared to a wild-type (WT) littermate (left). Weak activity is found in the forebrain and optic area only (arrowhead). Whole-mount X-gal staining. (F,G) Impaired heart looping in an E8.5 *Raldh2*–/– embryo (G), as seen by a lack of left-side lateralization of *Hand1*-expressing cells (arrow in F: wild-type littermate). Whole-mount *in situ* hybridization. b1–b3: branchial arches; ep: epicardium; em: extraembryonic membranes; fl: forelimb bud; fn: frontonasal mass; h: heart; hb: hindbrain; hl, hindlimb bud; L: left; R: right; s: somites. (A,B reprinted with permission from Niederreither et al.;[37] C–G reprinted with permission from Niederreither et al.[41])

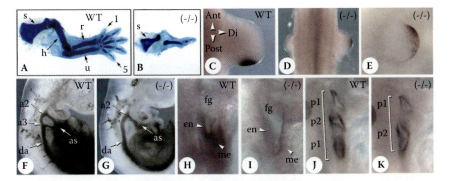

FIGURE 21.3 RA-rescue of *Raldh2*$^{-/-}$ knock-out embryos unveils novel developmental defects. (A,B) Abnormal forelimb development in an E14.5 *Raldh2*$^{-/-}$ fetus (B) after RA-rescue from E7.5 to 8.5, compare to WT littermate (A). Alcian blue cartilage staining. (C–E) At earlier stages, the rescued *Raldh2*$^{-/-}$ embryos (D,E) fail to activate *Sonic hedgehog* (*Shh*), or express it in a inappropriate distal/anterior location within the forelimb buds. (F,G) Impaired development of posterior branchial arches in a E9.5 RA-rescued *Raldh2*$^{-/-}$ embryo (G), as seen by the lack of development of third aortic arches. Intracardiac ink injections. (H,I) Impaired activation of the *Hoxa1* homeobox gene in the pharyngeal endoderm and mesoderm of a rescued E9.5 *Raldh2*$^{-/-}$ embryo (I). (J,K) Abnormal development of the posterior pharyngeal pouches in a rescued E9.5 *Raldh2*$^{-/-}$ embryo. The whole pharyngeal pouch region, visualized by expression of the *Pax9* gene, is reduced (compare brackets in J,K) and an abnormal, single "second pouch" is formed in the mutant (K). Ant: anterior; as: aortic sac; a2–a3: aortic arches; da: dorsal aorta; Di: distal; en: endoderm; fg: foregut; me: mesoderm; Post: posterior; p1–p3: pharyngeal pouches; r: radius; s: scapula; u: ulna; 1–5: digits. (A–E reprinted with permission from Niederreither et al.;[55] F–K reprinted with permission from Niederreither et al.[58])

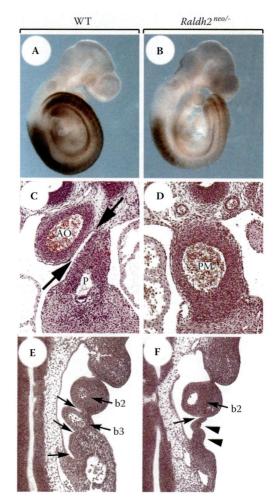

FIGURE 21.4 *Raldh2* engineered hypomorphic mutants display a DiGeorge syndrome-like phenotype. (A,B) Decreased *Raldh2* transcript levels in an E9.5 *Raldh2^neo/-^* embryo (B, compare with WT littermate, A). (C,D) Lack of septation of the outflow tract (persistant truncus arteriosus) in an E14.5 *Raldh2^-/-^* fetus (D, compare with the fully septated aorta and pulmonary trunk in a WT littermate, arrows in C). (E,F) Lack of development of the posterior branchial arches in an E9.5 *Raldh2^-/-^* embryo (F). Unlike its WT counterpart (E), the mutant embryo exhibits no visible third or fourth arches, although rudimentary branchial clefts are formed (arrowheads). AO: aorta; b2,b3: branchial arches; PT: pulmonary trunk; PTA: persistent truncus arteriosus; P2–P4: pharyngeal pouches. (Reprinted with permission from Vermot et al.[61])

17 Common Diseases Found in Inbred Strains of Laboratory Mice

John P. Sundberg and Tsutomu Ichiki

TABLE OF CONTENTS

Brother–sister matings of mice for 20 or more generations results in the creation of inbred strains.[1] As discussed in detail elsewhere in this handbook, the longer this breeding strategy continues the higher the degree of genetic homology within the strain. The Jackson Laboratory and other repositories maintain strains that have been inbred for hundreds of generations now. A combination of genetic polymorphisms or real mutations that already existed in the parental stocks become fixed over time in the new inbred strain as two recessive alleles that by themselves or in combination with another or possibly many such genes results in spontaneous background-specific diseases.[2] This can also be the result of mutations that occur spontaneously in the colony over time. The result is that each inbred strain has its own "disease profile"[3] and in many ways can be compared to an individual outbred person or domestic animal, or an isolated population, such as a breed of dogs, that is prone to developing certain types of unique diseases.

It is beyond the scope of this chapter to discuss all spontaneous background diseases in laboratory mice. These are well described in several books.[4–13] Methods to evaluate such animals are also described in detail elsewhere.[14,15] However, the need for good quality color images with descriptions has been a problem that is now being addressed by a variety of databases. The Jackson Laboratory Bioinformatics Program (www.informatics.jax.org) was a pioneer in this area.[16–18] Mouse Genome Informatics provides information of phenotypes based on specific gene mutations. Descriptions are supplemented with gross photographs of the live mice. We are developing ways to provide photomicrographs of classic lesions found in these mice as well. The Mouse Tumor Biology Database (http://tumor.informatics.jax.org)[19] combines clinical and genetic information with photomicrographs. This has been expanded to include a large inventory of commercially available antibodies useful for immunohistochemistry

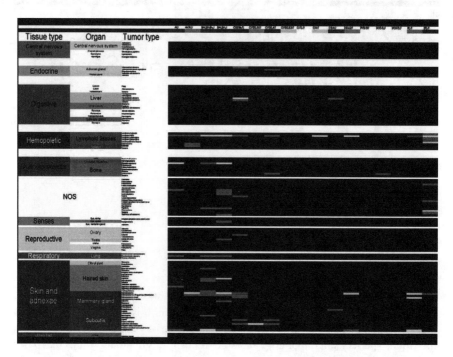

FIGURE 17.1 (See color insert following page 210.) Examples of the relative frequency of neoplastic disease types in major strains of inbred mice maintained at The Jackson Laboratory between 1987 and 2000. This table is a subdirectory of The Jackson Laboratory Colonies Disease Array Profile. Red indicates common lesions, yellow intermediate, and green rare types. More information can be found on the Mouse Phenome and Pathbase databases as described below and in other chapters in this book.

with information on methods and positive results in the form of photomicrographs.[20] The European Mouse Pathology Consortium, which includes Drs. J.P. Sundberg, C. Brayton, and J.M. Ward from the United States, is becoming an international repository of images that illustrate normal anatomy and lesions that occur spontaneously or as a result of genetic engineering in mice.[21,22] This database is online (www.pathbase.net) and described in detail in another chapter in this book.

When The Jackson Laboratory Pathology Program was developed by Dr. Sundberg in the mid-1980s, one of the first goals was to develop a relational database for medical recordkeeping.[23,24] This was refined over the years yielding a huge database to provide information on strain-specific diseases. Archives of paraffin blocks and frozen tumors were and continue to be heavily used by scientists both at The Jackson Laboratory and around the world. Summaries of these data are accessible online through the Mouse Phenome Database (http://aretha.jax.org/pub-cgi/phenome)[25,26] and Pathbase.net. The latter has links to images of the specific types of lesions. This is called The Jackson Laboratory Colonies Disease Array Profile (Figure 17.1).

TABLE 17.1
Relatively common diseases in C57BL/6J, BALB/cJ, 129P3/J, and C3H/HeJ inbred strains maintained at The Jackson Laboratory.[a]

Disease	C57BL/6J	BALB/cJ	129P3/J	C3H/HeJ
Cardiac mineralization	No	Epicardial	No	Myocardial
Retinal degeneration	No	No	No	Yes (*rd1/rd1*)
Mammary cancer	Rare	Papillary intraductal adenocarcinoma	No	Adenocarcinoma (MMTV dependent)
Lymphosarcoma	Rare	Common	No	Rare
Alopecia areata	No	No	No	20% >18 months old
Micro/anophthalmia	Yes	No	No	No
Polyarteritis nodosa	Yes	No	No	No
Internal hydrocephalus	Yes, >1:1000	Rare	No	Rare
Mural thrombi (heart)	No	Common	No	No
Myoepitheliomas	No	Common	No	No
Aspiration pneumonia	No	No	No	Yes, rare but unique to this strain
Hepatocellular adenoma	No	No	No	Yes
Testicular teratoma	No	No	Common	No
Leydig cell tumors	No	Rare	No	No
Hemangiomas	No	Rare	No	No
Pulmonary fibrosis	No	Rare	No	No
Rhabdomyosarcoma	No	Rare	No	No

[a] More information can be found on the Mouse Phenome and Pathbase databases.

Because it requires a full textbook to cover all of these diseases and to provide representative images, selected overviews on the four most commonly used inbred strains, C57BL/6J, BALB/cJ, 129P3/J, and C3H/HeJ, are provided (Table 17.1). Representative images of some of the more common lesions are shown (Figures 17.2 to 17.7). Details on the diseases can be found in the books referenced above or on the Web sites provided. This type of information is critical for anyone using inbred mice in their research, inasmuch as knowing diseases the mice naturally get helps differentiate these from lesions secondary to genetic engineering. Likewise, sudden changes in types of diseases in terms of frequency or appearance of new types of diseases can be significant for correct interpretation of research results.

REFERENCES

1. Silver, L. M., *Mouse Genetics. Concepts and Applications,* Oxford University Press, New York, 1995.

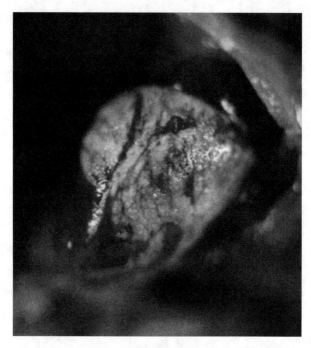

FIGURE 17.2 (See color insert following page 210.) Epicardial mineralization is a common finding in BALB strains and substrains. This is an example of severe epicardial mineralization in a 109-day-old female BALB/cByJ mouse.

2. Petkov, P. M., Ding, Y., Cassell, M. A., Zhang, W., Wagner, W., Sargent, E. E., Asquith, S., Crew, V., Johnson, K. A., Robinson, P., Scott, V. E., and Wiles, M. V., An efficient SNP system for mouse genome scanning and elucidating strain relationships, *Genome Res* 14, 1806–1811, 2004.
3. Mikaelian, I., Ichiki, T., Ward, J. M., and Sundberg, J. P., Diversity of spontaneous neoplasms in commonly used inbred strains and stocks of laboratory mice, in *The Laboratory Mouse*, Hedrich, H. J. ed. Academic Press, London, 2004, pp. 345–354.
4. Mohr, U., Dungworth, D. L., Capen, C. C., Carlton, W. W., Sundberg, J. P., and Ward, J. M., *Pathobiology of the Aging Mouse,* ILSI, Washington, DC, 1996.
5. Mohr, U., *International Classification of Rodent Tumors: The Mouse,* Springer Verlag, Berlin, 2001.
6. Maronpot, R. R., Boorman, G. A., and Gaul, B. W., *Pathology of the Mouse. Reference and Atlas,* Cache River Press, Vienna, IL, 1999.
7. Ward, J., Mahler, J., Maronpot, R., and Sundberg, J. P., *Pathology of Genetically Engineered Mice,* Iowa State University Press, Ames, IA, 2000.
8. Percy, D. H. and Barthold, S. W., *Pathology of Laboratory Rodents & Rabbits,* Iowa State University Press, Ames, IA, 2001.
9. Frith, C. H. and Ward, J. M., *Color Atlas of Neoplastic and Non-Neoplastic Lesions in Aging Mice,* Elsevier, Amsterdam, 1988.
10. Fredrickson, T. N., *Atlas of Mouse Hematopathology,* Harwood Academic, Amsterdam, 2000.

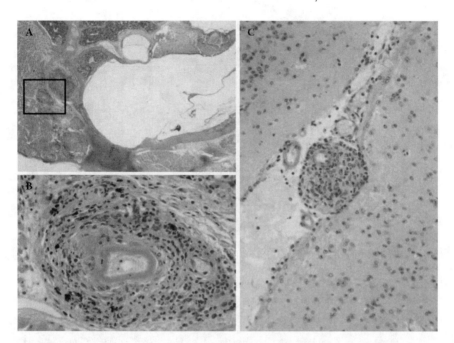

FIGURE 17.3 (See color insert following page 210.) Polyarteritis nodosa is a relatively common background disease in C57BL/6J mice. The neurological signs of head tilt are due to the arteritis of small- and medium-sized arteries going to the skull. In this case numerous affected arteries were found adjacent to the middle and outer ear (A). Higher magnification of boxed area in A reveals fibrinoid necrosis and surrounding inflammation (B). Cerebral arteries are also affected (C). These lesions were diagnosed in a 270-day-old female C57BL/6J mouse.

11. Smith, R. S., John, S. W. M., Nashina, P. M., and Sundberg, J. P., *Systematic Evaluation of the Mouse Eye. Anatomy, Pathology, and Biomethods,* CRC Press, Boca Raton, FL, 2002.

12. Sundberg, J. P., *Handbook of Mouse Mutations with Skin and Hair Abnormalities. Animal Models and Biomedical Tools,* CRC Press, Boca Raton, FL, 1994.

13. Bannasch, B. and Goessner, W., *Pathology of Neoplasia and Preneoplasia in Rodents,* Schattauer, Stuttgart, 1994.

14. Sundberg, J. P. and Ichiki, T., *Handbook on Genetically Engineered Mice,* CRC Press, Boca Raton, FL, 2005.

15. Sundberg, J. P. and Boggess, D., *Systematic Characterization of Mouse Mutations,* CRC Press, Boca Raton, FL, 2000.

16. Blake, J. A., Richardson, J. E., Bult, C., Kadin, J. A., Eppig, J. T., and Group, M. G. D., The Mouse Genome Database (MGD): The model organism database for the laboratory mouse, *Nucleic Acids Res* 30 (1), 113–115, 2002.

17. Blake, J. D., Richardson, J. E., Bult, C. J., Kadin, J. A., Eppig, J. T., and Group, M. G. D., MGD: The Mouse Genome Database, *Nucleic Acids Res* 31 (1), 193–195, 2003.

18. Bult, C. J., Blake, J. A., Richardson, J. E., Kadin, J. A., Eppig, J. T., Baldarelli, R. M., Barsanti, K., Baya, M., Beal, J. S., Boddy, W. J., Bradt, D. W.,

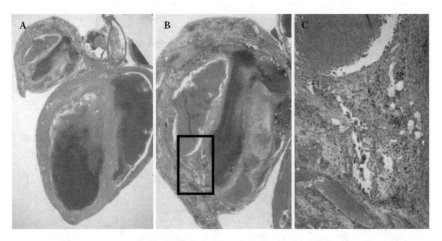

FIGURE 17.4 (See color insert following page 210.) Mural thrombosis is common in the hearts of BALB strains and substrains. This case is an example of a chronic mineralized left atrial mural thrombus in a 378-day-old female BALB/cJ mouse. Successive increases in magnification reveal the details of this lesion.

Burkart, D. L., Butler, N. E., Campbell, J., Corey, R., Corbani, L. E., Cousins, S., Dene, H., Drabkin, H. J., Frazer, K., Garippa, D. M., Glass, L. H., Goldsmith, C. W., Grant, P. L., King, B. L., Lennon-Pierce, M., Lewis, J., Lu, I., Lutz, C. M., Maltais, L. J., McKenzie, L. M., Miers, D., Modrusan, D., Ni, L., Ormsby, J. E., Qi, D., Ramachandran, S., Reddy, T. B., Reed, D. J., Sinclair, R., Shaw, D. R., Smith, C. L., Szauter, P., Taylor, B., VandenBorre, P., Walker, M., Washburn, L., Witham, I., Winslow, J., Zhu, Y., and Group, M. G. D., The Mouse Genome Database (MGD): Integrating biology with the genome, *Nucleic Acids Res* 32 (1), D476–D481, 2004.

19. Naf, D., Krupke, D. M., Sundberg, J. P., Eppig, J. T., and Bult, C. J., The mouse tumor biology database: A public resource for cancer genetics and pathology of the mouse, *Cancer Res* 62, 1235–1240, 2002.

20. Mikaelian, I., Nanney, L. B., Parman, K. S., Kusewitt, D., Ward, J. M., Naf, D., Krupke, D. M., Eppig, J. T., Bult, C. J., Seymour, R., Ichiki, T., and Sundberg, J. P., Antibodies that label paraffin-embedded mouse tissues: A collaborative endeavor, *Toxicol Pathol* 32, 1–11, 2004.

21. Schofield, P. N., Bard, J. B. L., Booth, C., Boniver, J., Covelli, V., Delvenne, P., Ellender, M., Engstrom, W., Goessner, W., Gruenberger, M., Hoefler, M., Hopewell, J., Mancuso, M.-T., Mothersill, C., Potten, C., Rozell, B., Soriola, H., Sundberg, J. P., and Ward, J. M., Pathbase: A database of mutant mouse pathology, *Nucleic Acids Res* 32, D512–515, 2004.

22. Schofield, P. N., Bard, J. B. L., Boniver, J., Covelli, V., Delvenne, P., Ellender, M., Engstrom, W., Goessner, W., Gruenberger, M., Hoefler, H., Hopewell, J. W., Mancuso, M., Mothersill, C., Quintanilla-Martinex, L., Rozell, B., Sariola, H., Sundberg, J. P., and Ward, A., Pathbase: A new reference resource and database for laboratory mouse pathology, *Radiation Protection Dosimetry* 112 (4), 525–528, 2004.

23. Sundberg, B. A. and Sundberg, J. P., A database system for small diagnostic laboratories, *Lab Anim* 19, 55–58, 1990.

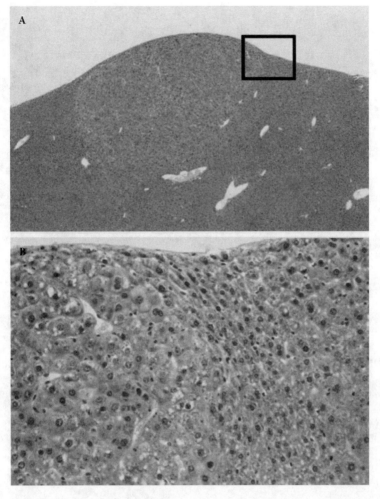

FIGURE 17.5 (See color insert following page 210.) Hepatocellular adenomas are common in old C3H/HeJ male mice. They appear as raised masses barely detectable at the gross level. Tinctorial qualities are similar to the adjacent normal parenchyma that they compress. This case illustrates a pale, well-circumscribed mass compressing the adjacent liver (A). Higher magnification of the boxed area in A illustrates the compression (B).

24. Sundberg, B. A. and Sundberg, J. P., Medical record keeping for project analysis, in *Systematic Approach to Evaluation of Mouse Mutations*, Sundberg, J. P. and Boggess, D. eds. CRC Press, Boca Raton, FL, 2000, pp. 47–55.

25. Paigen, K. and Eppig, J. T., A mouse phenome project, *Mamm Genome* 11 (9), 715–717, 2000.

26. Bogue, M., Mouse Phenome Project: Understanding human biology through mouse genetics and genomics, *J Appl Physiol* 95 (4), 1335–1337, 2003.

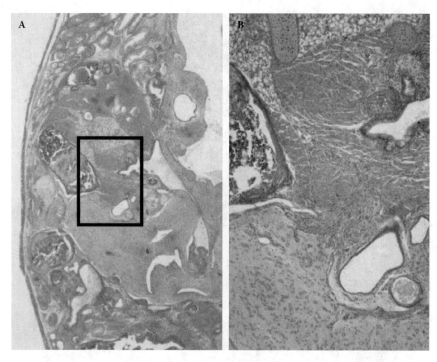

FIGURE 17.6 (See color insert following page 210.) Testicular teratomas are common unilateral masses found in most 129 strains and substrains. This is a case from a 28-day-old male 129-*Ifngr*^{tm1Agt}/J/*Ifngr*^{tm1Agt}/J mouse. The testicular cancer is a strain background characteristic and has nothing to do with the targeted mutation. Note the normal seminiferous tubules in the upper left area (A) and the majority of the testicle is compressed by a mixture of bone, skeletal muscle, tubules lined by a variety of epithelial cell types, and a large amount of neuropile. The higher magnification of the boxed area in A illustrates this mixture of cell types that are well differentiated and not invasive (B).

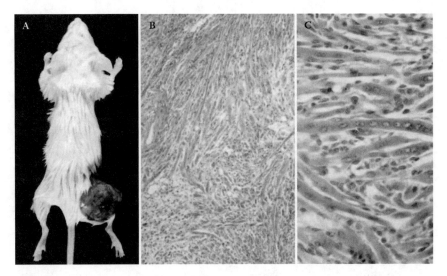

FIGURE 17.7 (See color insert following page 210.) Well-differentiated rhabdomyosar-comas are rare but found almost exclusively in BALB/cJ mice. This emaciated 118-day-old female BALB/cJ mouse has a large mass on its lower right leg. Removal of the skin revealed the irregular mottled mass (a). Histologically, these lesions consist of well-differentiated skeletal muscle (b,c) that is positive with Bodian, phosphotungstic acid hematoxylin, and iron hematoxylin histologic stains for cross striations and striated muscle actin isoforms by immunohistochemistry confirming that it is of skeletal muscle origin.

18 Colon Cancer and Polyposis Models

Makoto Mark Taketo

TABLE OF CONTENTS

INTRODUCTION

Colon cancer mortality takes the second to third highest position among all cancer deaths in most developed countries, and it is increasing in developing countries as well. Unlike lung cancer, where most cases should be prevented by eliminating tobacco smoke, there is no established environmental factor that can help eliminate colon cancer formation. Thus, it would be very important to develop workable mouse models of colon cancer and establish therapeutic and preventive strategies. Based on its genetic origin, colon cancer can be divided into two classes: polyposis colon cancer, and nonpolyposis colon cancer. Many genes, whose mutations are responsible for colon carcinogenesis, have been discovered through molecular genetic studies of hereditary cancer predisposition syndromes such as familial adenomatous polyposis (FAP), and hereditary nonpolyposis colon cancer (HNPCC). Induced mutations in mice of these genes have provided mouse models that are similar to human colon cancer and polyposis, even though they may not be identical to the human diseases. These mouse models have provided a large body of experimental evidence that helps understand the initiation, expansion, and progression of colon tumorigenesis. These mouse models have also turned out to be very useful tools to test chemotherapeutic and preventive agents. This chapter focuses on the mouse models for colon cancer and polyposis. I would also encourage readers to refer to another recent review on the pathology of mouse models for intestinal tumors.[1]

MOUSE MODELS FOR FAMILIAL ADENOMATOUS POLYPOSIS (FAP)

FAP is a hereditary disease with dominant inheritance that causes numerous colon polyps. Although the polyps are benign adenomas, some of the polyps will develop into malignant adenocarcinomas eventually, if left untreated. The adenomatous polyposis coli (*APC*) gene was identified on 5q as one of the genes commonly deleted in some FAP kindreds.[2,3] It encodes a huge protein of about 2850 amino acid residues, forms a complex with axin, and helps glycogen synthase kinase (GSK) 3β to phosphorylate N-terminal serine/threonine residues of β-catenin, accelerating its rapid degradation through ubiquitination.[4] If the *APC* gene is mutated, GSK3β cannot phosphorylate β-catenin. Unphosphorylated and therefore stabilized β-catenin accumulates in the cytoplasm, moves to the nucleus where it activates TCF/LEF transcription factors, inducing a set of new genes: *Wnt* target genes.[5] Activation of the Wnt pathway in the colonic epithelium appears to be one of the key events in the polyp initiation process.[6]

Apc^{Min} Mice, $Apc^{\Delta716}$ Mice, Apc^{1638N} Mice, and Other Apc Mutant Mice

The first mouse mutant in the *Apc* gene (the mouse homologue of human *APC* on mouse Chr 18) was discovered in a colony of randomly mutagenized mice.[7] This mutant, *Min* (Multiple intestinal neoplasia), was found to carry a truncation mutation at codon 850 of the *Apc* gene,[8] hence Apc^{Min}. In the C57BL/6 background, the heterozygote develops ~30 polyps in the small intestine. With the gene knock-out technology, several *Apc* mutations have been constructed. $Apc^{\Delta716}$ contains truncating mutation at codon 716, and Apc^{1638N} at codon 1638.[9,10] As in Apc^{Min} mice, both knock-out mutants develop polyps mainly in the small intestine. Histologically, all these *Apc* mutants form polyp adenomas indistinguishable from each other (Figure 18.1a–c). Interestingly, however, the polyp numbers are very different, even in the same C57BL/6J background. Namely, $Apc^{\Delta716}$ develops usually ~300 polyps, whereas Apc^{1638N} forms only ~3 polyps. Despite the numerous polyps developing in the small intestine in the $Apc^{\Delta716}$ (as well as in Apc^{Min}) mice, only a few polyps are formed in the colon, although the penetrance is

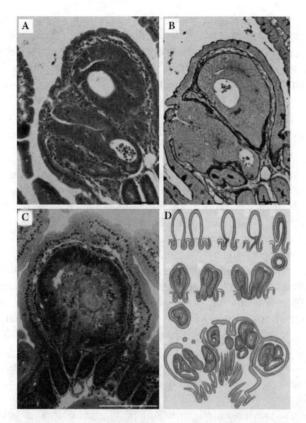

FIGURE 18.1 (See color insert following page 210.) Adenomatous polyps in the small intestine of the $Apc^{\Delta716}$ heterozygous mouse. (A) An early adenoma section stained with H&E.[6] (B) A silver staining for the basement membrane of a section adjacent to that in A.[6] (C) Toluidin blue staining of a thin section of a nascent polyp embedded in glycol-methacrylate resin.[10] (D) Schematic processes of polyp formation in the $Apc^{\Delta716}$ small intestine.[6] Scale bars: 100 µm.

complete. Because the polyp localization phenotype is opposite to that in human FAP, in which the polyps are formed primarily in the colon, and very rarely in the small intestine, it has been questioned whether *Apc* mutant mice can be accurate models for human FAP. We have presented an answer to this question in our recent paper; see below.[11]

Using $Apc^{\Delta716}$ mice, it was demonstrated that polyp formation is initiated by loss of heterozygosity (LOH) at the *Apc* locus in the proliferative zone cells,[10] followed by formation of an outpocket in the intestinal crypt (Figure 18.1d).[6] These results strongly suggested that the APC protein is essential for the proliferative zone cells to migrate along the crypt-villus axis, which is consistent with additional circumstantial evidence. Although aberrant crypt foci (ACF) have been proposed to be the precursor of the polyps in the human as well as in rodents, the term does not represent an histopathologic entity. In fact, azoxymethane-treated rat colon contains two types of microscopic lesions, and the one with β-catenin mutation and nuclear accumulation is not of the ACF appearance.[12] These microadenomas, lacking the appearance of ACF, are likely to be the precursors of colon cancer.[13]

Several other *Apc* gene knock-out mice have been reported. Their histopathology appears essentially the same as in *Apc^Min* or *Apc^Δ716*, with the only difference being the number of the intestinal polyps.

Modifier Genes of *Apc* Intestinal Polyposis

When the *Apc^Min* mutation is transferred to AKR mice, the intestinal polyp number decreases to only 1 to 3, due to a modifying locus *Mom1* (modifier of *Min*) on Chr 4.[14] Based on the results of a genetic analysis, this modifier was proposed to encode a form of secretory phospholipase A$_2$ (sPLA$_2$), but further analysis revealed a cluster of secretory phospholipases in this locus.[15] Because sPLA$_2$ cleaves long chain fatty acids from membrane phospholipids, it is difficult to explain the *Mom1* phenotype by the enzyme activity. Namely, sPLA$_2$ is wild-type in AKR, and mutated in C57BL/6. Thus, one would expect in the wild-type that an increased amount of AA is released and supplied to cyclooxygenases (COX) that produce tumor-promoting prostaglandins, which should lead to higher numbers of intestinal polyps in AKR; just the opposite phenotype from the experimental results above. In addition to *Mom1*, the mouse strain CAST carries a dominant modifier gene that confers significant protection against *Apc^Min* intestinal polyposis, although the gene remains to be identified and characterized.[15]

Introduction of a COX-2 (*Ptgs2*) gene mutation dramatically decreases the polyp number in *Apc* polyposis mice.[16] Likewise, a COX-1 (*Ptgs1*) mutation also reduces the polyp multiplicity in *Apc* polyposis mice[17] (see below).

Many other genes have been introduced into *Apc* polyposis mice, and some turned out to increase the number of polyps. For example, the DNA helicase gene mutation that is responsible for Bloom syndrome increases the number of intestinal polyps in *Apc^Min* mice.[18,19] On the other hand, the mutation for the telomerase RNA gene caused a complex phenotype in *Apc^Min* mice. The polyp numbers increased in several generations, and then decreased thereafter.[20]

Introduction of a homozygous mutation in p21, a cyclin inhibitor, into *Apc^1638N* mice increases tumor multiplicity about twice, and affects goblet cell differentiation.[21]

One of the matrix metalloproteinases matrilysin (*Mmp7*) is implicated in cancer invasion and metastasis, and its mutation reduces the polyp number in *Apc^Min* mice, indicating that the enzyme is involved in the formation of adenomatous polyps.[22]

Introduction of homozygous mutation for the inducible nitric oxide synthase (iNOS) gene reduced the polyp number in *Apc^Min* mice by less than half in the small intestine, and to about 10% of the control in the colon, suggesting a possibility in colonic polyp prevention with iNOS-selective inhibitors.[23]

Accumulating evidence indicates that the promoter regions of many cancer-related genes are silenced by methylation of the cytidine residues. However, the effects on intestinal polyposis by knock-out mutations in DNA methylation enzymes have been complex and confusing. Mutations in the DNA methyltransferase gene (*Dnmt1*), in conjunction with an enzyme inhibitor, reduced the polyp number in Apc^{Min} mice from over 100 to less than 2, suggesting suppressive effects of DNA hypomethylation on intestinal polyposis.[24] On the other hand, introduction of *Mbd2* homozygous mutation in Apc^{Min} mice also reduced the polyp numbers to ~1/10. Because *Mbd2* encodes the methyl-CpG binding repressor that recruits co-repressor complexes to methylated DNA, these results indicate that DNA methylation-mediated gene silencing is integral to the Apc^{Min} polyposis.[25]

The multidrug resistance 1 (*Mdr1*) gene promoter contains the TCF-LEF binding site, and is activated by Wnt signaling. *Mdr*-deficient Apc^{Min} mice ($Apc^{Min}/+ Mdr1a/b^{-/-}$) were constructed, and the polyp number was reduced to almost half of Apc^{Min} mice, although its clinical relevance remains to be investigated.[26]

As described above, all *Apc* mutant mice develop adenomatous polyps in the small intestine, rather than in the colon. However, recent studies show that an additional mutation in the *Cdx2* gene in $Apc^{\Delta716}$ mice reverses the polyp localization, shifting most polyps to the colon as in human FAP.[11] Interestingly, the dramatic increase in the colon polyp number is caused by the increased frequency of *Apc* LOH caused by chromosomal instability. The latter results from activation of the mTOR pathway and acceleration of the G_1 to S phase transition in the cell cycle.[11] These results present a new mechanism for chromosomal instability, and suggest a possibility of treatment and prevention of colon cancer with chromosomal instability.

MUTATIONS IN DNA MISMATCH REPAIR GENES

Mutations in DNA mismatch repair genes are involved in HNPCC, and show microsatellite instability (MSI), a mutator phenotype.[27,28] The proteins involved in DNA mismatch repair consist of three classes. Those essential for cell viability (MSH2 and MLH1), nonessential redundant genes (MSH3 and 6, PMS1, MLH3, and EXO1), and genes involved also in essential DNA replication processes (PCNA, RFC, RPA, DNA polymerase δ).[29] Although knock-out mutations have been introduced to many of these genes in the mouse, most intestinal tumors in these mice are adenomas, and malignant adenocarcinomas are very rare. Accordingly, these mice with single gene mutations are not practical models for colon cancer, but rather they are more suitable for analysis of early changes in the DNA mismatch repair lesions. Interestingly, however, compound mutants of these genes are more severe in their phenotypes, and some develop adenocarcinomas, although none of them are metastatic (see below).

Homozygous mutant mice in *Msh6*, one of the genes involved in DNA mismatch repair, develop lymphomas and intestinal adenomas, although they do not show MSI.[30] The intestinal adenomas appear to be caused by mutation in the *Apc* gene, suggesting that one of the targets of *Msh6* mutation is *Apc*. Homozygous mice in *Msh2* mutation, on the other hand, do not cause any intestinal tumors even after six months of age, despite their showing MSI.[31] Likewise, mutation in *Mlh1*, another gene involved in DNA mismatch repair, can cause GI adenomas (and carcinomas; see below) in both heterozygotes and homozygotes. When introduced into Apc^{1638N} mice, *Mlh1* mutation increased the polyp multiplicity 40- to 100-fold, due to substitution and frame shift mutations in the remaining *Apc* allele.[32,33]

When homozygous *Msh6* mutation is introduced into Apc^{1638N} mice, intestinal tumor multiplicity increases six- to sevenfold with base substitution mutations in the remaining *Apc* allele, although *Msh3* mutation introduced into Apc^{1638N} mice does not affect tumor multiplicity.[34] However, if *Msh3* mutation is added to Apc^{1638N} mice with *Msh6* mutation ($Apc^{+/1638N}$ $Msh3^{-/-}$ $Msh6^{-/-}$), intestinal tumor multiplicity increases because of truncation mutations in the remaining *Apc* allele due to base substitutions and frame shifts.[34] These data are consistent with the roles of *Msh6* and *Msh3* in repairing base substitution and frame shift mutations, respectively.

STABILIZING β-CATENIN MUTANT MICE

Because APC forms a complex with other proteins that mediates the Wnt signaling pathway, it is reasonable to ask whether mutations in other components of the complex can also cause polyps in the mouse intestines. In fact, stabilizing mutations in the serine/threonine residues of β-catenin have been identified in a subpopulation of colon cancer that do not carry *APC* mutations. To test such a possibility experimentally, conditional stabilizing β-catenin mutations have been introduced that are specifically expressed in the intestines. When expression of stabilized β-catenin is induced from calbindin promoter, mice developed only a few polyps in the small intestine.[35] In contrast, expression of Cre recombinase driven by cytokeratin 19 (K19) or fatty acid binding protein (FABP) gene promoter to introduce floxed stabilizing mutation in the β-catenin gene, caused formation of 700 to 3000 polyps in the small intestine.[36] These results confirm the role of Wnt signaling activation in polyp formation, and indicate that polyps are initiated essentially in the rapidly multiplying cells in the proliferative zone. The floxed β-catenin mutant mice have been used in several other organ systems, providing evidence for the roles of Wnt signaling in prostate tumorigenesis,[37] and embryonic and immune system development.[38,39]

MOUSE MODELS FOR GASTROINTESTINAL HAMARTOMA SYNDROMES

Hamartomas are benign tumors composed of indigenous components of the normal tissues, and therefore do not show any dysplastic histopathology. However, some hamartomas are reported to progress into malignant cancer, or hamartoma patients carry a higher risk of developing malignant tumors independent of the hamartomas.[40] Recently several genes have been identified whose mutations are responsible for the gastrointestinal hamartoma syndromes.[41] Induced mutations in some of the genes have turned out to cause hamartomas in the mouse. Below are several models of GI hamartoma syndromes.

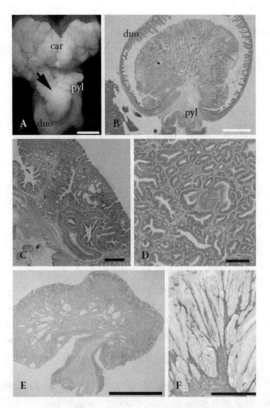

FIGURE 18.2 (See color insert following page 210.) Hamartomatous polyps in *Smad4*, *Cdx2*, and *Lkb1* heterozygotes, respectively. (A,B) A gastric polyp in a *Smad4+/−* mouse.[43] Dissection micrograph and low-power histology section (H&E) show a hamartoma developed in the pyloric region, protruding into the duodenum (arrow): car, gastric cardia; pyl, pylorus; duo, duodenum. (C,D) Histopathology of a hamartoma of the proximal colon in a *Cdx2+/−* mouse (H&E).[44] (E,F) Histopathology of a gastric hamartoma in an *Lkb1+/−* mouse.[48] Although E shows an H&E staining, the section in F was immunostained for α-smooth muscle actin. Note its arborizing pattern in the polyp stroma. Bars: A, 5 mm; B,E, 2 mm; C,D, 100 μm; and F, 500 μm.

The *SMAD4* (*DPC4*) gene encodes a cellular signaling molecule that forms a complex with *SMAD2/3* proteins, and mediates the TGF-β signaling as a nuclear transcription factor complex. In a subset of human juvenile polyposis, which develops gastrointestinal hamartomas, germ line mutations have been found in the *SMAD4* gene.[42] Consistent with this finding, the *Smad4* gene knock-out mice develop gastrointestinal hamartomas histologically similar to that in human juvenile polyposis (Figure 18.2a,b).[43] Interestingly, introduction of the *Smad4* mutations into the *Apc^{Δ716}* mice causes rapid progression of the intestinal polyps into malignant adenocarcinomas (see below).

CDX2 is a homeobox-containing transcription factor involved in gastrointestinal development and homeostasis. Although homozygous null mice are embryonic lethals, heterozygotes develop one to two hamartomas in the proximal colon (Figure 18.2c,d).[44] Although another paper claims that the tumor in the mutant is an adenocarcinoma, no convincing evidence has been presented in either histopathology or its disease course.[45] So far, no human hamartoma syndromes have been described in which the *CDX2* gene is mutated. Rather, reduced expression of *CDX2* is widely found in gastrointestinal cancers. In gastric cancer, loss of *CDX2* expression is strongly associated with poor postoperative survival.[46,47] Thus, it appears that *CDX2* plays a role similar to that of a tumor suppressor (see below).

Peutz-Jeghers' disease is another hamartoma syndrome, characterized by mucocutaneous pigmentation, gastrointestinal hamartomas, and predisposition to various cancers in multiple organs. Germ line mutations in the *LKB1* (*STK11*) gene appear to be responsible for the disease. Several groups constructed knockout mice for this gene and all found gastrointestinal hamartomas in the heterozygotes (Figure 18.2e,f), although homozygous embryos are lethal.[48,49] In addition, cancers of various organs are found in the *Lkb1* heterozygotes including hepatocellular carcinoma.[50]

Another important gene in hamartoma syndromes is *PTEN*, which encodes a protein/PI(3) phosphatase. Mutations in this gene have been found in Cowden disease and other tumor susceptibility syndromes. Although homozygous mutants are embryonic lethals, heterozygotes develop inflammatory tumors in various organs, including the colon.[51,52] Histologically, they appear as inflammatory polyps, rather than hamartomatous polyps.

Although homozygous mutation in PI 3-kinase γ, an upstream regulator of the Akt pathway, was reported to cause colon cancer, but not hamartomas,[53] the paper was retracted by the authors in 2003.

MOUSE MODELS FOR COLON CANCER

Although all *Apc* mutant mice develop adenomatous polyps, they do not progress into invasive or metastatic adenocarcinomas at a significant frequency. Because of the heavy tumor load in the small intestine, most *Apc* mutant mice die young (four to five months) due to anemia and cachexia, and some of them by intestinal intussusception. If additional mutations are introduced into these mice, however, the intestinal polyposis phenotypes are modified, and sometimes markedly malignant adenocarcinomas develop.

MOUSE MODELS FOR HEREDITARY NONPOLYPOSIS COLON CANCER

As described earlier, single homozygous mutations in the DNA mismatch repair genes rarely cause adenocarcinomas in the mouse. Interestingly, in compound mutant mice some of these genes can cause adenocarcinomas of the intestines. However, their histopathology is much milder than human colorectal cancer, and no metastasis to the liver or lymph node is observed.

ADDITIONAL MUTATIONS IN *APC^MIN* MICE, *APC^Δ716* MICE, OR *APC^1638N* MICE

For example, we have introduced the *Smad4* mutation (see above) into the *Apc^Δ716* polyposis mice, and constructed a model for malignant adenocarcinoma.[54] Although human homologues *SMAD4* and *APC* are on separate chromosomes, the mouse genes are both found on mouse Chr 18, about 30 cM apart. Because polyps are initiated by *Apc* LOH in *Apc^Δ716* intestines, and because this LOH is caused by loss of the entire Chr 18 due to recombination at the ribosomal DNA locus near the centromere, LOH of *Apc* also results in LOH at *Smad4*. Taking advantage of this fact, we have constructed mice that carried *Apc^Δ716* and *Smad4* mutations on the same chromosome in the *cis*-configuration. In the intestinal polyps, both *Apc* and *Smad4* loci are lost, resulting in homozygous mutant cells for both loci. It is important to note that the intestinal polyps in these mice progress rapidly into very invasive adenocarcinomas (Figure 18.3a–f).[54] Interestingly, however, these adenocarcinomas do not metastasize during the short lifespan of these mice. The histopathology is somewhat similar to that of the right-side colon cancer in humans that is often caused by mutations in the type II receptor for TGF-β. In fact, this model verifies tumor progression by sequential mutations in multiple genes. Moreover, some of these mice also develop adeno-carcinomas at the duodenal papilla of Vater, which is one of the complications in human FAP after the colectomy operation.

COLON CANCER MODELS WITH MUTATIONS IN OTHER GENES

Other colon cancer models have been reported that are caused by mutations in various other genes. For example, a knock-out mutation in the TGF-β1 gene (*Tgfb1*) introduced into *Rag2* mutant mice causes adenocarcinomas with strong local invasions.[55] On the other hand, it has been reported that the homozygous mutant in *Smad3*, encoding another cellular signaling molecule in the TGF-β pathway and a partner of SMAD4, develops colon cancer that metastasizes to the draining lymph nodes.[56] However, the penetrance of this phenotype appears to be low, and lymph node metastasis has not been found in all affected mice. Furthermore, the metastatic phenotype has not been observed in any similar knock-out mutants of *Smad4* constructed by other groups.[57]

With the use of the promoter for the villin gene whose expression is specific to the intestinal epithelium, transgenic mice have been constructed that express an activated mutant of K-*ras* (K-*ras^V12G*). Most transgenic mice develop single or multiple lesions, ranging from microadenomas to invasive adenocarcinomas, without mutations in *Apc*.[58] As in the *cis*-compound *Apc^Δ716Smad4* mutant mice, none of the adenocarcinomas in this model metastasize to distant loci, although the tumors are highly invasive locally.

Homozygous mutation in the *Muc2* gene that encodes the most abundant secreted gastrointestinal mucin causes adenomas and adenocarcinomas in the intestines.[59] Although the incidence and multiplicity are low, the adenocarcinoma is locally invasive without distant metastasis.

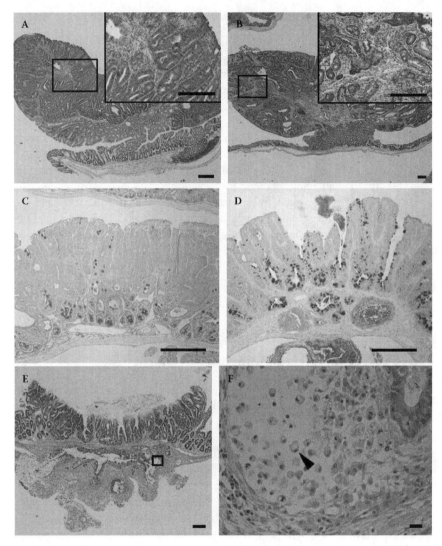

FIGURE 18.3 (See color insert following page 210.) Adenocarcinomas in the intestines of *cis-Apc$^{\Delta716}$Smad4* heterozygous mice.[54] (A,B) Representative colonic polyps in the *Apc$^{+/\Delta716}$* mouse (A; control) and *cis-Apc$^{+/\Delta716}$Smad4$^{+/-}$* mouse B at 14 weeks of age (H&E). Note from the scale bars that the polyp size is larger in B. (C,D) Small intestinal polyps in the *Apc$^{+/\Delta716}$* mouse (C; control) and *cis-Apc$^{+/\Delta716}$Smad4$^{+/-}$* mouse D at 20 weeks of age (Alucian blue). Note the invasive adenocarcinoma in the smooth muscle layer. (E,F) A more advanced adenocarcinoma in a 20-week-old *cis-Apc$^{+/\Delta716}$Smad4$^{+/-}$* mouse. Arrowhead in F shows one of the signet-ring cells in the boxed area in E. Bars: A–D, 250 μm; G, 125 mm; and F, 6.3 μm.

Mouse Models for Colon Cancer Associated with Inflammatory Bowel Disease

Colon cancer associated with inflammatory bowel disease is somewhat different from regular (i.e., noninflammatory) colon cancer. Malignant cancer develops after a long and sustained inflammation in the intestines. The etiology of the inflammation is mostly of autoimmune nature, and alterations in various immune cells and mediators have been reported. As models for this type of colon cancer, several mutant mouse strains have been constructed. For example, IL-10-deficient mice produce aberrant cytokines, especially IFNγ, and develop invasive adenocarcinoma in the colon in 60% of mice by six months of age.[60] Likewise, compound mutant mice for IL-2 and β2-microglobulin genes also develop ulcerative colitislike disease, and ~1/3 of them produces adenocarcinomas in the colon.[61] Interestingly, null mutant mice for one of the G proteins, Gαi2, also develop similar colitis and adenocarcinoma of the colon.[62] On the other hand, introduction of dominant negative N-cadherin causes inflammatory bowel disease and adenomas, but no carcinomas.[63]

MOUSE MODELS FOR TUMOR CHEMOTHERAPY AND CHEMOPREVENTION

Although colon cancer mouse models have greatly helped our understanding of the molecular mechanisms of tumorigenesis, another important application of these mice in biomedical research is preclinical evaluation of pharmaceutical and nutritional agents for treatment and prevention of cancer. Historically, rodent colon cancer induced by chemical carcinogens such as dimethylhydralazine or azoxymethane was used for evaluation of such drugs as nonsteroidal anti-inflammatory drugs (NSAIDs) on colon cancer prevention.[64,65] Although possible, these models are difficult to work with, simply because it takes a long time (1/2 to 1 year) to develop tumors at relatively low incidence and multiplicity. In contrast, genetically altered mice can develop large numbers of polyps suitable for statistical analyses in small numbers of animals at younger ages. It is also possible to introduce additional mutations in the drug target genes, as proof-of-principle experiments.

One such example is our study on the role of cyclooxygenase-2 (COX-2; *Ptgs2*) in intestinal polyposis using *Apc*$^{\Delta716}$ mice.[16] We first determined expression of COX-2 protein in intestinal polyps of various sizes, and found that COX-2 was expressed from a very early stage of polyp formation. Namely, polyps of 1 to 2 mm in diameter already express significant levels of COX-2 in the polyp stroma. We then introduced a COX-2 gene (*Ptgs2*) knock-out mutation into the *Apc*$^{\Delta716}$ mice, and discovered that both the number and size of polyps were reduced dramatically in the compound mutant mice, in a mutated gene dosage-dependent manner. To confirm the results with a pharmaceutical compound, we further dosed *Apc*$^{\Delta716}$ mice with COX-2 inhibitors, and demonstrated that the polyp number can be reduced in a dosage-dependent manner.[16,66] These results gave the rationale to treat human FAP patients with COX-2 inhibitors such as celecoxib or rofecoxib, and clinical trials confirmed the results of animal experiments, establishing COX-2 inhibitors as standard therapeutic agents for FAP as approved by the FDA.[67] Following these experiments, a number of reports have been published dosing the Apc mutant mice with various drugs or drug candidates. We have summarized such experiments in a recent review.[68] It has turned out that COX-2 plays a key role not only in benign polyposis, but also in malignant cancer of various organs.[69] It should be noted that the constitutive isozyme COX-1 (Ptgs2) also plays a significant role in polyposis. Introduction of a COX-1 mutation into Apc^{Min} mice reduces the number and size of intestinal polyps by ~80%, a similar effect by COX-2 mutation.[17] In fact, COX-1 and COX-2 cooperate in polyp formation by supplying PGE_2 that stimulates polyp angiogenesis.[70]

To assess the role of the (arachidonic acid) AA/COX-2/PGE_2 pathway further, we constructed several additional compound mutants of $Apc^{\Delta 716}$ mice with other genes in the pathway. Although cyclooxygenase is the rate-limiting enzyme in the pathway, its substrate AA is not freely available in $vivo$, but supplied on demand by the activity of phospholipases that cleaves AA from membrane phospholipids. One of the key enzymes in this activity is cytosolic phospholipase A_2 ($cPLA_2$). We introduced a knock-out gene for the enzyme into $Apc^{\Delta 716}$ mice, and demonstrated that polyp expansion is reduced significantly.[71] In the downstream of the pathway, on the other hand, the direct metabolite of AA by Cox is PGH_2 that is converted further to various prostanoids as PGA_2, D_2, $F_{2\alpha}$, E_2, I_2, and thromboxane A_2 (TXA_2). Among them PGE_2 appears to play the major role in polyposis by binding to four G-protein-coupled cell surface receptors, EP1, EP2, EP3, and EP4. To determine which particular EP receptor is involved in the polyposis phenotype, we introduced knock-out mutations for the receptors into $Apc^{\Delta 716}$ mice, respectively, and scored the polyp number and size.

The results clearly showed a significant suppression of the polyposis phenotype only in the compound mutant mice with the EP2 mutation.[72] Moreover, we found that induction of COX-2 itself, as well as other phenotypes associated with COX-2 induction such as angiogenesis and basement membrane changes, are also mediated by the PGE_2 and EP2 receptor. The former phenotype indicates that a positive feedback signal mediated through EP2 and intracellular cyclic AMP regulates COX-2 expression.[72]

It should be noted that several papers suggest that PGE_2 activates EP4 in epithelial cancer,[73] and that other prostanoids such as prostacyclin (PGI_2) and thromboxane A_2 (TxA_2) are involved in polyposis,[74,75] although these mechanisms remain to be confirmed by mutant mouse experiments.

Expression of COX-2 has been observed not only in adenomatous polyps, but also in hamartomatous polyps and malignant cancers as well. For example, COX-2 expression has been reported in Peutz-Jeghers gastrointestinal hamartomas, suggesting a possible therapeutic application of COX-2 inhibitors to the patients.[76] We have recently investigated expression of COX-2 in hamartomas of three models, namely, in mutant mice of $Smad4$, $Cdx2$, and $Lkb1$, respectively, and found significant expression levels in all models.[77] These results indicate that COX-2 induction is a general phenomenon common to most tumors, and expand the possibility to treat these diseases with COX-2 inhibitors to prevent malignant progression.

Although other possible targets have been proposed for the antitumor activities of NSAIDs,[78] these pathways remain to be investigated further by genetic means, that is, crossing of the target gene knock-out mice with the polyposis mice.

Another avenue of similar experiments is to study the effects of various diets and food additives, in attempts to identify the epidemiological effects on cancer initiation and progression. For example, we demonstrated that docosahexaenoic acid (DHA) reduces intestinal polyp development, although the effect is moderate and found only in female mice,[79] whereas feeding $Apc^{\Delta 716}$ mice with a high-fat diet increases polyp numbers significantly.[80] Likewise, a Western-style diet (high fat and low calcium) accelerates tumor formation in Apc^{1638N} mice.[81] It is also interesting to point out that calorie intake restriction by 40% reduces intestinal polyps in Apc^{Min} mice by ~60%, suggesting that dietary interventions can partially offset genetic susceptibility to intestinal carcinogenesis.[82]

In conclusion, many mouse models have been established that are useful for investigations of the initiation, expansion, and progression of colon cancer. They are also valuable tools to evaluate various pharmaceutical and biological agents for colon polyposis treatment and prevention of cancer. Yet, one of the key outstanding issues in this field of research is to establish practical models of colon cancer metastasis to the liver, lung, and lymph nodes. Such models should greatly help us find novel measures to overcome colon cancer in the 21st century.

ACKNOWLEDGMENT

I am grateful to the members of my laboratory who have contributed to the papers cited. The research programs in my laboratory have been supported by grants from MESSC, Japan; OPSR, Japan; University of Tokyo–Banyu Pharmaceutical Co. Joint Fund; Takeda Foundation; and Mitsubishi Foundation.

REFERENCES

1. Boivin, G. P., Washington, K., Yang, K., Ward, J. M., Pretlow, T. P., Russell, R., Besselsen, D. G., Godfrey, V. L., Doetschman, T., Dove, W. F., Pitot, H. C., Halberg, R. B., Itzkowitz, S. H., Groden, J., and Coffey, R. J., Pathology of mouse models of intestinal cancer: Consensus report and recommendations, *Gastroenterology* 124, 762–777, 2003.
2. Groden, J., Thliveris, A., Samowitz, W., Carlson, M., Gelbert, L., Albertsen, H., Joslyn, G., Stevens, J., Spirio, L., Robertson, M., Sargeant, L., Krapcho, K., Wolff, E., Burt, R., Hughes, J. P., Warrington, J., McPherson, J., Wasmuth, J., Le Paslier, D., Abderrahim, H., Cohen, D., Leppert, M., and White, R., Identification and characterization of the familial adenomatous polyposis coli gene, *Cell* 66, 589–600, 1991.

3. Kinzler, K. W., Nilbert, M. C., Vogelstein, B., Bryan, T. M., Levy, D. B., Smith, K. J., Preisinger, A. C., Hamilton, S. R., Hedge, P., Markham, A., Carlson, M., Joslyn, G., Groden, J., White, R., Miki, Y., Miyoshi, Y., Nishisho, I., and Nakamura, Y., Identification of a gene located at chromosome 5q21 that is mutated in colorectal cancers, *Science* 251, 1366–251, 1991.

4. Polakis, P., The oncogenic activation of beta-catenin, *Current Opinions Genetics Dev* 9, 15–21, 1999.

5. Korinek, V., Barker, N., Morin, P. J., van Wichen, D., de Weger, R., Kinzler, K. W., Vogelstein, B., and Clevers, H., Constitutive transcriptional activation by a β-catenin-Tcf complex in APC-/- colon carcinoma, *Science* 275, 1784–1787, 1997.

6. Oshima, H., Oshima, M., Kobayashi, M., Tsutsumi, M., and Taketo, M. M., Morphological and molecular processes of polyp formation in $Apc^{\Delta716}$ knockout mice, *Cancer Res* 57, 1644–1649, 1997.

7. Moser, A. R., Pitot, H. C., and Dove, W. F., A dominant mutation that predisposes to multiple intestinal neoplasia in the mouse, *Science* 247, 322–324, 1990.

8. Su, L.-K., Kinzler, K. W., Vogelstein, B., Preisinger, A. C., Moser, A. R., Luongo, C., Gould, K. A., and Dove, W. F., Multiple intestinal neoplasia caused by a mutation in the murine homolog of the APC gene, *Science* 256, 668–670, 1992.

9. Fodde, R., Edelmann, W., Yang, K., van Leeuwen, C., Carlson, C., Renault, B., Breukel, C., Alt, E., Lipkin, M., Khan, P. M., and Kucherlapati, R., A targeted chain-termination mutation in the mouse *Apc* gene results in multiple tumors, *Proc Natl Acad Sci USA* 91, 8969–8973, 1994.

10. Oshima, M., Oshima, H., Kitagawa, K., Kobayashi, M., Itakura, C., and Taketo, M., Loss of *Apc* heterozygosity and abnormal tissue building in nascent intestinal polyps in mice carrying a truncation *Apc* gene, *Proc Natl Acad Sci USA* 92, 4482–4486, 1995.

11. Aoki, K., Tamai, Y., Horiike, S., Oshima, M., and Taketo, M. M., Colonic polyposis caused by mTOR-mediated chromosomal instability in $Apc^{+/\Delta716}$ $Cdx2^{+/-}$ compound mutant mice, *Nat Gen* 35, 323–330, 2003.

12. Yamada, Y., Yoshimi, N., Hirose, Y., Kawabata, K., Matsunaga, K., Shimizu, M., Hara, A., and Mori, H., Frequent β-catenin gene mutations and accumulations of the protein in the putative preneoplastic lesions lacking macroscopic aberrant crypt foci appearance, in rat colon carcinogenesis, *Cancer Res* 60, 3323–3327, 2000.

13. Yamada, Y., Yoshimi, N., Hirose, Y., Matsunaga, K., Katayama, M., Sakata, K., Shimizu, M., Kuno, T., and Mori, H., Sequential analysis of morphological and biological properties of β-catenin-accumulated crypts, provable premalignant lesions independent of aberrant crypt foci in rat colon carcinogenesis, *Cancer Res* 61, 1874–1878, 2001.

14. MacPhee, M., Chepenik, K. P., Liddell, R. A., Nelson, K. K., Siracusa, L. D., and Buchberg, A. M., The secretory phospholipase A2 gene is a candidate for the *Mom1* locus, a major modifier of Apc^{Min}-induced intestinal neoplasia, *Cell* 81, 957–966, 1995.

15. Cormier, R. T., Bilger, A., Lillich, A. J., Halberg, R. B., Hong, K. H., Gould, K. A., Borenstein, N., Lander, E. S., and Dove, W. F., The *Mom1* AKR intestinal tumor resistance region consists of *Pla2g2a* and a locus distal to *D4Mit64*, *Oncogene* 19, 3182–3192, 2000.

16. Oshima, M., Dinchuk, J. E., Kargman, S. L., Oshima, H., Hancock, B., Kwong, E., Trzaskos, J. M., Evans, J. F., and Taketo, M. M., Suppression of intestinal poly-posis in $Apc^{\Delta716}$ knockout mice by inhibition of cyclooxygenase 2 (COX-2), *Cell* 87, 803–809, 1996.

17. Chulada, P. C., Thompson, M. B., Mahler, J. F., Doyle, C. M., Gaul, B. W., Lee, C., Tiano, H. F., Morham, S. G., Smithies, O., and Langenbach, R., Genetic disruption of *Ptgs-1*, as well as of *Ptgs-2*, reduces intestinal tumorigenesis in Min mice, *Cancer Res* 60, 4705–4708, 2000.

18. Luo, G., Santoro, I. M., McDaniel, L. D., Nishijima, I., Mills, M., Youssoufian, H., Vogel, H., Schultz, R. A., and Bradley, A., Cancer predisposition caused by elevated mitotic recombination in Bloom mice, *Nat Gen* 26, 424–429, 2000.

19. Heppner Goss, K., Risinger, M. A., Kordich, J. J., Sanz, M. M., Straughen, J. E., Losvek, L. E., Capobianco, A. J., German, J., Boivin, G. P., and Groden, J., Enhanced tumor formation in mice heterozygous for *Blm* mutation, *Science* 297, 2051–2053, 2002.

20. Rudolph, K. L., Millard, M., Bosenberg, M. W., and DePinho, R. A., Telomere dysfunction and evolution of intestinal carcinoma in mice and humans, *Nat Gen* 28, 155–159, 2001.

21. Yang, W. C., Mathew, J., Velcich, A., Edelmann, W., Kucherapati, R., Lipkin, M., Yang, K., and Augenlicht, L. H., Targeted inactivation of the $p21^{WAF/cip1}$ gene enhances Apc-initiated tumor formation and the tumor-promoting activity of a western-style high-risk diet by altering cell maturation in the intestinal mucosa, *Cancer Res* 61, 565–569, 2001.

22. Wilson, C. L., Heppner, K. J., P.A., L., Hogan, B. L. M., and Matrisian, L. M., Intestinal tumorigenesis is suppressed in mice lacking the metalloproteinase matrilysin, *Proc Natl Acad Sci USA* 94, 1402–1407, 1997.

23. Ahn, B. and Oshima, H., Suppression of intestinal polyposis in $Apc^{Min/+}$ mice by inhibiting nitric oxide production, *Cancer Res* 61, 8357–8360, 2001.

24. Laird, P. W., Jackson-Grusby, L., Faseli, A., Dickinson, S. L., Jung, W. E., Li, E., Weinberg, R. A., and Jaenisch, R., Suppression of intestinal neoplasia by DNA hypomethylation, *Cell* 81, 197–205, 1995.

25. Sansom, O. J., Berger, J., Bishop, S. M., Hendich, B., Bird, A., and Clarke, A. R., Deficiency of Mbd2 suppresses intestinal tumorigenesis, *Nat Genet* 34, 145–147, 2003.

26. Yamada, T., Mori, Y., Hayashi, R., Takada, M., Ino, Y., Naishiro, Y., Kondo, T., and Hirohashi, S., Suppression of intestinal polyposis in Mdr1-deficient $Apc^{Min/+}$ mice, *Cancer Res* 63, 895–901, 2003.

27. Modrich, P. and Lahue, R., Mismatch repair in replication fidelity, genetic recombination, and cancer biology, *Ann Rev Biochem* 65, 101–133, 1996.

28. Lynch, H. T. and de la Chapelle, A., Genetic susceptibility to non-polyposis colorectal cancer, *J Med Genet* 36, 801–818, 1999.

29. Kolodner, R. D., Putnam, C. D., and Myung, K., Maintenance of genome stability in *Saccharomyces cerevisiae*, *Science* 297, 552–557, 2002.

30. Edelmann, W., Yang, K., Umar, A., Heyer, J., Lau, K., Fan, K., Liedtke, W., Cohen, P. E., Kane, M. F., Lipford, J. R., Yu, N., Crouse, G. F., Pollard, J. W., T., K., Lipkin, M., Kolodner, R. D., and Kucherlapati, R., Mutation in the mismatch repair gene *Msh6* causes cancer susceptibility, *Cell* 91, 467–477, 1997.

31. Reitmair, A. H., Cai, J.-C., Bjerknes, M., Redston, M., Cheng, H., Pind, M. T. L., Hay, K., Mitri, A., Bapat, B. V., Mak, T. W., and Gallinger, S., MSH2 deficiency contributes to accelerated APC-mediated intestinal tumorigenesis, *Cancer Res* 56, 2922–2926, 1996.

32. Edelmann, W., Yang, K., Kuraguchi, M., Heyer, J., Lia, M., Kneitz, B., Fan, K., Brown, A. M. C., Lipkin, M., and Kucherapati, R., Tumorigenesis in *Mlh1* and *Mlh1/Apc*1638N mutant mice, *Cancer Res* 59, 1301–1307, 1999.

33. Kuraguchi, M., Edelmann, W., Yang, K., Lipkin, K., Kucherlapati, R., and Brown, A. M. C., Tumor-associated *Apc* mutations in *Mlh1*-/- *Apc*^1638N mice reveal a mutational signature of Mlh1 deficiency, *Oncogene* 19, 5755–5763, 2000.

34. Kuraguchi, M., Yang, K., Wong, E., Avdievich, E., Fan, K., Kolodner, R. D., Lipkin, M., Brown, A. M. C., Kucherlapati, R., and Edelmann, W., The distinct spectra of tumor-associated *Apc* mutations in mismatch repair-deficient *Apc*^1638N mice define the roles of MSH3 and MSH6 in DNA repair and intestinal tumorigenesis, *Cancer Res* 61, 7934–7942, 2001.

35. Romagnolo, B., Berrebi, D., Saadi-Keddoucci, S., Porteu, A., Pichard, A., Peuchmaur, M., Vandewalle, A., Kahn, A., and Perret, C., Intestinal dysplasia and adenoma in transgenic mice after overexpression of an activated β-catenin, *Cancer Res* 59, 3875–3879, 1999.

36. Harada, N., Tamai, Y., Ishikawa, T., Sauer, B., Takaku, K., Oshima, M., and Taketo, M. M., Intestinal polyposis in mice with a dominant stable mutation of the β-catenin gene, *EMBO J* 18, 5931–5942, 1999.

37. Gounari, F., Signoretti, S., Bronson, R., Klein, L., Sellers, W. R., Kum, J., Siermann, A., Taketo, M. M., von Boehmer, H., and Khazaie, K., Stabilization of β-catenin induces lesions reminiscent of prostatic intraepithelial neoplasia, but terminal squamous transdifferentiation of other secretory epithelia, *Oncogene* 4099, 4107, 2002.

38. Lickert, H., Domon, C., Huls, G., Wehrle, C., Duluc, I., Clevers, H., Meyer, B. I., Freund, J.-N., and Kemler, R., Wnt/β-catenin signaling regulates the expression of the homeobox gene *Cdx1* in embryonic intestine, *Development* 127, 3805–3813, 2000.

39. Gounari, F., Aifantis, I., Khazaie, K., Hoeflinger, S., Harada, N., Taketo, M. M., and von Boehmer, H., Somatic activation of β-catenin bypasses pre-TCR signaling and TCR selection in thymocyte development, *Nat Immunol* 2, 863–869, 2001.

40. Haggitt, R. C. and Reid, B. J., Hereditary gastrointestinal polyposis syndromes, *Am J Surgical Pathol* 10, 871–887, 1986.

41. Eng, C. and Peacocke, M., *PTEN* and inherited hamartoma-cancer syndromes, *Nat Gen* 19, 223, 1998.

42. Howe, J. R., Roth, S., Ringold, J. C., Summers, R. W., Jarvinen, H. J., Sistonen, P., Tomlinson, I. P., Houlston, R. S., Bevan, S., Mitros, F. A., Stone, E. M., and Aaltonen, L. A., Mutations in the *SMAD4/DPC4* gene in juvenile polyposis, *Science* 280, 1086–1088, 1998.

43. Takaku, K., Miyoshi, H., Matsunaga, A., Oshima, M., Sasaki, N., and Taketo, M. M., Gastric and duodenal polyps in *Smad4* (*Dpc4*) knockout mice, *Cancer Res* 59, 6113–6117, 1999.

44. Tamai, Y., Nakajima, R., Ishikawa, T., Takaku, K., Seldin, M. F., and Taketo, M. M., Colonic hamartoma development by anomalous duplication in *Cdx2* knock-out mice, *Cancer Res* 59, 2965–2970, 1999.

45. Chawengsaksophak, K., James, R., Hammond, V. E., Kontgen, F., and Beck, F., Homeosis and intestinal tumours in *Cdx2* mutant mice, *Nature* 386, 84–87, 1997.

46. Seno, H., Oshima, M., Taniguchi, M., Usami, K., Ishikawa, T., Chiba, T., and Taketo, M. M., CDX2 expression in the stomach with intestinal metaplasia and intestinal-type cancer: prognostic implications, *Int J Oncol* 21, 769–774, 2002.

47. Mizoshita, T., Tsukamoto, T., Nakanishi, H., Inada, K., Ogasawara, N., Joh, T., Itoh, M., Yamamura, Y., and Tatematsu, M., Expression of Cdx2 and the phenotype of advanced gastric cancers: Relationship with prognosis, *J Cancer Clin Oncol* 129, 727–734, 2003.

48. Miyoshi, H., Nakau, M., Ishikawa, T., Seldin, M. F., Oshima, M., and Taketo, M. M., Gastrointestinal hamartomatous polyposis in *Lkb1* heterozygous knockout mice, *Cancer Res* 62, 2261–2266, 2002.

49. Rossi, D. J., Ylikorkala, A., Korsisaari, N., Salovaara, R., Luukko, K., Launonen, V., Henkemeyer, M., Ristimäki, A., Aaltonen, L. A., and Mäkelä, T. P., Induction of cyclooxygenase-2 in a mouse model of Peutz-Jeghers polyposis, *Proc Natl Acad Sci USA* 99, 12327–12332, 2002.

50. Nakau, M., Miyoshi, H., Seldin, M. F., Imamura, M., Oshima, M., and Taketo, M. M., Hepatocellular carcinoma caused by loss of heterozygosity in *Lkb1* gene knockout mice, *Cancer Res* 62, 4549–4553, 2002.

51. di Cristofano, A., Pesce, B., Cordon-Cardo, C., and Pandolf, P. P., *Pten* is essential for embryonic development and tumor suppression, *Nat Gen* 19, 348–355, 1998.

52. Podsypanina, K., Ellenson, L. H., Nemes, A., Gu, J., Tamura, M., Yamada, K. M., Cordon-Cardo, C., Catoretti, T., Fisher, P. E., and Parsons, R., Mutation of *Pten/Mmac1* in mice causes neoplasia in multiple organ systems, *Proc Natl Acad Sci USA* 16, 1563–1568, 1999.

53. Sasaki, T., Irie-Sasaki, J., Horie, Y., Bachmaier, K., Fata, J. E., Li, M., Suzuki, A., Bouchard, D., Ho, A., Redston, M., Gallinger, S., Khokha, R., Mak, T. W., Hawkins, P. T., Stephens, L., Scherer, S. W., Tsao, M., and Penninger, J. M., Colorectal carcinomas in mice lacking the catalytic subunit of PI(3)Kγ, *Nature* 406, 897–902, 2000.

54. Takaku, K., Oshima, M., Miyoshi, H., Matsui, M., Seldin, M. F., and Taketo, M. M., Intestinal tumorigenesis in compound mutant mice of both *Dpc4* (*Smad4*) and *Apc* genes, *Cell* 92, 645–656, 1998.

55. Engle, S. J., Hoying, J. B., Boivin, G. P., Ormsby, I., Gartside, P. S., and Doetschman, T., Transforming growth factor β1 suppresses nonmetastatic colon cancer at an early stage of tumorigenesis, *Cancer Res* 59, 3379–3386, 1999.

56. Zhu, Y., Richardson, J. A., Parada, L. F., and Graff, J. M., *Smad3* mutant mice develop metastatic colorectal cancer, *Cell* 94, 703–714, 1998.

57. Yang, X., Letterio, J. J., Lechleider, R. J., Chen, L., Hayman, R., Gu, H., Roberts, A. B., and Deng, C., Targeted disruption of SMAD3 results in impaired mucosal immunity and diminished T cell responsiveness to TGF-β, *EMBO J* 18, 1280–1291, 1999.

58. Janssen, K.-P., El Mrjou, F., Pinto, D., Sastre, X., Rouillard, D., Fouquet, C., Soussi, T., Louvard, D., and Robine, S., Targeted expression of oncogenic K-ras in intestinal epithelium causes spontaneous tumorigenesis in mice, *Gastro-enterology* 2002.

59. Velcich, A., Yang, W., Heyer, J., Fragale, A., Nicholas, C., Viani, S., Kucherlapati, R., Lipkin, M., Yang, K., and Augenlicht, L., Colorectal cancer in mice genetically deficient in the mucin *Muc2*, *Science* 295, 1726–1729, 2002.

60. Berg, D. J., Davidson, N., Kühn, R., Müller, W., Menon, S., Holland, G., Thompson-Snipes, L., Leach, M. W., and Rennick, D., Enterocolitis and colon cancer in interleukin-10-deficient mice are associated with aberrant cytokine production and CD4+ TH1-like responses, *J Clin Invest* 98, 1010–1020, 1996.

61. Shah, S. A., Simpson, S. J., Brown, L. F., Comiskey, M., de Jong, Y. P., Allen, D., and Terhorst, C., Development of colonic adenocarcinomas in a mouse model of ulcerative colitis, *Inflamm Bowel Dis* 4, 196–202, 1998.

62. Rudolph, U., Finegold, M. J., Rich, S. S., Harriman, G. R., Srinivasan, Y., Brabet, P., Boulay, G., Bradley, A., and Birnbaumer, L., Ulcerative colitis and adenocarcinoma of the colon in Gα i2-deficient mice, *Nat Gen* 10, 143–150, 1995.

63. Hermiston, M. L. and Gordon, J. I., Inflammatory bowel disease and adenomas in mice expressing a dominant negative N-cadherin, *Science* 270, 1203–1207, 1995.

64. Taketo, M. M., Cyclooxygenase-2 inhibitors in tumorigenesis (Part I), *J Nat Cancer Inst* 90, 1529–1536, 1998.

65. Taketo, M. M., Cyclooxygenase-2 inhibitors in tumorigenesis (Part II), *J Nat Cancer Inst* 90, 1609–1620, 1998.

66. Oshima, M., Murai, N., Kargman, S., Arguello, M., Luk, P., Kwong, E., Taketo, M. M., and Evans, J. F., Chemoprevention of intestinal polyposis in the $Apc^{\Delta 716}$ mouse by rofecoxib, a specific cyclooxygenase-2 inhibitor, *Cancer Res* 61, 1733–1740, 2001.

67. Steinbach, G., Lynch, P. M., Phillips, R. K. S., Wallace, M. H., Hawk, E., Gordon, G. B., Wakabayashi, N., Sauders, B., Shen, Y., Fujimura, T., Su, L.-K., and Levin, B., The effect of celecoxib, a cyclooxygenase-2 inhibitor, in familial adenomatous polyposis, *N Eng J Med* 342, 1946–1952, 2000.

68. Oshima, M. and Taketo, M. M., COX selectivity and animal models for colon cancer, *Curr Pharm Des* 8, 102–1034, 2002.

69. Dannenberg, A. J. and Subbaramaiah, K., Targeting cyclooxygenase-2 in human neoplasia: Rationale and promise, *Cancer Cell* 4, 431–436, 2003.

70. Takeda, H., Sonoshita, M., Sugihara, K., Chulada, P. C., Langenbach, R., Oshima, M., and Taketo, M. M., Cooperation of cyclooxygenase 1 and cyclooxygenase 2 in intestinal polyposis, *Cancer Res* 63, 4872–4877, 2003.

71. Takaku, K., Sonoshita, M., Sasaki, N., Uozumi, N., Doi, Y., Shimizu, T., and Taketo, M. M., Suppression of intestinal polyposis in $Apc^{\Delta 716}$ knockout mice by an additional mutation in the cytosolic phospholipase A_2 gene, *J Biol Chem* 275, 34013–34016, 2001.

72. Sonoshita, M., Takaku, K., Sasaki, N., Sugimoto, Y., Ushikubi, F., Narumiya, S., Oshima, M., and Taketo, M. M., Acceleration of intestinal polyposis through prostaglandin receptor EP2 in $Apc^{\Delta 716}$ knockout mice, *Nature Med* 7, 1048–1051, 2001.

73. Sheng, H., Shao, J., Washington, M. K., and DuBois, R. N., Prostaglandin E2 increases growth and motility of colorectal carcinoma cells, *J Biol Chem* 276, 18075–18081, 2001.

74. Cutler, N. S., Graves-Deal, R., LaFleur, B. J., Gao, Z., Boman, B. M., Whitehead, R. H., Terry, E., Morrow, J. D., and Coffey, R. J., Stromal production of prostacyclin confers an antiapoptotic effect to colonic epithelial cells, *Cancer Res* 63, 1748–1751, 2003.

75. Daniel, T. O., Liu, H., Morrow, J. D., Crews, B. C., and Marnett, L. J., Thromboxane A_2 is a mediator of cyclooxygenase-2-dependent endothelial migration and angiogenesis, *Cancer Res* 59, 4574–4577, 1999.

76. de Leng, W. W., Westerman, A. M., de Rooij, F. W., Dekken, H. H., de Goeij, A. F., Gruber, S. B., Wilson, J. H., Offerhaus, G. J., Giardiello, F. M., and Keller, J. J., Cyclooxygenase 2 expression and molecular alterations in Peutz–Jeghers hamartomas and carcinomas, *Clin Cancer Res* 9, 3065–3072, 2003.

77. Takeda, H., Miyoshi, H., Tamai, Y., Oshima, M., and Taketo, M. M., Simultaneous expression of COX-2 and mPGES-1 in mouse gastrointestinal hamartomas, *Brit J Cancer* 90, 701–704, 2004.

78. DuBois, R. N., New agents for cancer prevention, *J Nat Cancer Inst* 94, 1732–1733, 2002.

79. Oshima, M., Takahashi, M., Oshima, H., Tsutsumi, M., Yazawa, K., Sugimura, T., Nishimura, S., Wakabayashi, K., and Taketo, M. M., Effects of docosahexaenoic acid (DHA) on intestinal polyp development in $Apc^{\Delta716}$ knock-out mice, *Carcinogenesis* 16, 2605–2607, 1995.

80. Hioki, K., Shivapurkar, N., Oshima, H., Alabaster, O., Oshima, M., and Taketo, M. M., Suppression of intestinal polyp development by low-fat and high-fiber diet in $Apc^{\Delta716}$ knock-out mice, *Carcinogenesis* 18, 1863–1865, 1997.

81. Yang, K., Edelman, W., Fan, K., Lau, K., Leung, D., Newmark, H., Kurcherlapati, R., and Lipkin, M., Dietary modulation of carcinoma development in a mouse model for human familial adenomatous polyposis, *Cancer Res* 58, 5713–5717, 1998.

82. Mai, V., Colbert, L. H., Berrigan, D., Perkins, S. N., Pfeiffer, R., Lavigne, J. A., Lanza, E., Heines, D. C., Schatzkin, A., and Hursting, S. D., Calorie restriction and diet composition modulate spontaneous intestinal tumorigenesis in Apc^{Min} mice through different mechanisms, *Cancer Res* 63, 1752–1755, 2003.

NOTE ADDED IN PROOF

Recently, cardiovascular side effects have been reported for some COX-2 inhibitors. Accordingly, clinical uses of these drugs as well as traditional NSAIDs need to be used with caution. (Ref. Solomon S. D., et al. Cardiovascular risk associated with celecoxib in a clinical trial for colorectal adenomaprevention. *New Engl. J. Med.* 352, 1071–1080, 2005. Bresalier, R. D., et al. Cardiovascular events associated with refecoxib in a colorectal adenomachemoprevention trial. *New Engl. J. Med.* 352, 1092–1102, 2005.)

19 Phenotypic Analysis of Mice with Steroid Deficiency

Noomen Ben El Hadj, Meng-Chun Hu, Hsueh-Ping Chu, Leo Chi-Kuang Wang, and Bon-chu Chung

TABLE OF CONTENTS

INTRODUCTION

Steroid hormones include glucocorticoids, mineralocorticoids, and sex hormones; they are major regulators of sexual differentiation, reproductive function, stress accommodation, and fluid and electrolyte balance.[1] Steroids are small soluble molecules that penetrate the cell's plasma membrane freely and bind to specific cytoplasmic or nuclear receptors. This binding of steroids to their receptors activates gene expression. Defects in steroid synthesis may lead to pathological disorders.[2] The first chemical reaction leading to the biosynthesis of all steroid hormones is the conversion of the abundant sterol precursor, cholesterol, to the C-21 steroid, pregnenolone. This reaction is catalyzed by cytochrome P450scc (cholesterol side chain cleavage enzyme, *Cyp11a1*), which plays a pivotal role in steroidogenesis because its product, pregnenolone, is considered to be an obligatory intermediate in the biosynthetic pathways through which steroid hormones are produced.[3] The *Cyp11a1* gene is expressed mainly in the adrenal cortex, testes, ovaries, and placenta, and to some extent in nerve and peripheral tissues.[4]

The hypothalamic-pituitary-adrenal (HPA) and the hypothalamic-pituitary-gonadal (HPG) axes are the main regulators of steroidogenesis. Gonadotropin-releasing hormone (GnRH) produced in the hypothalamus acts on the pituitary cells to secrete follicular stimulating hormone (FSH) and luteinizing hormone (LH), which stimulate gonadal steroidogenesis. Similarly, hypothalamic corti-cotropin releasing factor (CRF) acts on the pituitary cells to secrete adrenocorti-cotropin hormone (ACTH), which in turn activates adrenal steroidogenesis. Only aldosterone is regulated by the renin-angiotensin system.[5] Steroid hormones are produced at different concentrations or ratios to control the feedback of the HP axis in regulating the secretion of CRF, ACTH, GnRH, FSH, and LH either positively or negatively.[6,7] Dysfunction in any step of the HPG or HPA axes resulting in disruption of the endocrine balance often leads to a vast number of pathologies. Indeed, spontaneous or engineered mutations in the genes encoding steroidogenic enzymes, pituitary and hypothalamic peptides, and their receptors result in major metabolic disorders and defect of normal development of both internal and external male and female structures.[8]

In mammals, male accessory sex organ (MASO) development requires the differentiation of the bipotential gonad into a hormonally functional testis. The fetal testis secretes both steroid (testosterone, T) and nonsteroid (Müllerian inhib-iting substance, MIS) hormones. These two hormones trigger the development of MASO and the regression of the female phenotype. MIS causes regression of the Müllerian ducts, which in its absence would give rise to the fallopian tubes, uterus, and the upper part of the vagina as is observed in female embryos.[9] T induces the Wolffian ducts (WD) to differentiate into epididymis (Ep), vas deferens (VD), and seminal vesicles (SV). The metabolite of testosterone, 5 alpha-dihydrotestosterone, is the active ligand in the urogenital sinus (UGS) and is responsible for its differentiation into the ampullary gland, prostate (Pr), urethra, bulbourethral glands (BUG), and penis (P).

The presence of clinical cases with different degrees of genital tract malfor-mations indicates the effects of steroid hormones in male accessory sex organ differentiation and the necessity to maintain physiologic ratios between different kinds of steroids.[10,11] Congenital adrenal hyperplasia is a group of autosomal recessive disruptions of adrenal steroidogenesis due to a genetic disorder in one of the steroidogenic enzymes, most frequently mutations in the 21-hydroxylase gene. It is the most common cause of androgen-mediated virilization of the UGS in females.[12,13] Patients with a mutation in *Cyp11a1* (CYP11A1) have been described.[14,15] In a previous report we showed that targeted disruption of the *Cyp11a1* gene is compatible with term gestation and gives a severe phenotype in which the newborn mice die shortly after birth.[16] Here we used corticosteroid replacement to keep *Cyp11a1*-deficient mice alive for a longer time in order to

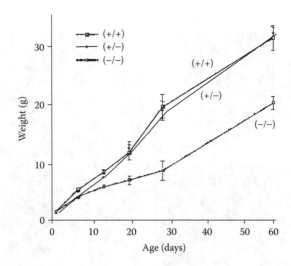

FIGURE 19.1 Growth retardation of *Cyp11a1* null mice. Growth curves are assessed in wild-type (open squares), heterozygous (filled squares), and *Cyp11a1* null (open circles) male mice. These mice (4/genotype) were injected daily with corticoids in the evening and their body weights were measured weekly. Growth rate of *Cyp11a1* null mice is severely reduced.

assess the developmental effects of *Cyp11a1* deficiency, particularly on reproductive function. Absence of P450scc activity leads to typical symptoms of steroidogenesis disruption with sex reversal, genital tract infantilism, gametogenesis impairment, and growth retardation.

RESULTS

Cyp11a1 Null Mice Rescued with Corticosteroids Exhibit Growth Retardation

We previously reported that *Cyp11a1*-deficient newborn mice die shortly after birth from hypoaldosteronism.[16] In order to maintain these mice alive and to study their phenotype we rescued them with mineralocorticoids and glucocorticoids, which are compulsory for survival. Although *Cyp11a1*-deficient newborn mice have normal body weight at birth (comparable to wild-type and heterozygous pups: 1.47 ± 0.1 +/+; 1.5 ± 0.12 +/−; 1.38 ± 0.15 −/−), they grow more slowly than the wild-type littermates. At two months their body weights were almost 2/3 of those of wild-type mice (30.2 ± 2.05 +/+; 31.1 ± 0.89 +/−; 19.48 ± 1.02 −/−) (Figure 19.1). Wild-type and heterozygous mice have a similar growth rate. In spite of substitutive corticotherapy and salt addition in their drinking water, *Cyp11a1* null mice show muscle atrophy, decreased food intake, weakness, lethargy, and dehydration, and die in most cases with hypoglycemic syndrome (shaking, anorexia, dizziness, atonia).

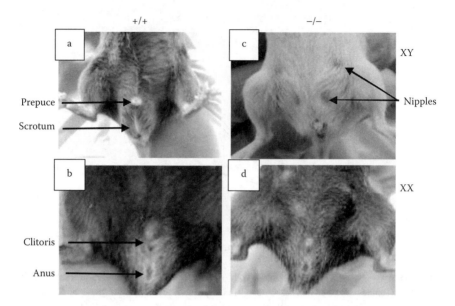

FIGURE 19.2 Male to female sex reversal and cryptorchidism of *Cyp11a1* null mice. (a) XY wild-type mice (2 months) have normal male external genitalia with complete testis descent into the scrotum as well as a large distance between the anus and prepuce. (b) Normal female mice have short anus–prepuce distance and persistence of nipples in their abdomen. (c) *Cyp11a1*-deficient XY mice have female phenotype with nipples and short anus–prepuce distance. The scrotum does not develop and their testes remain in abdominal position. (d) *Cyp11a1*-deficient XX mice have normal female phenotype.

SEX REVERSAL AND CRYPTORCHIDISM OF *CYP11A1* NULL MALE MICE

Because sex steroids are required for the differentiation of male phenotype in mammals,[17] we examined carefully the sexual characteristics of *Cyp11a1*-deficient mice in different developmental stages. The defect of male external genitalia differentiation is noticeable as early as 2 days after birth. Indeed, *Cyp11a1*-deficient pups having the *Sry* gene (XY genetic male) present a sex-reversed genitalia with small prepuce and short ano-genital distance (data not shown). Between 7 and 10 days, *Cyp11a1*-deficient XY genetic males start to develop nipples exactly as in XX females. In normal male development, at around 15 days the testes descend completely inside the scrotum, but in *Cyp11a1*-deficient males the scrotum was not formed yet. At 2 months XY wild-type mice have normal male external genitalia with large distance between the anus and prepuce and complete descent of testes in the scrotum (Figure 19.2a). However *Cyp11a1*-deficient XY mice still show nipples in their abdomen and short anus–prepuce distance compared to XY wild-type individuals (Figure 19.2c). Moreover, these mice are cryptorchid, either because their testes fail to migrate in the absence of sex hormone signaling or because the scrotum did not develop. Although XY mice have major external genital defects, XX *Cyp11a1*-deficient mice show normal external female genitalia when compared to the XX wild-type female mice (Figure 19.2d). Nipples develop normally in XX *Cyp11a1*-deficient mice.

INTERNAL REPRODUCTIVE TRACT PHENOTYPE OF *CYP11A1* NULL MICE

Defect of Male Accessory Sex Organ Differentiation in *Cyp11a1* Null Mice

The male reproductive system in the mouse includes testes, derivatives of the Wolffian duct (WD), and derivatives of the urogenital sinus (UGS). Dissection of *Cyp11a1*-deficient male urogenital tracts at different developmental stages revealed that the most dramatically affected genital organs are the derivatives of the UGS. In normal conditions UGS differentiates into prostatic glands (Pr) and bulbourethral glands (BUG), and the Wolffian duct differentiates into ducts efferens, epididymis (Ep), vas deferens (VD), seminal vesicles (SV), and ampulla. This differentiation, mediated by androgens, starts in late stages of gestation (embryonic day 14 to 18 in mice) and continues throughout the post-natal life. Only after puberty do these organs reach their adult size and shape.[18] Because sex steroids are not required for survival, we can follow internal genital development of *Cyp11a1* null mice by maintaining them alive with only corti-costeroid treatment.

Careful examination of the Wolffian duct derivatives in the second day after birth shows normal epididymis and vas deferens, but seminal vesicles cannot be found. In the wild-type seminal vesicles appear as two thin tubules of 2 to 3 mm of length at the intersection of the vas deferens. At two weeks all the derivatives of the Wolffian duct including Ep, VD, and SV are differentiated in the wild-type. The two VD are fused at the neck of the bladder (B). In the ampullary region SV and prostatic glands are converged (Figure 19.3a). *Cyp11a1*-deficient littermates do not develop SV and their Ep and VD are thinner. Besides, the two VD of *Cyp11a1* null mice do not fuse on the neck of the B and extend to the lower part of UGS (Figure 19.3b).

At four weeks when the SV of wild-type males are well developed, *Cyp11a1* null mice show a vestige tiny primordia of SV and regressing Ep and VD. During the first week after birth we cannot recognize the anatomical defects of *Cyp11a1* null mice UGS derivatives because their differentiation starts after birth. At two weeks, whereas Pr, BUG, U, and penis are formed in wild-type, the *Cyp11a1*-deficient male UGS differentiate into a female phenotype with a vaginalike structure enclosing the U. Prostatic glands, BUG, and penis are totally absent (Figure 19.3b,d). Because the ampulla is not formed in *Cyp11a1* null mice, the VD do not fuse with the UGS and extended parallel to the U where they degen-erate as thin fibers near the preputial glands (Figure 19.3d). The preputial gland, which is a modified sebaceous gland derived from epidermis, exists in *Cyp11a1* null males, but its size is reduced (Figure 19.3c,d).

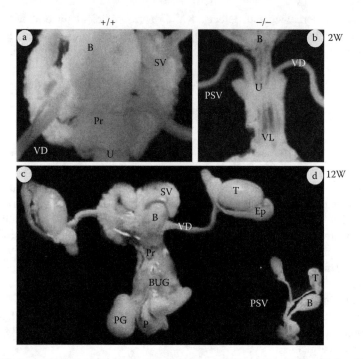

FIGURE 19.3 Defect of male accessory sex organ differentiation in *Cyp11a1* null mice. (a) Wild-type urogenital tracts of 2-week mice include two vas deferens (VD) fused at the neck of the bladder (B). In the ampullary region, VD, seminal vesicles (SV), and prostatic glands (Pr) converge, giving rise to the urethra (U). (b) *Cyp11a1*-deficient urogenital tracts lack SV and Pr. Their VD do not fuse at the neck of the bladder and therefore the ampulla is not formed. VD extend to the lower part of the urogenital sinus, which differentiates into a vaginalike (VL) structure. (c) 12-week male genital tracts include testis, Ep, bulbouretral glands (BUG), and penis (P). (d) *Cyp11a1*-deficient mice showing smaller male internal genitalia of the testis. PSV: primordial of seminal vesicle, U: urethra.

Genital Tract Infantilism of *Cyp11a1* Null Female Mice

In all mammals, in the absence of hormonally functional testis or in the presence of ovary, the urogenital tract develops a female phenotype. Indeed, the absence of androgens leads to the regression of the Wolffian duct and the absence of anti-Müllerian substance allows the differentiation of the Müllerian duct into fallopian tubes, uterus, and upper vagina.[19] Unlike males, no developmental defect was noted in the *Cyp11a1*-deficient female genitalia. During the first 2 weeks after birth, *Cyp11a1*-deficient females show comparable size and morphology of ovary (Ov), uterine horns (UH), vagina (V), and clitoral glands (CG) to those of the wild-type females. At 4 weeks, when puberty starts, internal genitalia of wild-type mice are well developed, whereas *Cyp11a1* null genital tracts are hypoplastic: the ovaries, uterine horns, and vagina are smaller than those of the wild-type. The size of the clitoral glands is also reduced. After puberty, at 12 weeks age, *Cyp11a1*-deficient

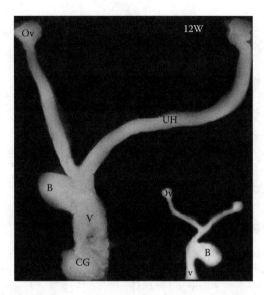

FIGURE 19.4 Underdevelopment of the female urogenital tract of *Cyp11a1* null mice. (A). Female urogenital tracts from wild-type (left) and *Cyp11a1*-deficient (right) mice showing different developmental stages. At 12 weeks (12W), *Cyp11a1* null genital tracts are hypoplastic: the ovaries (Ov), uterine horns (UH), vagina (V), and clitoral glands (CG) are smaller than those of the wild-type. B: bladder.

females still show an infantile urogenital tract. Their ovaries have much reduced size compared to wild-type ovaries. Their vaginas and uteri do not grow and seem to be regressing in size (Figure 19.4).

HISTOLOGICAL MODIFICATIONS IN GENITAL TRACT OF *CYP11A1* NULL MICE

Sex hormones are known to be involved in growth and functional differentiation of the male and female ductal epithelia;[20] we examined the histology of Wolffian duct derivatives in *Cyp11a1* null mice. Before puberty, comparable histology of the epididymis and vas deferens between wild-type and *Cyp11a1* null mice are observed (data not shown). Yet in older *Cyp11a1* null mice (30 days), the histophysiological aspects of the ducts are markedly affected.

Microscopic examination of the wild-type epididymis (Ep), site of sperm maturation and storage, reveals a convoluted tubule with pseudostratified columnar epithelium lined by long microvilli (arrow in Figure 19.5a). In *Cyp11a1*-deficient Ep, there is a dramatic reduction in epithelial height accompanied by total absence of microvilli and reduction of the tubule lumen (Figure 19.5b). Although its general architecture is conserved, the epithelium of *Cyp11a1*-deficient epididymis is aberrant and lacks microvilli. It cannot exert its physiological functions (sperm maturation, phagocytosis, protein synthesis, storage, etc.). Finally, the VD, which transports sperm to the ampullary region, is markedly atrophic in *Cyp11a1*-deficient mice but shows different layers of the muscularis

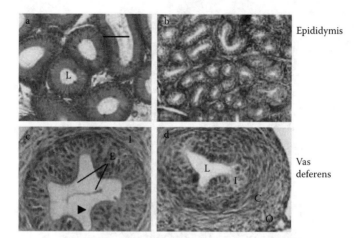

Epididymis

Vas deferens

FIGURE 19.5 Histological defects of male genital tract of 30-day *Cyp11a1* null mice. (a) Longitudinal sections of wild-type epididymis show a pseudostratified columnar epithelium lined by long microvilli (arrow). The lumen (L) of the tubule is well formed. (b) In *Cyp11a1*-deficient mice, the lumen of the tubule is reduced and the microvilli are absent. (c) The wild-type vas deferens is well developed and its epithelium (E) is full of blebs and microvilli (arrow). (d) *Cyp11a1*-deficient vas deferens is markedly atrophic and its epithelium lacks blebs and microvilli. I: inner longitudinal layer, C: middle circular layer, O: outer longitudinal layer.

as in wild-type mice. These muscular layers are thinner in *Cyp11a1*-deficient VD and lined by an epithelium lacking blebs and microvilli (arrow in Figure 19.5c).

DISCUSSION

Much of the experimental research of the functions of sex steroid receptors has depended upon *in vitro* systems as well as *in vivo* methods that require surgical castration or the pharmacological administration of hormone antagonists. However, recently developed techniques that allow for manipulation of the mouse genome have been utilized to generate gene-targeted animals that lack steroidogenic genelike *Cyp11a1*. These animals, combined with the naturally existing *Tfm* mice which lack a functional androgen receptor, now provide *in vivo* models for further study of the various actions of the sex steroids and their receptors.

Alfred Jost pioneered the field of reproductive endocrinology with his seminal observation that two hormones produced by the testes are required for the male embryo to develop a normal internal reproductive tract. Testosterone induces the Wolffian ducts to differentiate into epididymis, vas deferens, and seminal vesicles. Müllerian inhibiting substance (MIS) causes regression of the Müllerian ducts, which in its absence would normally develop into fallopian tubes, uterus, and upper vagina as is observed in female embryos.

In all mammalian species testicular androgens control three aspects of male phenotypic development: conversion of the Wolffian ducts into epididymis, vas deferens, and seminal vesicles; formation of the male urethra and prostate; and formation of the phallus. Studies in animals and humans with mutations in the genes that encode the androgen receptor or steroid 5 alpha-reductase indicate that testosterone virilizes Wolffian ducts and that 5-reduced androgens are essential for formation of the prostate and phallus.[21] Wolffian duct virilization has been thought to be under androgen effect. Our knock-out mice show that the duct precursor is originally formed, but it lacks some of the accessory organs while others remain differentiated in the absence of sex steroids.

Our results show that sex steroids, estrogens in particular, are necessary for folliculogenesis. Indeed, estrogen promotes follicular development by potentiating follicular development, steroidogenesis, granulosa cell expression of gonadotropin receptors, and gap junction formation by granulosa cells, and by inhibiting granulosa cell apoptosis. These events lead to ovulation. Our knock-out mouse studies show that female hormones are not required before puberty, but are required for reproductive functions after puberty.

ACKNOWLEDGMENTS

We would like to thank Shu-Jan Chou for excellent technical assistance, and Transgenic Core Facility at Academia Sinica for the generation of knock-out mouse lines. This work was supported by grant NSC 91-2311-B-001-085 from National Science Council, and by Academia Sinica, Republic of China, AS1IMB1PP.

REFERENCES

1. Hu, M.C. et al., Regulation of steroidogenesis in transgenic mice and zebrafish, *Mol. Cell. Endocrinol.*, 171, 9, 2001.
2. Miller, W.L. and Levine, L.S., Molecular and Clinical advances in congenital adrenal hyperplasia, *J. Ped.*, 111, 1, 1988.
3. Chung, B.C., Guo, I.C., and Chou, S.J., Transcriptional regulation of the *CYP11A1* and ferredoxin genes, *Steroids*, 62, 37, 1997.
4. Chung, B.C. et al., Human cholesterol side-chain cleavage enzyme, P450scc: cDNA cloning, assignment of the gene to chromosome 15, and expression in the placenta, *Proc. Natl. Acad. Sci. U. S. A.*, 83, 8962, 1986.
5. Bader, M. et al., Tissue renin-angiotensin systems: new insights from experimental animal models in hypertension research, *J. Mol. Med.*, 79, 76, 2001.
6. Wierman, M.E., Bruder, J.M., and Kepa, J.K., Regulation of gonadotropin-releasing hormone (*GnRH*) gene expression in hypothalamic neuronal cells, *Cell. Mol. Neurobiol.*, 15, 79, 1995.
7. Pacak, K., Stressor-specific activation of the hypothalamic-pituitary-adrenocortical axis, *Physiol. Res.*, 49 Suppl. 1, S11, 2000.
8. Brinkmann, A.O., Molecular basis of androgen insensitivity, *Mol. Cell. Endocrinol.*, 179, 105, 2001.

9. Behringer, R.R., The *in vivo* roles of mullerian-inhibiting substance, *Curr. Top. Dev. Biol.*, 29, 171, 1994.

10. Hiort, O., Holterhus, P.M., and Nitsche, E.M., Physiology and pathophysiology of androgen action, *Baillieres Clin. Endocrinol. Metab.*, 12, 115, 1998.

11. Kater, C.E. and Biglieri, E.G., Disorders of steroid 17 alpha-hydroxylase deficiency, *Endocrinol. Metab. Clin. North Am.*, 23, 341, 1994.

12. Ganesan, A. et al., Congenital adrenal hyperplasia: preliminary observations of the urethra in 9 cases, *J. Urol.*, 167, 275, 2002.

13. White, P.C. and Speiser, P.W., Congenital adrenal hyperplasia due to 21-hydroxylase deficiency, *Endocr. Rev.*, 21, 245, 2000.

14. Katsumata, N. et al., Compound heterozygous mutations in the cholesterol side-chain cleavage enzyme gene (*CYP11A*) cause congenital adrenal insufficiency in humans, *J. Clin. Endocrinol. Metab.*, 87, 3808, 2002.

15. Tajima, T. et al., Heterozygous mutation in the cholesterol side chain cleavage enzyme (*p450scc*) gene in a patient with 46,XY sex reversal and adrenal insufficiency, *J. Clin. Endocrinol. Metab.*, 86, 3820, 2001.

16. Hu, M.C. et al., Steroid deficiency syndromes in mice with targeted disruption of Cyp11a1, *Mol. Endocrinol.*, 16, 1943, 2002.

17. Couse, J.F. and Korach, K.S., Exploring the role of sex steroids through studies of receptor deficient mice, *J. Mol. Med.*, 76, 497, 1998.

18. Cunha, G.R. et al., Normal and abnormal development of the male urogenital tract. Role of androgens, mesenchymal-epithelial interactions, and growth factors, *J. Androl.*, 13, 465, 1992.

19. Jost, A., A new look at the mechanisms controlling sex differentiation in mammals, *Johns Hopkins Med. J.*, 130, 38, 1972.

20. Nielsen, M. et al., Ontogeny of oestrogen receptor alpha in gonads and sex ducts of fetal and newborn mice, *J. Reprod. Fertil.*, 118, 195, 2000.

21. Russell, D.W. and Wilson, J.D., Steroid 5 alpha-reductase: two genes/two enzymes, *Annu. Rev. Biochem.*, 63, 25, 1994.

20 External Genitalia Development: A Model System to Study Organogenesis

*K. Suzuki, H. Nishida, S. Ohta, Y. Satoh,
Y. Xu, Y. Zhang, Y. Wada, Y. Ogino, N. Nakagata,
T. Ohba, and G. Yamada*

TABLE OF CONTENTS

Developmental biology has a long history as an experimental biology. Clearly, experimental biology and medicine originate, in a sense, from descriptive research. Ever since such descriptive research was first studied, numerous functional analyses became possible and were performed. Efficient manipulation of genes at the whole-body level (gene manipulation at the level of whole body of mammals) only became possible after many years of this period until the development of transgenic technology and ES cell (embryonic stem cell) technology.

Retrospectively, analytical science, such as biochemistry, was also becoming possible at that time which was also coincident with the generation of interdisciplinary science with chemistry and biology to create biochemistry. Since then, it became possible to efficiently isolate and clone various genes leading to the establishment of molecular biology. Such technology exists as a background for the creation of "reproductive engineering" because molecular biological techniques are necessary to manipulate genes before introduction into the whole body. It is also necessary for the analysis of introduced genes in "transgenic animals".

After several decades, it is now becoming "normal" to analyze the role of genes *in vivo*. The next period was spent isolating many genes from genomic DNA (chromosomal genes including relevant exons and introns) and for cloning cDNAs. Around that time, isolating putative gene fragments was, in a sense, enough to publish significant reports. Subsequently, gene sequences *per se* were already registered in the sequence databases by genome projects in various species. Researchers kept trying to analyze gene functions of the so-called "developmental control genes", such as homeobox genes, *in vivo* in the mammalian (mouse) whole body.[1] Since then, gene introduction into fertilized eggs became possible to create transgenic mice (Tg mice). This enables "gain-of-function" studies with a gene of interest driven by a promoter of choice to achieve ectopic transgene expression. To analyze gene functions by gain-of-function strategies, injection of the cloned DNA, such as the growth hormone expression vector, yielded the so-called "giant mice" and drew much attention. This was just the beginning of Pandora's box opening, reporting the importance of manipulation in the whole mouse body.

Following such vast progress of molecular biology and embryonic manipulation, transgenic technology and targeted mutagenesis (gene knock-out, KO, technology) have become routine and are widely utilized today. A good example of showing the research trends utilizing engineered mice is to examine scientific meetings in the corresponding field. A meeting named Mouse Molecular Genetics was held annually since the late 1980s. In the first years, most of the presentations focused on Tg mouse technology and phenotype analyses. Most work is now devoted to KO mouse analyses and applications toward developing mouse models of human diseases.

There is no doubt about the role of KO research in elucidating various gene functions which are necessary in many physiological processes, such as hormone actions, digestion, development, and so on. After the declaration of the completion of genome sequencing (genome project), it is becoming even more necessary to perform such genetically engineered mouse studies because vast amounts of genome information is now available to the biomedical research community.

EXTERNAL GENITALIA DEVELOPMENT: A MODEL FOR ORGANOGENESIS?

Numerous studies revealed aspects of developmental organogenesis. Mammalian external genitalia (Figure 20.1) display highly specialized anatomical structures suitable for efficient internal fertilization and show marked morphological variations between species. Such morphologic variations of external genitalia reflect various modes of copulatory behaviors and characteristics of various species. However, very little research has been undertaken to investigate external genitalia development.[2–4]

In humans, a congenital anomaly displaying developmental defects both in appendages such as limbs and external genitalia were reported.[3,5,6] In fact, such human congenital disorders did display and indicate possible shared aspects of developmental programs between different budding organs such as limbs and external genitalia appendages. Careful clinical examinations of such patients indicated a common hypoplasia among outgrowing/budding appendages such as limbs or external genitalia of some patients, although insights on their developmental backgrounds remained unelucidated at that time.

The genital tubercle (hereafter described as GT), an anlage of the external genitalia, differentiates into a penis in the male and a clitoris in the female (Figure 20.2). During the mouse embryonic development the structural difference between the male and female genital ridge is first recognized at embryonic day 12.5 (E12.5).[7,8]

In the field of molecular developmental biology, researchers also noticed "shared" expression patterns in different outgrowing appendages such as similar expression patterns of several developmental controlling genes in limbs and external genitalia.

Pioneering studies suggested that GT development shows developmental similarities with limb development, considering that both structures exhibit prominent outgrowth before terminal differentiation of the organ in embryos.[2,9]

Would it be possible to learn some lessons from analyses of developmental controlling genes involved in limb development? There were numerous achievements during the history of molecular and developmental analyses on vertebrate limb development. One of the key findings in limb developmental biology was the finding of a key developmental region identified on the basis of its unique function and marker gene expression: the finding of the apical ectodermal ridge (AER).

Vertebrate limb development depends on the establishment and maintenance of the AER, a specialized ectodermal structure at the distal tip of the limb bud.[10–12] The AER, a transient specialized distal epithelium, is essential for vertebrate embryonic limb outgrowth along the proximodistal (PD) axis. It is well recognized that proliferation of limb mesenchymal cells is controlled by the AER. Several key findings were discovered in relation with growth factor functions related to AER. Such findings offered fundamental cues for understanding

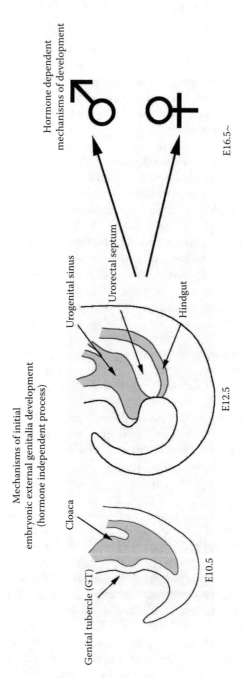

FIGURE 20.1 An illustration of mouse external genitalia development.

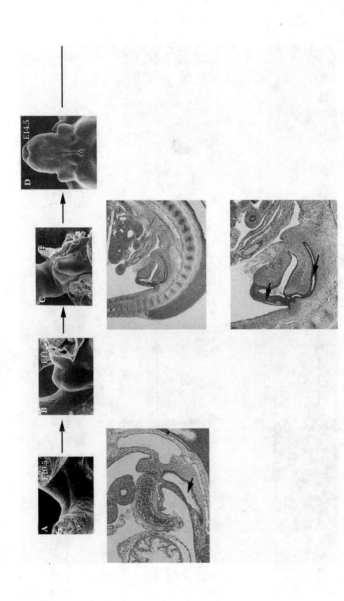

FIGURE 20.2A (See color insert following page 210.) Mouse genital tubercle (GT) development during embryogenesis; illustration of its development by SEM and sagittal sections of embryos for early hormone independent stages. A slight sign of genital tubercle outgrowth is observed adjacent to the cloaca (A,E) around E10.5 during mouse development. Later, GT outgrowth becomes more prominent and the cloaca is divided into urogenital sinus and hindgut by urorectal septum at E12.5 (F,G).

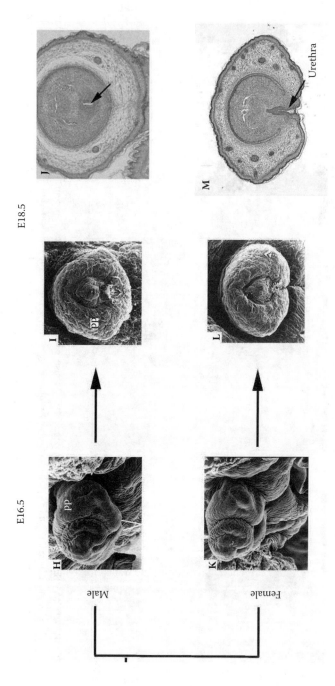

FIGURE 20.2B (See color insert following page 210.) A characteristic point of the developmental processes for external genitalia formation lies in the hormone-dependent phase of development after E16.5 in mouse development (lower part, H–M). Several corresponding sagittal sections are shown for (E–G). pp: prepuce.

the role of AER in limb appendage outgrowth. The sonic hedgehog (SHH)/ fibroblast growth factor (FGF) feedback signaling loop, for example, that operates between the polarizing region and the AER, may coordinate growth and patterning for the limb.[13–15] By the developmental and functional analyses of such a signaling region, numerous growth factors and transcriptional factors are expressed in such regions.

EMERGENCE OF GROWTH FACTORS: FIBROBLAST GROWTH FACTORS (FGFS) AS A BEGINNING OF GROWTH FACTOR ANALYSIS OF EXTERNAL GENITALIA FORMATION

Fibroblast growth factors (FGFs) are essential signaling molecules for embryogenesis.[16–22] One of the FGFs, FGF8, is important for growth and patterning of embryonic limbs.[11,23–25] It is possible that loss-of-function studies of the *Fgf8* failed to identify limb phenotypes due to compensation by other *Fgf(s)*.[26]

As indicated in the introduction, several modern molecular developmental techniques were established to analyze such complex actions of genetic cascades during organogenesis. The general gene knock-out technology often failed to identify gene functions if such genes possessed several functions in early and late embryonic stages because simple introduction of the mutation often leads to early lethal phenotypes of the homozygotes, and thus failed in identifying their functions at later stages.

In fact, more recent compound conditional KO mouse studies revealed complex action of a growth factor cascade during limb morphogenesis and also some information on external genitalia morphogenesis.[26–28]

Recently, we investigated the molecular mechanisms of the GT formation and showed that the distal urethral epithelium (DUE), the region marked by *Fgf8* expression (Figure 20.3), regulates the outgrowth of mouse GT. Based on several functional assays, we proposed that the DUE of the developing GT could function as a putative signaling center orchestrating the embryonic development of external genitalia.[29–31] In addition to several *Fgfs*, expression of multiple regulatory genes involved in the formation of other appendages were detected in the developing GT. More recent study also revealed factors other than *Fgfs*, such as *Bmp7*, are expressed in the DUE region indicating the complexity of such growth factor cascades.[31,32]

Such redundant modes of gene expression in the transient signaling epithelia prompted us to speculate about putative shared mechanisms for organogenesis. Such similarities suggest the existence of shared molecular developmental programs for vertebrate appendages. In this context, the role of regulatory factors and pathways, such as for *Fgfs*, *Bmps*, *Wnt*, and *Sonic hedgehog* (*Shh*), would require further analysis. They may display similar as well as divergent roles in regulating the morphogenesis of various organs.

Expression pattern of *Fgf8* during embryonic limb and external genitalia formation

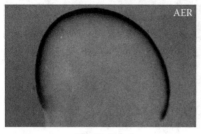 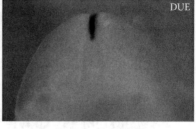

Hindlimb   Genital tubercle (GT)

FIGURE 20.3 Expression pattern of *Fgf8* in AER: apical ectodermal ridge, the key signaling epithelia orchestrating the limb appendage formation and in DUE (distal urethral epithelium in genital tubercle appendage formation). The AER was shown to function as signaling epithelia during limb formation. The DUE was suggested to possess a character of signaling epithelia for external genitalia development.[29]

IMPLICATIONS FOR UNDERSTANDING HUMAN BIRTH DEFECTS BASED ON MOLECULAR EMBRYOLOGY FOR EXTERNAL GENITALIA DEVELOPMENT

It would be useful to explore the mechanisms of external genitalia development integrated with recent findings on the molecular regulatory programs controlling GT morphogenesis considering the recent increase of some human birth defects, such as hypospadias[33–35] and cryptorchidism.[36–38] In fact, there are several reports indicating a possible increase of the frequency of the hypospadias during newborn infants worldwide. Depending on the regions or countries, the frequency of the hypospadias is reported as dramatically increasing. Thus, understanding its mechanisms is essential for society as well as for science. A detailed description of mouse GT formation will constitute a solid basis for further studies on mammalian external genitalia.

ROLE OF *FGF10* DURING EXTERNAL GENITALIA FORMATION: THE FIRST CASE OF GENE KNOCK-OUT MOUSE MODEL FOR EXTERNAL GENITALIA DEVELOPMENT

The first incidence demonstrating the importance of gene KO mice for the analysis of external genitalia formation was the case of *Fgf10* KO mice. Following initial outgrowth, the GT differentiates into a penis or clitoris with the preputial glands, urethrae, prepuces, and the corpus cavernosum glans. Following mesenchymal differentiation of the GT, *Fgf10* was expressed in the mesenchyme adjacent to the midline urethral plate at E13.5, when there is no gross sexual dimorphism.[29] It was shown that the external genitalia of *Fgf10* mutant mice have striking defects in urethra formation.[29] Such dysmorphogenesis was clearly evident by SEM analysis.[39] Characteristic morphological defects were found in the ventral (lower) side of the glans penis and the clitoridis, especially in the prepuce. This was the first case demonstrating a gene function by knock-out mice for external genitalia formation.

In normal females at E18.5, morphogenesis of the urethra and prepuce in the clitoris is largely similar to that of the penis, though the incorporation of the tubular urethral epithelium is incomplete. In mutants, the tubular urethra was not formed at the glans penis or clitoris, resulting in a failure of the prepuce to fuse at the ventral (lower) midline of the glands.

IDENTIFICATION OF AN ESSENTIAL EPITHELIAL GENE: THE ROLE OF *SHH* IN URETHRAL PLATE EPITHELIA V.S. MESENCHYMAL *FGF10*

Another essential gene potentially interacting with the *Fgf* pathway was revealed. Sonic hedgehog was originally identified in *Drosophila* as Hedgehog (*Hh*) and it was shown that *Hh* has fundamental functions in *Drosophila* organogenesis. Later the *Hh* homologue was identified in mammals, and one of them, *Shh*, was shown to possess essential functions in organogenesis of the brain, neural tube, lung, digestive tract, limb, and skin development. It was shown that *Shh* was expressed in the urethral plate epithelium[29,40] and *Shh* mutant mice revealed complete GT agenesis suggesting its vital importance during initiation of external genitalia formation.[40, 41]

The unique character of the analysis of external genitalia formation lies in utilizing both gene knock-out mouse (*in vivo*) and organ culture (*in vitro*) systems.

Functional analysis by organ culture revealed that *Shh* could regulate mesenchymaly expressed genes, patched 1 (*Ptch1*), bone morphogenetic protein 4 (*Bmp4*), *Hoxd13*, and fibroblast growth factor 10 (*Fgf10*) and the genital tubercle (GT). Regulation of mesenchymal *Fgf10* expression by the epithelial *Shh* was indicated for late GT development. By explant culture experiments, SHH protein-induced *Bmp4* and *Fgf10* expression in GT mesenchyme and inhibition of *Shh* signaling leads to downregulation of mesenchymal genes, such as *Ptch1* and *Fgf10*. The reduction of mesenchymal *Fgf10* expression may underlie hypoplasia consistent with the previous data that *Fgf10* mutation induces severe ventral GT dysmorphogenesis.[29]

These results suggest a dual mode of *Shh* function, first by the regulation of initiating GT outgrowth, and second, by subsequent GT differentiation. This was a unique observation that *Shh* has several phases of functions in early and late stages of external genitalia development.

GENES NECESSARY FOR SETTING INITIAL DEVELOPMENTAL FIELD: GENETIC CASCADES NECESSARY FOR EXTERNAL GENITALIA OUTGROWTH INDUCTION

Activities of *Shh* for both GT outgrowth and differentiation were also demonstrated. *Shh* KO mice displayed complete GT agenesis. Our search for the possible cause leading to such agenesis revealed several key features before genital bud formation in such mutant mice. Furthermore, the regulation of apoptosis during GT formation was studied. Increased cell death and reduced cell proliferation in the *Shh* mutant mice genital "field" were shown.[40] A search for alterations of *Shh* downstream gene expression identified a dramatic shift of *Bmp4* expression from the mesenchyme to the epithelium of the *Shh* mutant before GT outgrowth. This was one of the first findings of genetic cascade for initial onset of external genitalia formation among several essential growth factor genes.

THE ROLE OF DEVELOPING EPITHELIA IN EPITHELIAL–MESENCHYMAL INTERACTION: THE DEVELOPMENT OF EXTERNAL GENITALIA AND EMBRYONIC GUT

It was stated that urethral plate/tube formation was attributed partly to endodermal morphogenesis, although very little was reported on the underlying molecular mechanisms.[42–44]

As stated in this review, numerous molecular developmental analyses of organogenesis were done, making it possible to compare the molecular cascade between different organogenesis processes. It was mentioned that some outgrowing (budding) organs might possess shared functions of molecular cascades as stated above.

Although the gross histological structures are different in some organogenesis processes, it would be still intriguing to monitor the similarity of molecular cascades between different epithelial–mesenchymal interactions. The morphogenetic processes of gut formation could be referred to for comparison. *Shh* was recently identified as a crucial regulator emanating from the inner epithelium to the outer gut mesenchyme, and thus was described as a radially acting regulatory molecule.[45–47] *Shh* signaling from the urethral epithelium in the developing GT might be regarded as similar to the regulatory mode of *Shh* function during gut development as mesenchymal gene expression; for example, *Fgf10* expression could be controlled. However, one could also point to the differences in which GT development includes dynamic epithelial and mesenchymal tissue arrangements, for example, urethral tube formation, which forms initially as a ventral groove, unlike regulating differentiation radially.

More recently our analyses revealed some conserved and also divergent mechanisms of copulatory organ morphogenesis as revealed by studying the fish copulatory organ, gonopodium, formation (Ogino et al.).[48] It was found that a *Shh* homologue was expressed in the developing fish copulatory organ (gonopodium: the modified anal fin derived structure serving as a copulatory organ for some fish) in its distal region (see Figure 20.4). Our analyses, including the functional assay using a *Shh* inhibitor substrate, revealed the importance of *Shh* for such gonopodia suggesting its importance in divergent animal species ranging from fish to mammals. Our studies also indicated its involvement in epithelial–mesenchymal interaction in gonopodium formation and also in

FIGURE 20.4 Androgen-dependent sexual dimorphism in Western mosquitofish (*Gambusia affinis*) male. The distal portion of the gonopodium (GP) is composed of third, fourth, and fifth fin rays serving as a copulatory organ. For copulation, the GP swings forward and sperm bundles, spermatozeugmata, are directly transported into the female urogenital sinus.

mammalian external genitalia formation. Further studies are required to clarify to what extent such mechanisms are conserved in divergent animal species.

ROLE OF BONE MORPHOGENETIC PROTEIN (*BMP*) GENES IN EXTERNAL GENITALIA DEVELOPMENT

Bone morphogenetic protein genes (*Bmps*) and their antagonists were spatio-temporally expressed during GT development.[31] Exogenously applied BMP increased apoptosis of GT and inhibited its outgrowth. It was shown that the distal urethral epithelium (DUE) was marked by *Fgf8* expression, and can control the GT outgrowth. Exogenously applied BMP4 downregulated the expression of *Fgf8* and *Wnt5a*, concomitant with increased apoptosis and decreased cell proliferation of the GT mesenchyme.

Furthermore, *Noggin* mutants and *Bmpr1a* conditional mutant mice displayed hypoplasia and hyperplasia of the external genitalia, respectively.[31] *Noggin* mutant mice exhibited downregulation of *Wnt5a* and *Fgf8* expression with decreased cell proliferation. Consistent with such findings, *Wnt5a* mutant mice displayed GT agenesis with decreased cell proliferation. By contrast, *Bmpr1a* mutant mice displayed decreased apoptosis and augmented *Fgf8* expression in the DUE associated with GT hyperplasia.

These results suggest that some of the *Bmps* could negatively affect proxi-modistally oriented outgrowth of GT with regulatory functions on cell proliferation and apoptosis in the external genitalia appendage. In the mouse, *Bmp4* is frequently expressed adjacent to *Fgf*-expressing regions and is involved in the regulation of cell proliferation and differentiation.[49] Several *Bmps* are expressed in the mouse dorsal forebrain and facial primordial.[49,50,51] Ectopic application of

BMP4 reduces the expression of both *Shh* and *Fgf8*[52] and represses anterior neural gene expression promoting apoptosis in mouse forebrain explants.[53] *Bmp4* and *Fgf10* are often expressed in adjacent domains during organogenesis.[54] During limb morphogenesis, BMP-FGF crosstalk has been suggested as functioning during, for example, the AER formation or differentiation of the interdigit apoptotic zone. Inactivation of BMP signaling results in the loss of *Fgf8* expression in the AER of the limb.[54,55] Apart from limbs, developmental budding processes are often influenced by the interaction of positive growth regulators (e.g., the FGF family or SHH) and negative growth regulators (such as BMP2, BMP4, and BMP7).[57,58] Further work on the orchestration of *Bmp* activities in relation to the distal signaling region such as AER and DUE in different appendages is required.

DO SIGNALING EPITHELIA EXPLAIN ALL ASPECTS OF EXTERNAL GENITALIA DEVELOPMENT? THE ROLE OF DUE

Although elucidation of the distal urethral epithelium developmental mechanisms, including the functional analysis of DUE, are essential for analyzing developmental mechanisms of external genitalia formation, it would be worth noticing that signaling epithelia may function only in a limited phase of some developmental programs. The notion of signaling epithelia, such as AER and enamel knot, did include the concept that they function only in a limited phase of developmental programs including the rapid appearance and disappearance of those regions for early patterning of organogenesis.

Normal *Fgf8* expression in the DUE remains until E14.0,[29] whereas GT outgrowth continues. In fact, GT hyperplasia was observed in newborn *Bmpr1a* KO mice.[31] This might reflect DUE-less-dependent GT outgrowth for late-stage GTs. In this respect previous findings suggesting the importance of normal GT ectoderm for proper GT development would be intriguing, albeit based on broad ectoderm–mesenchymal recombination assays available at that time.[59]

External genital morphogenesis has drawn very little attention so far from a molecular developmental viewpoint. However, as discussed in this review, external genitalia exhibit intriguing biological and medical mechanisms common to other organ systems. Therefore, it is necessary to further analyze organogenesis of these structures as a mouse model.

ACKNOWLEDGMENT

We thank Drs. Juan Carlos Izpisua Belmonte, Cheng-Ming Chuong, E.M. De Robertis, Richard R. Behringer, Andrew McMahon, Karen M. Lyons, Brigid Hogan, John McLachlan, Rolf Zeller, Eduardo Rosa-Molinar, Urlich Ruether, Blanche Capel, Patrick P.L. Tam, Cathy Mendelsohn, Matthew Scott, and Cliff Tabin for help; and Shiho Kitagawa for assistance. This research was supported in part by a Grant-in-Aid for the 21st Century COE Research from the Ministry of Education,

Culture, Sports, Science and Technology; by a Grant-in-Aid for Scientific Research on Priority Areas (A), General Promotion of Cancer Research in Japan; and by a Grant for Child Health and Development (14C-1) from the Ministry of Health, Labor and Welfare.

REFERENCES

1. Kessel, M. and Gruss, P. Murine developmental control genes. *Science* 249, 374–9 (1990).
2. Dolle, P., Izpisua-Belmonte, J. C., Brown, J. M., Tickle, C., and Duboule, D. *HOX-4* genes and the morphogenesis of mammalian genitalia. *Genes Dev* 5, 1767–7 (1991).
3. Kondo, T., Zakany, J., Innis, J. W., and Duboule, D. Of fingers, toes and penises. *Nature* 390, 29 (1997).
4. Warot, X., Fromental-Ramain, C., Fraulob, V., Chambon, P., and Dolle, P. Gene dosage-dependent effects of the *Hoxa-13* and *Hoxd-13* mutations on morphogenesis of the terminal parts of the digestive and urogenital tracts. *Development* 124, 4781–91 (1997).
5. Goff, D. J. and Tabin, C. J. *Hox* mutations au naturel. *Nat Genet* 13, 256–8 (1996).
6. Mortlock, D. P. and Innis, J. W. Mutation of *HOXA13* in hand-foot-genital syndrome. *Nat Genet* 15, 179–80 (1997).
7. Brennan, J. et al. *Sry* and the testis: Molecular pathways of organogenesis. *J Exp Zool* 281, 494–500 (1998).
8. Ikeda, Y., Shen, W. H., Ingraham, H. A., and Parker, K. L. Developmental expression of mouse steroidogenic factor-1, an essential regulator of the steroid hydroxylases. *Mol Endocrinol* 8, 654–62 (1994).
9. Yamaguchi, T. P., Bradley, A., McMahon, A. P., and Jones, S. A Wnt5a pathway underlies outgrowth of multiple structures in the vertebrate embryo. *Development* 126, 1211–23 (1999).
10. Duboule, D. The function of *Hox* genes in the morphogenesis of the vertebrate limb. *Ann Genet* 36, 24–9 (1993).
11. Johnson, R. L. and Tabin, C. J. Molecular models for vertebrate limb development. *Cell* 90, 979–90 (1997).
12. Tickle, C. and Eichele, G. Vertebrate limb development. *Ann Rev Cell Biol* 10, 121–52 (1994).
13. Haramis, A. G., Brown, J. M., and Zeller, R. The limb deformity mutation disrupts the *SHH/FGF-4* feedback loop and regulation of 5' *HoxD* genes during limb pattern formation. *Development* 121, 4237–45 (1995).
14. Niswander, L., Jeffrey, S., Martin, G. R., and Tickle, C. A positive feedback loop coordinates growth and patterning in the vertebrate limb. *Nature* 371, 609–12 (1994).
15. Zuniga, A., Quillet, R., Perrin-Schmitt, F., and Zeller, R. Mouse Twist is required for fibroblast growth factor-mediated epithelial-mesenchymal signalling and cell survival during limb morphogenesis. *Mech Dev* 114, 51–9 (2002).
16. Bellusci, S., Grindley, J., Emoto, H., Itoh, N., and Hogan, B. L. Fibroblast growth factor 10 (FGF10) and branching morphogenesis in the embryonic mouse lung. *Development* 124, 4867–78 (1997).

17. Cohn, M. J., Izpisua-Belmonte, J. C., Abud, H., Heath, J. K., and Tickle, C. Fibroblast growth factors induce additional limb development from the flank of chick embryos. *Cell* 80, 739–46 (1995).

18. Martin, G. R. The roles of FGFs in the early development of vertebrate limbs. *Genes Dev* 12, 1571–86 (1998).

19. Ornitz, D. M. FGFs, heparan sulfate and FGFRs: Complex interactions essential for development. *Bioessays* 22, 108–12 (2000).

20. Wall, N. A. and Hogan, B. L. Expression of bone morphogenetic protein-4 (BMP-4), bone morphogenetic protein-7 (BMP-7), fibroblast growth factor-8 (FGF-8) and sonic hedgehog (SHH) during branchial arch development in the chick. *Mech Dev* 53, 383–92 (1995).

21. Yamasaki, M., Miyake, A., Tagashira, S., and Itoh, N. Structure and expression of the rat mRNA encoding a novel member of the fibroblast growth factor family. *J Biol Chem* 271, 15918–21 (1996).

22. Yonei-Tamura, S. et al. FGF7 and FGF10 directly induce the apical ectodermal ridge in chick embryos. *Dev Biol* 211, 133–43 (1999).

23. Crossley, P. H. and Martin, G. R. The mouse *Fgf8* gene encodes a family of polypeptides and is expressed in regions that direct outgrowth and patterning in the developing embryo. *Development* 121, 439–51 (1995).

24. Crossley, P. H., Minowada, G., MacArthur, C. A., and Martin, G. R. Roles for FGF8 in the induction, initiation, and maintenance of chick limb development. *Cell* 84, 127–36 (1996).

25. Ohuchi, H. et al. Involvement of androgen-induced growth factor (*FGF-8*) gene in mouse embryogenesis and morphogenesis. *Biochem Biophys Res Commun* 204, 882–8. (1994).

26. Moon, A. M. and Capecchi, M. R. Fgf8 is required for outgrowth and patterning of the limbs. *Nat Genet* 26, 455–9 (2000).

27. Frank, D. U. et al. An Fgf8 mouse mutant phenocopies human 22q11 deletion syndrome. *Development* 129, 4591–603 (2002).

28. Sun, X., Mariani, F. V., and Martin, G. R. Functions of FGF signalling from the apical ectodermal ridge in limb development. *Nature* 418, 501–8 (2002).

29. Haraguchi, R. et al. Molecular analysis of external genitalia formation: The role of fibroblast growth factor (Fgf) genes during genital tubercle formation. *Development* 127, 2471–9 (2000).

30. Ogino, Y. et al. External genitalia formation: Role of fibroblast growth factor, retinoic acid signaling, and distal urethral epithelium. *Ann NY Acad Sci* 948, 13–31 (2001).

31. Suzuki, K. et al. Regulation of outgrowth and apoptosis for the terminal appendage: External genitalia development by concerted actions of BMP signaling [corrected]. *Development* 130, 6209–20 (2003).

32. Morgan, E. A., Nguyen, S. B., Scott, V., and Stadler, H. S. Loss of Bmp7 and Fgf8 signaling in *Hoxa13*-mutant mice causes hypospadia. *Development* 130, 3095–109 (2003).

33. Baskin, L. S. Hypospadias and urethral development. *J Urol* 163, 951–6 (2000).

34. Kurzrock, E. A., Jegatheesan, P., Cunha, G. R., and Baskin, L. S. Urethral development in the fetal rabbit and induction of hypospadias: A model for human development. *J Urol* 164, 1786–92 (2000).

35. van der Werff, J. F. and Ultee, J. Long-term follow-up of hypospadias repair. *Br J Plast Surg* 53, 588–92 (2000).

36. Jegou, B., Laws, A. O., and de Kretser, D. M. The effect of cryptorchidism and subsequent orchidopexy on testicular function in adult rats. *J Reprod Fertil* 69, 137–45 (1983).

37. Weidner, I. S., Moller, H., Jensen, T. K., and Skakkebaek, N. E. Cryptorchidism and hypospadias in sons of gardeners and farmers. *Environ Health Perspect* 106, 793–6 (1998).

38. Weidner, I. S., Moller, H., Jensen, T. K., and Skakkebaek, N. E. Risk factors for cryptorchidism and hypospadias. *J Urol* 161, 1606–9 (1999).

39. Suzuki, K. et al. Embryonic development of mouse external genitalia: insights into a unique mode of organogenesis. *Evol Dev* 4, 133–41 (2002).

40. Haraguchi, R. et al. Unique functions of Sonic hedgehog signaling during external genitalia development. *Development* 128, 4241–50 (2001).

41. Perriton, C. L., Powles, N., Chiang, C., Maconochie, M. K., and Cohn, M. J. Sonic hedgehog signaling from the urethral epithelium controls external genital development. *Dev Biol* 247, 26–46 (2002).

42. Hayward, S. W. et al. Interactions between adult human prostatic epithelium and rat urogenital sinus mesenchyme in a tissue recombination model. *Differentiation* 63, 131–40 (1998).

43. Kurzrock, E. A., Baskin, L. S., Li, Y., and Cunha, G. R. Epithelial-mesenchymal interactions in development of the mouse fetal genital tubercle. *Cells Tissues Organs* 164, 125–30 (1999).

44. Kurzrock, E. A., Baskin, L. S., and Cunha, G. R. Ontogeny of the male urethra: theory of endodermal differentiation. *Differentiation* 64, 115–22 (1999).

45. Roberts, D. J. et al. Sonic hedgehog is an endodermal signal inducing *Bmp-4* and *Hox* genes during induction and regionalization of the chick hindgut. *Development* 121, 3163–74 (1995).

46. Roberts, D. J., Smith, D. M., Goff, D. J., and Tabin, C. J. Epithelial–mesenchymal signaling during the regionalization of the chick gut. *Development* 125, 2791–801 (1998).

47. Sukegawa, A. et al. The concentric structure of the developing gut is regulated by Sonic hedgehog derived from endodermal epithelium. *Development* 127, 1971–80 (2000).

48. Ogino, Y., Katoh, H., and Yamada, G. Androgen dependent development of a modified anal fin, gonopodium, as a model to understand the mechanism of secondary sexual character expression in vertebrates. *FEBS Lett.* 575(1–3), 119–126 (2004).

49. Hogan, B. L. Bone morphogenetic proteins: Multifunctional regulators of vertebrate development. *Genes Dev* 10, 1580–94 (1996).

50. Barlow, A. J. and Francis-West, P. H. Ectopic application of recombinant BMP-2 and BMP-4 can change patterning of developing chick facial primordia. *Development* 124, 391–8 (1997).

51. Furuta, Y., Piston, D. W., and Hogan, B. L. Bone morphogenetic proteins (BMPs) as regulators of dorsal forebrain development. *Development* 124, 2203–12 (1997).

52. Ohkubo, Y., Chiang, C., and Rubenstein, J. L. Coordinate regulation and synergistic actions of BMP4, SHH and FGF8 in the rostral prosencephalon regulate morphogenesis of the telencephalic and optic vesicles. *Neuroscience* 111, 1–17 (2002).

53. Graham, A., Francis-West, P., Brickell, P., and Lumsden, A. The signalling molecule BMP4 mediates apoptosis in the rhombencephalic neural crest. *Nature* 372, 684–6 (1994).

54. Weaver, M., Dunn, N. R., and Hogan, B. L. Bmp4 and Fgf10 play opposing roles during lung bud morphogenesis. *Development* 127, 2695–704 (2000).

55. Ahn, K., Mishina, Y., Hanks, M. C., Behringer, R. R., and Crenshaw, E. B., 3rd. BMPR-IA signaling is required for the formation of the apical ectodermal ridge and dorsal-ventral patterning of the limb. *Development* 128, 4449–61 (2001).

56. Pizette, S., Abate-Shen, C., and Niswander, L. BMP controls proximodistal outgrowth, via induction of the apical ectodermal ridge, and dorsoventral patterning in the vertebrate limb. *Development* 128, 4463–74 (2001).

57. Hogan, B. L. Morphogenesis. *Cell* 96, 225–33 (1999).

58. Jung, H. S. et al. Local inhibitory action of BMPs and their relationships with activators in feather formation: Implications for periodic patterning. *Dev Biol* 196, 11–23 (1998).

59. Murakami, R. and Mizuno, T. Proximal-distal sequence of development of the skeletal tissues in the penis of rat and the inductive effect of epithelium. *J Embryol Exp Morphol* 92, 133–43 (1986).

21 Genetic Approaches to Investigate Retinoic Acid Functions in Mouse Development

Pascal Dollé, Karen Niederreither,
Julien Vermot, Vanessa Ribes,
Jabier Gallego-Llamas, and Isabelle Le Roux

TABLE OF CONTENTS

INTRODUCTION

The lipophilic vitamin A (retinol) and its derivatives, the retinoids, have long been known to be essential for proper embryonic and fetal development. Maternal vitamin A deficiency (VAD) in mice or rats leads to a well-defined spectrum of fetal abnormalities affecting many organ systems.[1] Most severe VAD conditions can lead to earlier embryonic defects in rodents, although such experiments require a retinoid supplementation to be performed during early pregnancy, in

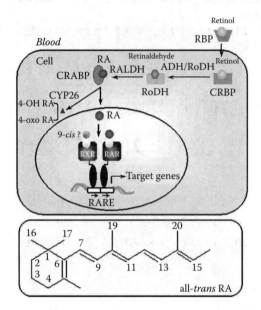

FIGURE 21.1 Scheme of the intracellular pathways involving retinoids and chemical formula of all-*trans* retinoic acid (below). CRBP: cellular retinol binding protein; CRABP: cellular retinoic acid binding protein; RARE: retinoic acid response element; RBP: retinol binding protein; RoDH: retinol dehydrogenase.

order to prevent lack of embryo implantation and infertility (e.g., Reference 2). Another vertebrate model that has been extensively studied is the VAD quail embryo.[3] Studies performed in this model reported VAD-induced defects in early heart, hindbrain, limb bud, or somite development (for review, see Reference 4).

With the exception of the eye protoreceptor protein rhodopsin, which uses retinaldehyde as a chromophore, most of the biological functions of vitamin A are mediated by its acidic derivative retinoic acid (RA) (Figure 21.1). RA acts as a ligand for a subfamily of nuclear receptors, the RA receptors (RAR) α, β, and γ. These receptors are ligand-inducible transcription factors that function as heterodimers with members of another subfamily, the "retinoid X" receptors (RXR) α, β, and γ. The RXRs are heterodimerization partners for several other nuclear receptors, including the thyroid hormone (TR), vitamin D (VDR), or peroxisome proliferator-activated (PPAR) receptors.

Both the RARs and RXRs exist as amino terminal isoform variants generated by alternative splicing or promoter usage of the corresponding genes (for review, see Reference 5). Isoforms of a given receptor may differ in their transcriptional activation properties, but share the same DNA-binding domain. Experimentally, the RXRs can specifically bind and be activated by a RA stereoisomer, 9-*cis* RA, whereas RARs can accommodate both all-*trans* and 9-*cis* RA. However, and in contrast to all-*trans* RA, the 9-*cis* isomer was not detected endogenously in developing or adult rodent tissues (e.g., Reference 6). It is therefore unclear whether 9-*cis* RA is significantly involved in biological processes, or if RXRs act primarily as unliganded partners *in vivo*. There may be several hundred direct RA-target genes (containing RAR/RXR binding sites, termed "RA response elements" or RAREs) in mammalian genomes, encoding very diverse protein families (for review, see Reference 7).

The present chapter presents an overview of the genetic approches that were used in the mouse to unravel developmental functions of proteins involved in production or transduction of the RA signal. It then focuses on recent work of our group that aimed at characterizing the roles of RALDH2, a critical enzyme required for RA synthesis during development.

RA RECEPTORS HAVE PLEIOTROPIC AND MOSTLY REDUNDANT DEVELOPMENTAL FUNCTIONS

A first genetic approach to study *in vivo* RA functions, initiated some 15 years ago, was to generate mice with targeted disruptions of the RAR or RXR genes. Both null alleles and isoform-specific loss-of-function mutations were produced (for review, see ref. 8). It turned out, somewhat surprisingly (given the fact that most of these genes, including their major isoforms, exhibit differential expression patterns during development),[9] that these receptors are to a large extent functionally redundant. Thus, individual RAR null mutations give rise to limited sets of phenotypic defects in specific organ systems (e.g., RARα null male mutants undergo postnatal testis degeneration).[10] The RXRγ mutants are viable and show no detectable defect, whereas RXRβ mutants have a selective abnormality of the testicular Sertoli cells.[11] RXRα is the only retinoid receptor whose null mutation is lethal before birth, due to a defect in myocardial differentiation and subsequent heart failure.[12]

Various combinations of RAR/RXR double mutants were generated by intercrosses, possibly to unravel redundant developmental functions (for review, see ref. 8). Most of these genetic combinations are developmentally lethal. When analyzed histologically, a complex pleiotropic set of defects is seen in the RAR compound mutants.[13,14] These include all abnormalities previously described in the rodent fetal VAD syndrome, as well as a number of novel abnormalities. Most of the defects are common to several mutant genotypes, although they tend to be more severe or more penetrant in a given mutant combination (for instance, severe craniofacial skeletal defects and limb patterning defects are especially found in RARα/RARγ double mutants).

RAR/RXR compound mutants were also generated (see 8,15, and refs. therein). Their phenotypic analysis implicated RXRα as the main RAR heterodimeric partner for RAR-mediated developmental functions, as most of the defects found in RAR single or double mutants were recapitulated in various RAR/RXRα compound mutants.[15] Alleles were also generated to selectively disrupt the AF1 or AF2 domains of RXRα.[16,17] These domains code for two transcriptional activation functions, which are, respectively, ligand independent (AF1) or inducible (AF2). Analysis of these mutants (and the corresponding RAR compound mutants) revealed a differential contribution of the RXR transcriptional activation functions for developmental processes, and provided genetic evidence for a ligand-dependent (AF2-mediated) RXR activity in some of these processes (see refs. 16 and 17 for further discussion).

Altogether, this work confirmed that RARs and RXRs transduce the pleiotropic functions of RA as a developmental signaling molecule. These studies also provided evidence for RAR-independent functions of RXRs, that is, functions that would involve other nuclear receptor heterodimerization partner(s). Due to several reasons (phenotypic heterogeneity, variable stages of lethality, and need to generate large numbers of mice), the compound RAR/RXR mutants do not represent an easy system for investigating the mechanisms underlying the phenotypic defects, and few molecular studies (e.g. refs. 18 and 19) have been performed in these mutants.

RETINOIC ACID SYNTHESIS IS ENZYMATICALLY REGULATED IN EMBRYONIC TISSUES

The observation of a functional redundancy between RA receptors supported the idea that regulation of the synthesis and production of RA as an active ligand may be a key step in determining its biological functions. A number of studies involving direct detection by HPLC (see 20, and refs. therein) or indirect detection via RA-responsive transgenes (see below) have demonstrated that this retinoid (and some of its derivatives) exhibits differential distribution and levels among embryonic structures. It was first proposed that, due to its diffusible properties and its uneven distribution in embryonic structures such as the limb bud, this molecule may behave as a morphogen, that is, may elicit distinct morphogenetic responses according to local concentration gradients.[21] As discussed hereafter, there is clear *in vivo* evidence suggesting that RA acts as a diffusible signal to elicit responses in regions or cell layer(s) distinct from those that synthesize it; there is, however, no firm evidence for a role as a true morphogen.

RA is generated intracellularly from retinol by two oxidative reactions that transform its alcohol moiety into an aldehyde (retinaldehyde) and carboxylic acid (RA) (Figure 21.1). Mammalian embryos receive retinol of maternal origin via transplacental transfer. Retinol binding protein (RBP), a protein involved postnatally in the plasmatic transport of retinol, also appears to be involved in the transfer of retinol to the embryo at early developmental stages. Indeed, this protein is specifically produced by the embryonic visceral yolk sac endoderm, the layer that mediates maternal–embryonic exchanges before establishment of a functional placenta.[22] The functional significance of its expression in the yolk sac is not established, as *RBP*[−/−] knock-out mice are viable and only display an abnormal vision phenotype.[23]

Several groups sought to determine whether specific enzyme(s) are involved in the generation of RA in embryonic or adult tissues.* Some members of the

---

* In addition to the enzymes described hereafter, two types of proteins act as specific cytosolic binding proteins for retinol (the cellular retinol binding proteins, CRBPs) and RA (the cellular RA binding proteins, CRBAPS), respectively. The function(s) of these proteins in retinoid metabolism and signal transduction is not clearly elucidated, and current hypotheses are not elaborated herein due to space constraints.

alcohol dehydrogenase (ADH) family, especially ADH1, ADH3, and ADH4, can efficiently metabolize retinol into retinaldehyde *in vitro* (see 24,25, and refs. therein). ADH3 is a ubiquitously expressed enzyme, whereas ADH1 and ADH4 exhibit differential temporal and spatial expression profiles in the mouse embryo. Despite this fact, the respective null mutant mice are viable and show no phenotypic abnormality.[25,26] These mutants are impaired in their ability to produce retinaldehyde following exogenous administration of retinol, demonstrating that the corresponding enzymes can act *in vivo* as retinol dehydrogenases. Furthermore, *Adh4⁻/⁻* embryos exhibit an increased rate of lethality when produced from VAD mothers, indicating that this enzyme participates in retinol metabolism during development, and that its activity becomes critical under limiting conditions.[27] Altogether, the current evidence does not allow assignment of a specific role for a given ADH enzyme in tissue-specific production of RA during development. There is no further genetic evidence for specific developmental functions of short-chain dehydrogenase/reductases (SDR), another class of enzymes that can act in retinol oxidation (reviewed in ref. 28).

DIFFERENTIAL EXPRESSION PATTERNS OF MURINE RETINALDEHYDE DEHYDROGENASES

Three members of the aldehyde dehydrogenase (ALDH) family have been shown to have a high catalytic activity on retinaldehyde. Two of these, the retinaldehyde dehydrogenases (RALDH) 1 and 3, were first characterized as human ALDH isozymes and were subsequently found to correspond to two RA-synthesizing activities present in the dorsal and ventral embryonic mouse retina.[29,30] RALDH2 was identified through a PCR cloning strategy as a novel enzyme that may account for RA production in early embryonic trunk structures and ventral retina.[31] RALDH2 and RALDH3 are at least tenfold more efficient than RALDH1 in metabolizing retinaldehyde.[32,33]

All three *Raldh* genes show distinct patterns of expression during mouse development. *Raldh1* expression starts at about midgestation (E9.5). This enzyme is strongly expressed in the dorsal—and to a lesser extent the ventral—retina, as well as other eye structures such as the lens epithelium. It is also expressed in various other developing organs such as the lung, liver, intestine, and kidney.[34,35] Surprisingly, *Raldh1⁻/⁻* knock-out mouse mutants are viable and do not display any known morphological or physiological defect.[36] This could be because its lack of function can be compensated by the presence of other RALDH enzymes in certain structures, such as the developing eye. The lack of a developmental phenotype could also reflect the fact that this enzyme is only marginally in endogenous RA synthesis and that its organ-specific activity, which persists during adult life, may be linked to the metabolism of other aldehyde(s).

The *Raldh2* enzyme is mostly expressed during development. It is induced during gastrulation in the posterior embryonic mesoderm, and eventually exhibits spatially restricted expression patterns in many mesodermal derivatives, such as the somites, the cervical and posterior branchial arch mesenchyme, the posteriormost

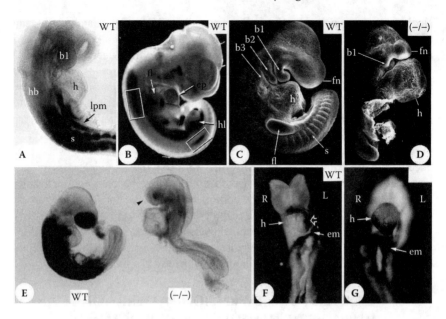

FIGURE 21.2 (See color insert following page 210.) The RALDH2 enzyme is required for retinoic acid synthesis during early embryogenesis. (A,B) Spatially restricted expression patterns of the *Raldh2* gene in E8.5 and E10.E mouse embryos, respectively. Boxed areas indicate *Raldh2*-expressing motor neuron populations in the brachial and lumbar spinal cord. Whole-mount *in situ* hybridizations with digoxigenin-labeled riboprobes. (C,D) Scanning electron micrographs of E9.5 wild-type and *Raldh2-/-* knock-out embryos, respectively. (E) Deficient activity of a RA-responsive *lacZ* reporter transgene in an E8.5 *Raldh2-/-* embryo (right) compared to a wild-type (WT) littermate (left). Weak activity is found in the forebrain and optic area only (arrowhead). Whole-mount X-gal staining. (F,G) Impaired heart looping in an E8.5 *Raldh2-/-* embryo (G), as seen by a lack of left-side lateralization of *Hand1*-expressing cells (arrow in F: wild-type littermate). Whole-mount *in situ* hybridization. b1–b3: branchial arches; ep: epicardium; em: extraembryonic membranes; fl: forelimb bud; fn: frontonasal mass; h: heart; hb: hindbrain; hl, hindlimb bud; L: left; R: right; s: somites. (A,B reprinted with permission from Niederreither et al.;[37] C–G reprinted with permission from Niederreither et al.[41])

heart tube mesoderm, or the flank lateral plate mesenchyme (Figure 21.2A,B).[37] These sites of expression correlate with most of the early embryonic regions that are known to be RA-rich (see above). *Raldh2* is also transiently expressed in the rostral forebrain neuroepithelium during formation of the eye vesicle by ~E8.5,[38] and in discrete brachial and lumbar motor neuron pools within the developing spinal cord (Figure 21.2B).[31] During late embryonic and fetal development, *Raldh2* is expressed in several tissues, such as the pleura and pericardium/epicardium, the kidney cortex, and specific limb regions (Figure 21.2B).[37]

Raldh3 expression, on the other hand, could account for most of the RA synthesis in embryonic craniofacial structures. Indeed, this enzyme is highly expressed from E8.5 in frontonasal ectoderm, and eventually its expression becomes restricted to the developing lens and nasal placodes.[29,30] Although expression strongly persists in the nasal epithelium, it eventually disappears from the lens epithelium and appears in the developing ventral retina.[38] Few other embryonic structures express *Raldh3*; these include the mesonephros (primitive kidney) and later, the renal excretory ducts.[29,30,35] *Raldh3* knock-out mutant mice die at birth due to abnormal morphogenesis of the nasal cavities, including atresia of the choanae that prevents breathing at birth.[39] They also display mild ocular abnormalities, especially a shortening of the ventral retina. No abnormality of the kidney or excretory ducts has been detected. The relatively limited extent and severity of these abnormalities suggest a possible redundancy with RALDH1 or RALDH2 function(s), which is currently being tested by genetic crosses.

A fourth retinaldehyde dehydrogenase (RALDH4) has recently been characterized.[40] This enzyme differs from the previously known RALDH in that it has a high catalytic activity on 9-*cis* RA. Its expression appears to be restricted to adult liver and kidney, and no additional site of expression was found during embryogenesis (ref. 40; our unpublished data). Its *in vivo* function remains to be established.

RALDH2 IS MAINLY RESPONSIBLE FOR RA SYNTHESIS AT EARLY EMBRYONIC STAGES

Raldh2-targeted null mutations were found to be early embryonic lethal.[41,42] The null mutant embryos die at E9.5–10.5 and exhibit complex morphogenetic defects. Morphologically, their most striking features are an abnormal, bulging heart tube, a shortening of the embryonic anteroposterior axis with small and compact somites, and a mild frontonasal truncation with apparent lack of posterior (second to sixth) branchial arches (Figure 21.2c,d). The *Raldh2* null allele was crossed with a RA-responsive transgenic line (RARE-hsp-*lacZ*; ref. 43), and it was shown that transgenic null mutant embryos display almost no activation of the reporter transgene, except in the craniofacial region which expresses *Raldh3* (Figure 21.2e). This clearly established that the *Raldh2*$^{-/-}$ phenotype results from a deficiency in endogenous embryonic RA.[41]

Extensive molecular studies were performed on the *Raldh2*$^{-/-}$ mutants that were needed to understand the basis of the abnormal heart phenotype.[44] Analysis of genes involved in early heart regionalization, such as *Hand1*, indicated that the abnormal *Raldh2*$^{-/-}$ heart morphology reflects a lack of left–right (L–R) looping morphogenesis of the linear heart tube (Figure 21.2f,g). Expression of left-side-specific determinants such as *nodal* or *Pitx2* was not altered, indicating that the looping defect was not linked to abnormal embryonic L–R axis (*situs*) determination. Additional molecular markers enabled assessment of how chamber patterning and differentiation could proceed in the absence of RA synthesis. Notably, expression of *Tbx5*, a major determinant of posterior heart chambers (atrium and sinus venosus), was spatially reduced in *Raldh2*$^{-/-}$ embryos. This correlated with poor histological development of these chambers in the mutant embryos. These findings demonstrated a requirement of embryonic RA for proper development of posterior heart structures, which was previously suggested by analyses of the pattern of activity of a RA-reporter transgene in the mouse heart and the effects of excess RA in the chick embryonic heart tube.[45,46]

Distinct analyses characterized the mechanism underlying the abnormal head phenotype in *Raldh2*$^{-/-}$ embryos.[47] A segmentation and patterning defect was thus uncovered in the hindbrain of the mutant embryos. Transient segmentation into eight rhombomeres (r) is a developmental characteristic of this region of the brain stem, which will be critical for proper patterning of the cranial nerves and their nuclei, and for establishing segmental pathways of neural crest cell migration towards the branchial region. In *Raldh2*$^{-/-}$ embryos, the hindbrain segmentation defect appears to result from early changes in the distribution of molecular determinants of anterior rhombomere identity (such as *Krox20* and *Hoxb1*), which extend abnormally throughout the posterior hindbrain at the expense of determinants of posterior (r5–r8) rhombomeric identity. In other words, the posterior hindbrain of *Raldh2*$^{-/-}$ mutants appears to be "transformed" towards a more anterior identity. As a result, posterior hindbrain neural crest cells are severely deficient and fail to colonize the posterior (second to sixth) branchial arch region. RA can therefore be considered as a diffusible "posteriorizing" signal, produced mesodermally to regulate the expression of posterior rhombomeric determinants in the hindbrain neuroepithelium** (see also refs. 18,19,48).

RA-RESCUE OF NULL MUTANT EMBRYOS UNVEILS ADDITIONAL RALDH2 DEVELOPMENTAL FUNCTIONS

As the early lethality of *Raldh2*$^{-/-}$ null mutants precludes an analysis of later functions of this enzyme, we initiated several approaches and developed new genetic tools for this purpose. As a first approach, we attempted to extend the viability of the *Raldh2*$^{-/-}$ embryos by rescuing their early phenotypic defects through maternally administered RA. Maternal RA reaching the embryo via transplacental transfer can have potent effects, and is actually a teratogen when given in excess. A retinoic acid embryopathy has been reported in the 1980s in infants whose mothers were exposed to RA during pregnancy[49] and rodent models were extensively used to study the effects of RA on various developmental processes (such as hindbrain or axial skeletal patterning (e.g., refs. 50,51). We administered RA at subteratogenic doses to *Raldh2*$^{+/-}$ pregnant mice and observed that, when given over a critical period (E7.5 to 8.5, although extended administration until E9.5 and 10.5 further increases the percentage of *Raldh2*$^{-/-}$ fetuses recovered), exogenous RA can improve heart morphogenesis and rescue the

---

** Another class of enzymes, the cytochrome P450s CYP26A1, B1, and C1, specifically act on RA to metabolize it into polar compounds, and thus trigger its catabolism (see Figure 21.1). Their expression patterns suggest they may be important for finely tuning embryonic RA levels, particularly during rhombomere and branchial arch development. *Cyp26A1* null mutant mice display a rhombomeric phenotype somewhat opposite to that of *Raldh2* mutants, that is, a partial "posteriorization" of anterior rhombomeres.[53,54]

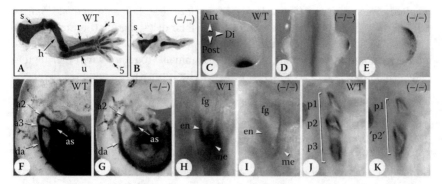

FIGURE 21.3 (See color insert following page 210.) RA-rescue of *Raldh2*–/– knock-out embryos unveils novel developmental defects. (A,B) Abnormal forelimb development in an E14.5 *Raldh2*–/– fetus (B) after RA-rescue from E7.5 to 8.5, compare to WT littermate (A). Alucian blue cartilage staining. (C–E) At earlier stages, the rescued *Raldh2*–/– embryos (D,E) fail to activate Sonic hedgehog (*Shh*), or express it in an inappropriate distal/anterior location within the forelimb buds. (F,G) Impaired development of posterior branchial arches in an E9.5 RA-rescued *Raldh2*–/– embryo (G), as seen by the lack of development of third aortic arches. Intracardiac ink injections. (H,I) Impaired activation of the *Hoxa1* homeobox gene in the pharyngeal endoderm and mesoderm of a rescued E9.5 *Raldh2*–/– embryo (I). (J,K) Abnormal development of the posterior pharyngeal pouches in a rescued E9.5 *Raldh2*–/– embryo. The whole pharyngeal pouch region, visualized by expression of the *Pax9* gene, is reduced (compare brackets in J,K) and an abnormal, single "second pouch" is formed in the mutant (K). Ant: anterior; as: aortic sac; a2–a3: aortic arches; da: dorsal aorta; Di: distal; en: endoderm; fg: foregut; me: mesoderm; Post: posterior; p1–p3: pharyngeal pouches; r: radius; s: scapula; u: ulna; 1–5: digits. (A–E reprinted with permission from Niederreither et al.;[55] F–K reprinted with permission from Niederreither et al.[58])

viability of null mutants until fetal stages. Our initial trials were performed by providing RA via oral gavage[41] (see also ref. 42). For further experiments, to avoid any deleterious effect due to bolus administration, we established conditions of administration within the maternal food and only used doses that were non-teratogenic in littermate control embryos.[52]

RA-rescue of the *Raldh2*–/– embryos thus helped to uncover a set of novel specific abnormalities. The most striking defect observed at fetal stages was the presence of markedly hypoplastic forelimbs (Figure 21.3a,b).[55] By modulating the stages and doses administered, we showed that exogenous RA can efficiently rescue forelimb growth, if provided to *Raldh2*–/– mutants at least until E10.5. Interestingly, though, even the most efficiently rescued forelimbs showed alterations of the anteroposterior (AP) patterning of skeletal elements (such as mirror-image polydactyly). Molecular analysis showed that the mutant forelimb buds failed to develop a functional "polarizing region". This signaling center, normally located in the postero-distal limb bud mesenchyme and characterized by the production of the Sonic hedgehog (*Shh*) signaling molecule (Figure 21.3c), coordinates both the growth and patterning of the developing limbs by acting on ectodermally derived growth factors and mesodermally expressed transcriptional regulators, respectively. The *Raldh2*[-/-] forelimb buds failed to activate *Shh* or exhibited weak, ectopic *Shh* expression along their distal margin (Figure 21.3d,e). These data implicate embryonic RA, normally produced by RALDH2 in the flank mesenchyme, as an important upstream determinant (together with other regulators such as dHAND (refs. 56,57) of the forelimb polarizing region.[55]

Another remarkable abnormality in the rescued *Raldh2*[-/-] mutants is impaired development of the posterior (third to sixth) branchial arches, and the corresponding aortic arteries (Figure 21.3f,g).[58] The second arches develop normally in the RA-rescued *Raldh2*[-/-] embryos (Figure 21.3g), whereas they do not form in the unrescued mutants (Figure 21.2d). Thus, there is differential sensitivity to RA along the embryonic branchial region, such that the posteriormost arches require locally high RA levels that can only be achieved via RALDH2 function, whereas the developing second arch requires lower RA levels that can be achieved in mutants via maternal administration. Altered gene expression was found in all components of the posterior branchial region (ectoderm, mesenchyme, and foregut endoderm) of the rescued *Raldh2*[-/-] embryos (Figure 21.3h–j), indicating that RA normally produced by RALDH2 in the mesoderm acts as a diffusible molecule to regulate gene expression in adjacent tissue layers. In addition, the migratory pathways of postotic hindbrain neural crest cells were perturbed in the rescued *Raldh2*[-/-] embryos.[58] This provides a rationale for a cardiovascular defect consistently seen at later stages in the rescued mutants, that is, a lack of septation of their outflow tract region into aorta and pulmonary artery (persistent truncus arteriosus; see below). Indeed, a subpopulation of hindbrain neural crest cells (the "cardiac" neural crest), that normally migrate along the posterior aortic arches to reach the outflow tract region, are important for the septation process.[59] Development of the posterior cranial nerves is also severely impaired in the rescued *Raldh2*[-/-] mutants. In particular, lack of vagal nerve outgrowth leads to aganglionosis of the developing stomach and gut wall, a condition that mimics human Hirschsprung's disease.[58]

DECREASED RALDH2-MEDIATED RA SYNTHESIS LEADS TO A DIGEORGE SYNDROMELIKE PHENOTYPE

There are several examples of engineered mouse lines in which a *neo* selectable marker was inserted within introns, and that appear to behave as hypomorphic mutants. The molecular mechanism responsible for the hypomorphic effect has not been formally described, although it is thought that the presence of cryptic splice sites within the *neo* sequence leads to aberrant splicing or destabilization of the transcript product.[60] A *Raldh2* construct containing an intronic *neo* sequence was generated as a tool for conditional somatic mutagenesis (see below), and mice harboring this novel allele were produced (as homozygotes or heterozygotes for the *Raldh2*[neo] and the null allele) and analyzed.

At embryonic stages, these mutants were shown to exhibit an overall decrease in *Raldh2* transcript (Figure 21.4a,b) and protein (not shown) levels. Unlike the null mutants, the *Raldh2*[neo/-] mice die during late fetal development or shortly after birth, whereas a small fraction of the homozygous $Raldh2^{neo/neo}$ mutants are viable.[61] The cause of death appears to be a lack of septation of the heart outflow tract (persistent truncus arteriosus; Figure 21.4c,d), which is lethal at birth because it does not allow separate pulmonary and systemic arterial systems to function. Aberrant patterns of the ascending aorta and large arterial vessels were also seen in these mutants. Histological and skeletal analyses performed at fetal stages showed additional abnormalities of the thymus, parathyroid glands, and laryngeal/tracheal cartilages, which all derive from the posteriormost branchial arches.[61] No other organ system appeared to be affected. Analyses performed at early embryonic stages revealed a selective hypoplasia of the third to sixth branchial arch region (Figure 21.4e,f) and a range of local molecular alterations that were clearly reminiscent of, although somewhat less severe than, those found in the RA-rescued $Raldh2^{-/-}$ embryos (see above).

It was rather unexpected to see that a global decrease in RALDH2 activity leads to such a restricted set of embryonic defects. Outflow tract and large vessel abnormalities, as well as thymus/parathyroid agenesis or ectopia, are found in various compound RAR mutant mice (see above), but these defects belong to a complex and pleiotropic spectrum of abnormalities. The hypomorphic $Raldh2$ mutants revealed a particular sensitivity of the developing posterior branchial apparatus to a decrease in endogenous RA levels. Other developing regions and organ systems seem to be more resistant to fluctuations of RA levels, as none of the other defects characteristic of RAR compound mutants (above) or $Raldh2^{-/-}$ null mutants (above) are recapitulated in the hypomorphic mutants, even in a milder form.

Interestingly, the set of defects seen in the $Raldh2^{neo}$ mutants phenocopy the characteristic abnormalities of the human DiGeorge syndrome (DGS). This dominant syndrome is caused in most cases by a microdeletion of chomosome 22, leading to haploinsufficiency of (at least) the $TBX1$ gene.[62] As the DGS is highly variable in its expressivity, it is tempting to speculate that conditions leading to altered embryonic RA signaling may represent the "modifiers" of the human DGS. Furthermore, impaired RA signaling may on its own be responsible for DGS-like abnormalities, as such defects are the first ones to appear in mice where levels of embryonic RA are experimentally decreased. In humans, both genetic and environmental causes (poor nutrition, anorexia, vitamin A deficiency) may be involved, perhaps in conjunction, in the generation of such abnormalities. Alcohol consumption during early pregnancy may also be implicated, as ethanol can act as a substrate for some of the enzymes (the alcohol dehydrogenases) involved in RA synthesis.[63] By competing with endogenous retinol, it may thus potentiate any condition leading to impaired RA signaling.

CONDITIONAL MUTAGENESIS: AN ALTERNATIVE APPROACH TO INVESTIGATE TISSUE-SPECIFIC RALDH2 FUNCTIONS

To study additional RALDH2 functions, we generated a novel allele that allows for tissue-specific somatic inactivation. In this allele, *loxP* sites (target sites for

WT   *Raldh2*^{neo/-}

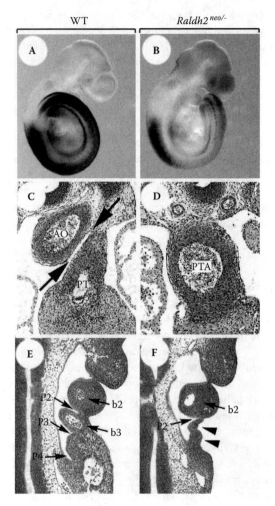

FIGURE 21.4 (See color insert following page 210.) *Raldh2* engineered hypomorphic mutants display a DiGeorge syndromelike phenotype. (A,B) Decreased *Raldh2* transcript levels in an E9.5 *Raldh2*^{neo/-} embryo (B, compare with WT littermate, A). (C,D) Lack of septation of the outflow tract (persistant truncus arteriosus) in an E14.5 *Raldh2*^{-/-} fetus (D, compare with the fully septated aorta and pulmonary trunk in a WT littermate, arrows in C). (E,F) Lack of development of the posterior branchial arches in an E9.5 *Raldh2*^{-/-} embryo (F). Unlike its WT counterpart (E), the mutant embryo exhibits no visible third or fourth arches, although rudimentary branchial clefts are formed (arrowheads). AO: aorta; b2,b3: branchial arches; PT: pulmonary trunk; PTA: persistent truncus arteriosus; P2–P4: pharyngeal pouches. (Reprinted with permission from Vermot et al.[61])

the bacterial cre recombinase) were inserted on each side of the *Raldh2* exon that encodes most of the catalytic domain. Cre-mediated recombination will lead to the excision of the exon, generating a loss-of-function allele. We produced mice carrying the unexcised (floxed) allele, which are fully viable, and showed that this allele behaves as a null after excision by crossing the floxed mutants with transgenic mice expressing cre in a ubiquitous manner. Progeny from this cross exhibit an early lethal embryonic phenotype similar to that of conventional null mutants (our unpublished data).

We will use the floxed allele to generate tissue-specific loss of functions of RALDH2. Such a strategy should be particularly suitable for investigating the role of RALDH2-mediated RA synthesis during motor neuron specification within the spinal cord. *Raldh2* is expressed in specific pools of ventral neuron precursors at brachial and lumbar levels of the embryonic spinal cord, and experimental work in the chick has implicated RA as a determinant for a subset of motor neurons that will innervate some of the limb muscles.[64] Conditional *Raldh2* mutagenesis using a cre recombinase expressed in the developing spinal cord will generate a mouse model for investigating this hypothesis. More generally, both the conditional mutant(s) and the available *Raldh2* null and hypomorphic mutants can be used as model systems for further studies to identify the genetic pathways regulated by RA in various developmental processes, using transcriptome-based approaches.

ACKNOWLEDGMENTS

We would like to thank Professor P. Chambon for initiating this work and for his constant support.

Work in the authors' laboratory is supported by funds from the CNRS, the INSERM, the Ministère de la Recherche, the Association pour la Recherche sur le Cancer, the Fondation pour la Recherche Médicale, the Institut Universitaire de France, and the European Union.

REFERENCES

1. Wilson, J. G., Roth, C. B., and Warkany, J. An analysis of the syndrome of malformations induced by maternal vitamin A deficiency. Effects of restoration of vitamin A at various times during gestation. *Am J Anat* 92, 189–217 (1953).

2. White, J. C. et al. Vitamin A deficiency results in the dose-dependent acquisition of anterior character and shortening of the caudal hindbrain of the rat embryo. *Dev Biol* 220, 263–84 (2000).

3. Dersch, H. and Zile, M. H. Induction of normal cardiovascular development in the vitamin A-deprived quail embryo by natural retinoids. *Dev Biol* 160, 424–33 (1993).

4. Zile, M. H. Function of vitamin A in vertebrate embryonic development. *J Nutr* 131, 705–8 (2001).

5. Chambon, P. A decade of molecular biology of retinoic acid receptors. *Faseb J* 10, 940–54 (1996).

6. Mic, F. A. et al. Retinoid activation of retinoic acid receptor but not retinoid X receptor is sufficient to rescue lethal defect in retinoic acid synthesis. *Proc Natl Acad Sci U S A* 100, 7135–40 (2003).

7. McCaffery, P. and Dräger, U. C. Regulation of retinoic acid signaling in the embryonic nervous system: A master differentiation factor. *Cytokine Growth Factor Rev* 11, 233–49 (2000).

8. Mark, M. et al. A genetic dissection of the retinoid signalling pathway in the mouse. *Proc Nutr Soc* 58, 609–13 (1999).

9. Mollard, R. et al. Tissue-specific expression of retinoic acid receptor isoform transcripts in the mouse embryo. *Mech Dev* 94, 223–32 (2000).

10. Lufkin, T. et al. High postnatal lethality and testis degeneration in retinoic acid receptor alpha mutant mice. *Proc Natl Acad Sci U S A* 90, 7225–9 (1993).

11. Kastner, P. et al. Abnormal spermatogenesis in RXR beta mutant mice. *Genes Dev* 10, 80–92 (1996).

12. Kastner, P. et al. Genetic analysis of RXR alpha developmental function: Convergence of RXR and RAR signaling pathways in heart and eye morphogenesis. *Cell* 78, 987–1003 (1994).

13. Lohnes, D. et al. Function of the retinoic acid receptors (RARs) during development (I). Craniofacial and skeletal abnormalities in RAR double mutants. *Development* 120, 2723–48 (1994).

14. Mendelsohn, C. et al. Function of the retinoic acid receptors (RARs) during development (II). Multiple abnormalities at various stages of organogenesis in RAR double mutants. *Development* 120, 2749–71 (1994).

15. Kastner, P. et al. Vitamin A deficiency and mutations of RXRalpha, RXRbeta and RARalpha lead to early differentiation of embryonic ventricular cardiomyocytes. *Development* 124, 4749–58 (1997).

16. Mascrez, B. et al. The RXR alpha ligand-dependent activation function 2 (AF-2) is important for mouse development. *Development* 125, 4691–707 (1998).

17. Mascrez, B. et al. Differential contributions of AF-1 and AF-2 activities to the developmental functions of RXR alpha. *Development* 128, 2049–62 (2001).

18. Dupé, V. et al. Key roles of retinoic acid receptors alpha and beta in the patterning of the caudal hindbrain, pharyngeal arches and otocyst in the mouse. *Development* 126, 5051–9 (1999).

19. Wendling, O. et al. Roles of retinoic acid receptors in early embryonic morphogenesis and hindbrain patterning. *Development* 128, 2031–8 (2001).

20. Maden, M. et al. The distribution of endogenous retinoic acid in the chick embryo: Implications for developmental mechanisms. *Development* 125, 4133–44 (1998).

21. Thaller, C. and Eichele, G. Identification and spatial distribution of retinoids in the developing chick limb bud. *Nature* 327, 625–8 (1987).

22. Ward, S. J. et al. A retinol-binding protein receptor-mediated mechanism for uptake of vitamin A to postimplantation rat embryos. *Biol Reprod* 57, 751–5 (1997).

23. Vogel, S. et al. Retinol-binding protein-deficient mice: Biochemical basis for impaired vision. *Biochemistry* 41, 15360–8 (2002).

24. Molotkov, A., Fan, X. and Duester, G. Excessive vitamin A toxicity in mice genetically deficient in either alcohol dehydrogenase Adh1 or Adh3. *Eur J Biochem* 269, 2607-12 (2002).

25. Molotkov, A. and Duester, G. Genetic evidence that retinaldehyde dehydrogenase Raldh1 (Aldh1a1) functions downstream of alcohol dehydrogenase *Adh1* in metabolism of retinol to retinoic acid. *J Biol Chem* 278, 36085–90 (2003).

26. Deltour, L., Foglio, M. H., and Duester, G. Metabolic deficiencies in alcohol dehydrogenase *Adh1*, *Adh3*, and *Adh4* null mutant mice. Overlapping roles of Adh1 and Adh4 in ethanol clearance and metabolism of retinol to retinoic acid. *J Biol Chem* 274, 16796–801 (1999).

27. Deltour, L., Foglio, M. H., and Duester, G. Impaired retinol utilization in *Adh4* alcohol dehydrogenase mutant mice. *Dev Genet* 25, 1–10 (1999).

28. Napoli, J. L. 17 beta-Hydroxysteroid dehydrogenase type 9 and other short-chain dehydrogenases/reductases that catalyze retinoid, 17 beta- and 3 alpha-hydroxy-steroid metabolism. *Mol Cell Endocrinol* 171, 103–9 (2001).

29. Li, H. et al. A retinoic acid synthesizing enzyme in ventral retina and telencephalon of the embryonic mouse. *Mech Dev* 95, 283–9 (2000).

30. Mic, F. A. et al. RALDH3, a retinaldehyde dehydrogenase that generates retinoic acid, is expressed in the ventral retina, otic vesicle and olfactory pit during mouse development. *Mech Dev* 97, 227–30 (2000).

31. Zhao, D. et al. Molecular identification of a major retinoic-acid-synthesizing enzyme, a retinaldehyde-specific dehydrogenase. *Eur J Biochem* 240, 15–22 (1996).

32. Gagnon, I., Duester, G., and Bhat, P. V. Kinetic analysis of mouse retinal dehydrogenase type-2 (RALDH2) for retinal substrates. *Biochim Biophys Acta* 1596, 156–62 (2002).

33. Grun, F. et al. Aldehyde dehydrogenase 6, a cytosolic retinaldehyde dehydrogenase prominently expressed in sensory neuroepithelia during development. *J Biol Chem* 275, 41210–8 (2000).

34. Haselbeck, R. J., Hoffmann, I., and Duester, G. Distinct functions for Aldh1 and Raldh2 in the control of ligand production for embryonic retinoid signaling pathways. *Dev Genet* 25, 353–64 (1999).

35. Niederreither, K. et al. Differential expression of retinoic acid-synthesizing (RALDH) enzymes during fetal development and organ differentiation in the mouse. *Mech Dev* 110, 165–71 (2002).

36. Fan, X. et al. Targeted disruption of *Aldh1a1* (*Raldh1*) provides evidence for a complex mechanism of retinoic acid synthesis in the developing retina. *Mol Cell Biol* 23, 4637–48 (2003).

37. Niederreither, K. et al. Restricted expression and retinoic acid-induced down-regulation of the retinaldehyde dehydrogenase type 2 (*RALDH-2*) gene during mouse development. *Mech Dev* 62, 67–78 (1997).

38. Wagner, E., McCaffery, P., and Dräger, U. C. Retinoic acid in the formation of the dorsoventral retina and its central projections. *Dev Biol* 222, 460–70 (2000).

39. Dupé, V. et al. A newborn lethal defect due to inactivation of retinaldehyde dehydrogenase type 3 is prevented by maternal retinoic acid treatment. *Proc Natl Acad Sci U S A* 100, 14036–41 (2003).

40. Lin, M. et al. Mouse retinal dehydrogenase 4 (*RALDH4*), molecular cloning, cellular expression, and activity in 9-cis-retinoic acid biosynthesis in intact cells. *J Biol Chem* 278, 9856–61 (2003).

41. Niederreither, K. et al. Embryonic retinoic acid synthesis is essential for early mouse post-implantation development. *Nat Genet* 21, 444–8 (1999).

42. Mic, F. A. et al. Novel retinoic acid generating activities in the neural tube and heart identified by conditional rescue of *Raldh2* null mutant mice. *Development* 129, 2271–82 (2002).

43. Rossant, J. et al. Expression of a retinoic acid response element–*hsplacZ* transgene defines specific domains of transcriptional activity during mouse embryogenesis. *Genes Dev* 5, 1333–44 (1991).

44. Niederreither, K. et al. Embryonic retinoic acid synthesis is essential for heart morphogenesis in the mouse. *Development* 128, 1019–31 (2001).

45. Moss, J. B. et al. Dynamic patterns of retinoic acid synthesis and response in the developing mammalian heart. *Dev Biol* 199, 55–71 (1998).

46. Osmond, M. K. et al. The effects of retinoic acid on heart formation in the early chick embryo. *Development* 113, 1405–17 (1991).
47. Niederreither, K. et al. Retinoic acid synthesis and hindbrain patterning in the mouse embryo. *Development* 127, 75–85 (2000).
48. Gould, A., Itasaki, N., and Krumlauf, R. Initiation of rhombomeric *Hoxb4* expression requires induction by somites and a retinoid pathway. *Neuron* 21, 39–51 (1998).
49. Lammer, E. J. et al. Retinoic acid embryopathy. *N Engl J Med* 313, 837–41 (1985).
50. Wood, H., Pall, G., and Morriss-Kay, G. Exposure to retinoic acid before or after the onset of somitogenesis reveals separate effects on rhombomeric segmentation and 3' *HoxB* gene expression domains. *Development* 120, 2279–85 (1994).
51. Kessel, M. and Gruss, P. Homeotic transformations of murine vertebrae and concomitant alteration of *Hox* codes induced by retinoic acid. *Cell* 67, 89–104 (1991).
52. Niederreither, K. et al. Retinaldehyde dehydrogenase 2 (RALDH2)-independent patterns of retinoic acid synthesis in the mouse embryo. *Proc Natl Acad Sci U S A* 99, 16111–6 (2002).
53. Sakai, Y. et al. The retinoic acid-inactivating enzyme CYP26 is essential for establishing an uneven distribution of retinoic acid along the anterio-posterior axis within the mouse embryo. *Genes Dev* 15, 213–25 (2001).
54. Abu-Abed, S. et al. The retinoic acid-metabolizing enzyme, CYP26A1, is essential for normal hindbrain patterning, vertebral identity, and development of posterior structures. *Genes Dev* 15, 226–40 (2001).
55. Niederreither, K. et al. Embryonic retinoic acid synthesis is required for forelimb growth and anteroposterior patterning in the mouse. *Development* 129, 3563–74 (2002).
56. Fernandez-Teran, M. et al. Role of dHAND in the anterior-posterior polarization of the limb bud: Implications for the Sonic hedgehog pathway. *Development* 127, 2133–42 (2000).
57. Charite, J., McFadden, D. G., and Olson, E. N. The bHLH transcription factor dHAND controls Sonic hedgehog expression and establishment of the zone of polarizing activity during limb development. *Development* 127, 2461–70 (2000).
58. Niederreither, K. et al. The regional pattern of retinoic acid synthesis by RALDH2 is essential for the development of posterior pharyngeal arches and the enteric nervous system. *Development* 130, 2525–34 (2003).
59. Farrell, M. et al. A novel role for cardiac neural crest in heart development. *Trends Cardiovasc Med* 9, 214–20 (1999).
60. Meyers, E. N., Lewandoski, M., and Martin, G. R. An *Fgf8* mutant allelic series generated by cre- and flp-mediated recombination. *Nat Genet* 18, 136–41 (1998).
61. Vermot, J. et al. Decreased embryonic retinoic acid synthesis results in a DiGeorge syndrome phenotype in newborn mice. *Proc Natl Acad Sci U S A* 100, 1763–8 (2003).
62. Baldini, A. DiGeorge syndrome: The use of model organisms to dissect complex genetics. *Hum Mol Genet* 11, 2363–9 (2002).
63. Deltour, L., Ang, H. L., and Duester, G. Ethanol inhibition of retinoic acid synthesis as a potential mechanism for fetal alcohol syndrome. *Faseb J* 10, 1050–7 (1996).
64. Sockanathan, S. and Jessell, T. M. Motor neuron-derived retinoid signaling specifies the subtype identity of spinal motor neurons. *Cell* 94, 503–14 (1998).

22 Mouse Models For Developmental Biology: Functional Analysis of Ror and Wnt Signaling

Isao Oishi, Akinori Yoda, Shuichi Kani, and Yasuhiro Minami

TABLE OF CONTENTS

INTRODUCTION

Receptor tyrosine kinases (RTKs) play pivotal roles in developmental morphogenesis by regulating growth, differentiation, motility, adhesion, and death of many types of cells.[1] The Ror-family RTKs are a recently identified family of orphan RTKs, characterized by the presence of extracellular Frizzled-like cysteine-rich domains and Kringle domains, both of which are believed to play important roles in protein–protein interactions.[2,3] The cytoplasmic regions of the Ror-family proteins possess canonical tyrosine kinase domains, highly related to those of the Trk-family RTKs. The Ror-family RTKs are evolutionarily well conserved among *Caenorhabditis elegans*,[4] *Aplysia*,[5] *Drosophila melanogaster*,[6,7] *Xenopus*,[8] mice,[9] and humans.[10] In mammals, there exist the two members of this family, Ror1 and Ror2 (Figure 22.1). The developmental expression patterns of *Ror1* and *Ror2* during mouse embryogenesis were assessed by *in situ* hybridization.[11,12] During early embryogenesis, remarkable expressions of *Ror1* and *Ror2* were observed in

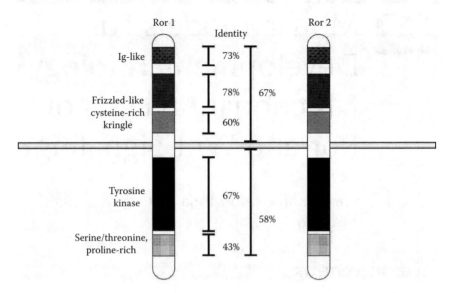

FIGURE 22.1 Structural features of mouse *Ror1* and *Ror2*. Amino acid identities (%) between the corresponding domains (i.e., the Ig-like domain, cysteine-rich domain, Kringle domain, tyrosine kinase domain, and serine/threonine, proline-rich domain) and regions (i.e., the extracellular and intracellular regions) of mouse *Ror1* and *Ror2* are compared.

the developing face such as the frontonasal process and pharyngeal arches that are derived from cephalic neural crest cells. In addition, both *Ror1* and *Ror2* are highly expressed in the developing limb mesenchyme at early stages. In the developing limbs, the expression of *Ror1* is restricted to the proximal regions of the limb buds, and *Ror2* expression is extended throughout the limbs. At later stages, both *Ror1* and *Ror2* are also expressed in the developing heart and lung. In the heart they are expressed in the myocardium and interventricular septum, but not in the epicardium, and in the lung they are expressed in primitive alveoli.

Members of the Wnt-family proteins have also been implicated in a variety of developmental morphogenetic processes.[13] Previous studies indicated that Wnt-family proteins can be classified into at least two subfamilies:[14] one is the Wnt1 class (e.g., Wnt1, Wnt3a, Wnt8) which activates the canonical Wnt/β-catenin pathway to regulate cell proliferation and cell fate,[15,16] and the other is the Wnt5a class (e.g., Wnt5a, Wnt11) which activates a noncanonical Wnt pathway to regulate cell polarity in *Drosophila* and convergent extension movements in *Xenopus* and zebrafish.[17–21] Recently, it was reported that Wnt5a is capable of activating c-Jun N-terminal kinase (JNK) in mammalian cells and regulating convergent extension movements in *Xenopus*.[22] Among the *Wnt* family genes in mammals, mouse *Wnt5a* exhibits a remarkably similar developmental expression pattern to mouse *Ror2*. Like *Ror2, Wnt5a* is expressed in the developing face, limbs and tail, lungs, and genitals.[23,24]

In this chapter, we describe the comparison of developmental phenotypes of *Ror2* and *Wnt5a* mutant mice and then provide biochemical and biological evidence indicating physical and functional interactions between Ror2 and Wnt5a. We also discuss the functions of Ror2 in the noncanonical Wnt signaling.

PHENOTYPES OF THE *ROR1* AND *ROR2* MUTANT MICE

To understand the functional roles of *Ror1* and *Ror2* during development, mice lacking either of these genes were established.[25–27] *Ror1* mutant mice newborns show a gross appearance very similar to that observed in the wild-type mice and any apparent abnormalities were not found in the skeletal and cardiac systems, although the *Ror1* mutant mice exhibit forced respiration and cyanosis and die within 24 hours after birth without nursing. *Ror2* mutant newborns also show forced respiration and cyanosis, and die within 6 hours after birth. In contrast to *Ror1* mutant mice, *Ror2* mutant mice exhibit dwarfism with short limbs and tail, and facial anomalies. Histological analysis of the lungs from *Ror1* and *Ror2* mutant newborns revealed that the mutant mice failed to expand their alveolar sacs, and it is assumed that both mutant mice died due to pulmonary dysfunctions.

Interestingly, the *Ror2* mutant mice, but not *Ror1* mutant mice, possess a ventricular septal defect (VSD). In addition to their cardiac anomaly, *Ror2* mutant mice possess abnormalities in their skeletal systems. They have craniofacial anomalies, including hypoplasia of mandibula, and fore- and hind limb anomalies, characterized by mesomeric bone dysplasia (significant or complete loss of the radius, ulna, tibia, and fibula). Skeletal defects are also found in axial bones, including fusion of adjoining ribs. Histological examination of bones showed gross abnormalities in anlage and growth plate organization in *Ror2* mutant mice, suggesting that *Ror2* gene products are required for the proper proliferation, differentiation, motility, and function of chondrocytes. Furthermore, hypoplasia of the genital tubercle (outgrowth reduction) is observed in *Ror2* mutant mice. Because the expression patterns of *Ror1* and *Ror2* are similar during development, the lack of apparent abnormalities in *Ror1* mutant mice may be attributable to the functional redundancy between Ror1 and Ror2.

To test a possible genetic interaction between *Ror1* and *Ror2* during development, we established and analyzed the *Ror1/Ror2*-double mutant mice.[27] *Ror1/Ror2*-double mutant mice exhibit perinatal lethality and enhanced *Ror2* mutant phenotypes. Interestingly, *Ror1/Ror2*-double mutant mice exhibit a drastic enhancement of the skeletal phenotypes observed in *Ror2* mutant mice. Compared with *Ror2* mutants, more severe hypoplasia of the maxilla and mandible, and malformation of vertebra and ribs are found in the double mutant mice. In the *Ror1/Ror2*-double mutant mice, the proximal long bones, humerus and femur, are affected in addition to the distal long bones. Intriguingly, *Ror1/Ror2*-double mutant mice showed a sternal defect (sternal agenesis) and dysplasia of the symphysis of the pubic bones, skeletal phenotypes that were not observed in *Ror2* mutant mice. Furthermore, *Ror1/Ror2*-double mutant mice exhibit the complete transposition of the great artery in addition to VSD, a phenotype also not observed in *Ror2* mutant mice. Collectively, these findings indicate that *Ror1* and *Ror2* interact genetically during the development of the skeletal and cardiovascular systems.

THE *ROR2* MUTANT MICE AS A MODEL FOR HUMAN GENETIC DISORDERS

Interestingly, it was reported that mutations within the *Ror2* gene account for autosomal dominant Brachydactyly type B (BDB)[28,29] and recessive Robinow syndrome.[30,31] BDB is a dominant skeletal disorder characterized by hypoplasia/aplasia of distal phalanges and nails. It has been shown that BDB is caused by mutations in *Ror2*, which result in the C-terminal truncation of Ror2 protein either before or after the kinase domain. On the other hand, Robinow syndrome is a multisystemic disease characterized by skeletal dysplasia with mesomelic limb shortening and hemivertebrae, segmental defects of the spine, brachydactyly, genital hypoplasia, and a dysmorphic facial appearance.[32] Interestingly, developmental pathology of *Ror2* mutant mice explains many of the developmental malformations found in patients with Robinow syndrome.[31,33] These results suggest that Robinow syndrome is caused by loss-of-function mutation of *Ror2*. Indeed, the mutations found in *Ror2* in Robinow syndrome are missense, nonsense, and frameshift mutations in both intracellular and extracellular domains, resulting in reduced or eliminated functions. To date, any mutations in human *Ror1* have not been reported. Further study is required to clarify a possible relationship of *Ror1* mutation(s) with human genetic disorders.

COMPARISON OF PHENOTYPES OF *ROR2* AND *WNT5A* MUTANT MICE

Because both Ror1 and Ror2 possess the cysteine-rich domains (CRDs) within their extracellular regions, closely resembling the Wnt-binding sites of the Frizzled (Fz) proteins,[34,35] it can be assumed that Ror-family RTKs may interact with a member(s) of the Wnt family of proteins. Interestingly, it was found that overall phenotypes of *Ror2* mutant mice are very similar to those of *Wnt5a* mutant mice originally reported by Yamaguchi et al.[23] Both *Ror2* and *Wnt5a* mutant newborns exhibit dwarfism, facial abnormalities, short limbs and tails, and respiratory dysfunction, and die shortly after birth (Table 22.1). In the heart, as with *Ror2* mutant mice, *Wnt5a* mutant mice exhibit VSD. In addition, *Wnt5a* mutant mice show the complete transposition of the great arteries, a phenotype reminiscent of *Ror1/Ror2*-double mutant mice. In the lungs, both *Ror2* and *Wnt5a* mutant mice exhibit abnormalities with foreshortened trachea along the P–D axis and a reduced number of cartilage rings[24,36] In the genitals, both mutant mice exhibit outgrowth defects. Furthermore, both mutant mice exhibit similar skeletal abnormalities, including hypoplasia of the maxilla and mandible and dysplasia of limb long bones and vertebrae. Although several developmental defects are more profound in *Wnt5a* mutant mice compared to *Ror2* mutant mice, remarkably similar phenotypes of *Ror2* and *Wnt5a* mutant mice indicate a possible functional interaction of Ror2 with Wnt5a during mouse development.

TABLE 22.1

Defects in the *Ror1, Ror2, Ror1/Ror2,* and *Wnt5a* Mutant Mice

	Ror1	Ror2	Ror1/Ror2	Wnt5a
Phenotypes Observed in All Mutant Mice				
Neonatal lethality	+	+	+	+
Respiratory dysfunction	+	+	+	+
Phenotypes Observed in the Ror2, Ror1/Ror2, and Wnt5a, but not Ror1 Mutant Mice				
Dysplasia of the distal long bones	–	+	+	+
Ventricular septal defect (VSD)	–	+	+	+
Dwarfism	–	+	++	++
Short limbs and tail	–	+	++	++
Facial anomalies	–	+	++	++
Dysplasia of the proximal long bones	–	+	++	++
Hypoplasia of the maxilla and mandible	–	+	++	++
Phenotypes Observed in the Ror1/Ror2 and Wnt5a Mutant Mice				
Transposition of the great arteries	–	–	+	+

PHYSICAL AND FUNCTIONAL INTERACTIONS BETWEEN ROR2 AND WNT5A

We have recently shown that Ror2 and Wnt5a can interact both physically and functionally.[36] *In vitro* binding assay revealed that Wnt5a selectively binds to Ror2, presumably via its extracellular CRD. Furthermore, when expressed ectopically in mammalian cells, Ror2 associates physically with rat Frizzled2 and human Frizzled5, a putative receptor for Wnt5a, but not with mouse Frizzled8, suggesting the possible function of Ror2 as a component of receptor for Wnt5a (Figure 22.2). To test this, the effects of Wnt5a and Ror2 on JNK activities in NIH3T3 cells and on convergent extension movements in *Xenopus* were examined. Expression of Wnt5a or Ror2 alone in NIH3T3 cells results in the activation of JNK, suggesting that Ror2 is also involved in the JNK pathway. Furthermore, co-expression of Wnt5a and Ror2 in the cells has an additive effect on JNK activity. In *Xenopus*, convergent extension movements stimulated by BVg1 are inhibited by injection of *Wnt5a* mRNA or *Ror2* mRNA alone. Interestingly, co-injection of *Wnt5a* and *Ror2* mRNAs synergistically inhibits convergent extension movements, indicating that Wnt5a and Ror2 interact functionally. In addition, Hikasa et al. have shown that *Xenopus* Ror2 can interact with Xwnt11, a member of the Wnt5a class, and that inhibition of convergent extension movements is reversed by a dominant negative form of Cdc42 that has been suggested to mediate the noncanonical Wnt pathway.[8] Taken together, these results suggest that Ror2 and a Wnt protein of Wnt5a class interact both physically and functionally, and that Ror2 may act as a receptor for a Wnt of Wnt5a class to activate noncanonical Wnt signaling.

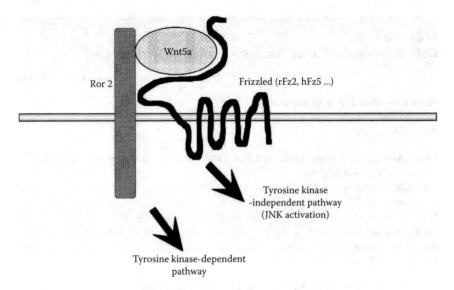

FIGURE 22.2 Model of Ror2-mediated signaling following Wnt5a stimulation. Ror2 possesses tyrosine kinase-dependent and -independent functions.

TYROSINE KINASE-DEPENDENT AND -INDEPENDENT FUNCTIONS OF ROR2

Intriguingly, the tyrosine kinase-deficient Ror2 mutant can also mediate activation of the JNK pathway in NIH3T3 cells and inhibition of convergent extension movements in *Xenopus*, suggesting that Ror2 possesses a tyrosine kinase-independent function.[8,36] In this respect, it is important to note that the *C. elegans* orthologue of Ror2, CAM-1, possesses both tyrosine kinase-dependent and -independent functions during development.[2–4] CAM-1 regulates cell migration, asymmetric cell divisions, and axon outgrowth during the developmental processes; however, tyrosine kinase activity of CAM-1 is not required for cell migration. Although at present tyrosine kinase-dependent functions of Ror2 are largely unknown, we have recently found that a protein serine/threonine kinase, casein kinase Iε (CKIε), can interact with Ror2 both physically and functionally, and that CKIε can be involved in tyrosine kinase-dependent functions of Ror2 (Kani et al.[37]). Further study will be required to clarify the biological significance of tyrosine kinase-dependent and -independent functions of Ror-family RTKs.

ACKNOWLEDGMENTS

Work in the authors' laboratory was supported in part by a Grant-in-Aid for Scientific Research from JSPS, a Research Grant for Comprehensive Research on Aging and Health, a Research Grant for Cardiovascular Diseases, and a Research Grant for Pediatric Research from the Ministry of Health and Welfare of Japan, and by the Mochida Memorial Foundation for Medical and Pharmaceutical Research, Japan Cardiovascular Research Foundation, Nippon Boehringer Ingelheim, Co., Ltd., and Daiichi Pharmaceutical Co., Ltd.

REFERENCES

1. Schlessinger, J., Cell signaling by receptor tyrosine kinases, *Cell*, 103, 211, 2000.
2. Forrester, W. C., The Ror receptor tyrosine kinase family, *Cell. Mol. Life Sci.*, 59, 83, 2002.
3. Yoda, A., Oishi, I., and Minami, Y., Expression and function of the Ror-family receptor tyrosine kinases during development: Lessons from genetic analyses of nematodes, mice, and humans, *J. Recept. Signal Transduct. Res.*, 23, 1, 2003.
4. Forrester, W. C. et al., A *C. elegans* Ror receptor tyrosine kinase regulates cell motility and asymmetric cell division, *Nature*, 400, 881, 1999.
5. McKay, S. E. et al., *Aplysia ror* forms clusters on the surface of identified neuroendocrine cells, *Mol. Cell. Neurosci.*, 17, 821, 2001.
6. Wilson, C., Goberdhan, D. C., and Steller, H., *Dror*, a potential neurotrophic receptor gene, encodes a *Drosophila* homolog of the vertebrate Ror family of Trk-related receptor tyrosine kinases, *Proc. Natl. Acad. Sci. USA*, 90, 7109, 1993.
7. Oishi, I. et al., A novel *Drosophila* receptor tyrosine kinase expressed specifically in the nervous system. Unique structural features and implication in developmental signaling, *J. Biol. Chem.*, 272, 11916, 1997.
8. Hikasa, H. et al., The *Xenopus* receptor tyrosine kinase Xror2 modulates morphogenetic movements of the axial mesoderm and neuroectoderm via Wnt signaling, *Development*, 129, 5227, 2002.
9. Oishi, I. et al., Spatio-temporally regulated expression of receptor tyrosine kinases, mRor1, mRor2, during mouse development: Implications in development and function of the nervous system, *Genes Cells*, 4, 41, 1999.
10. Masiakowski, P. and Carroll, R. D., A novel family of cell surface receptors with tyrosine kinase-like domain, *J. Biol. Chem.*, 267, 26181, 1992.
11. Matsuda, T. et al., Expression of the receptor tyrosine kinase genes, *Ror1* and *Ror2*, during mouse development, *Mech. Dev.*, 105, 153, 2001.
12. Al-Shawi, R. et al., Expression of the *Ror1* and *Ror2* receptor tyrosine kinase genes during mouse development, *Dev. Genes Evol.*, 211, 161, 2001.
13. Wodarz, A. and Nusse, R., Mechanisms of Wnt signaling in development, *Ann. Rev. Cell Dev. Biol.*, 14, 59, 1998.
14. Kuhl, M. et al., The Wnt/Ca2+ pathway: A new vertebrate Wnt signaling pathway takes shape, *Trends Genet.*, 16, 279, 2000.
15. Cadigan, K. M. and Nusse, R., Wnt signaling: A common theme in animal development, *Genes Dev.*, 11, 3286, 1997.
16. Sokol, S. Y., Wnt signaling and dorso-ventral axis specification in vertebrates, *Curr. Opin. Genet. Dev.*, 9, 405, 1999.
17. Moon, R. T. et al., Xwnt-5A: A maternal Wnt that affects morphogenetic movements after overexpression in embryos of *Xenopus laevis*, *Development*, 119, 97, 1993.
18. Heisenberg, C. P. et al., Silberblick/Wnt11 mediates convergent extension movements during zebrafish gastrulation, *Nature*, 405, 76, 2000.

19. Sokol, S., A role for Wnts in morpho-genesis and tissue polarity, *Nat. Cell Biol.*, 2, E124, 2000.

20. Tada, M. and Smith, J. C., *Xwnt11* is a target of *Xenopus* Brachyury: Regulation of gastrulation movements via Dishevelled, but not through the canonical Wnt pathway, *Development*, 127, 2227, 2000.

21. Wallingford, J. B. et al., Dishevelled controls cell polarity during Xenopus gastrulation, *Nature*, 405, 81, 2000.

22. Yamanaka, H. et al., JNK functions in the non-canonical Wnt pathway to regulate convergent extension movements in vertebrates, *EMBO Rep.*, 3, 69, 2002.

23. Yamaguchi, T. P. et al., A Wnt5a pathway underlies outgrowth of multiple structures in the vertebrate embryo, *Development*, 126, 1211, 1999.

24. Li, C. et al., Wnt5a participates in distal lung morphogenesis, *Dev. Biol.*, 248, 68, 2002.

25. Takeuchi, S. et al., Mouse Ror2 receptor tyrosine kinase is required for the heart development and limb formation, *Genes Cells*, 5, 71, 2000.

26. DeChiara, T. M. et al., *Ror2*, encoding a receptor-like tyrosine kinase, is required for cartilage and growth plate development, *Nat. Genet.*, 24, 271, 2000.

27. Nomi, M. et al., Loss of *mRor1* enhances the heart and skeletal abnormalities in *mRor2*-deficient mice: Redundant and pleiotropic functions of mRor1 and mRor2 receptor tyrosine kinases, *Mol. Cell. Biol.*, 21, 8329, 2001.

28. Oldridge, M. et al., Dominant mutations in *ROR2*, encoding an orphan receptor tyrosine kinase, cause brachydactyly type B, *Nat. Genet.*, 24, 275, 2000.

29. Schwabe, G. C. et al., Distinct mutations in the receptor tyrosine kinase gene *ROR2* cause brachydactyly type B, *Am. J. Hum. Genet.*, 67, 822, 2000.

30. Afzal, A. R. et al., Recessive Robinow syndrome, allelic to dominant brachydactyly type B, is caused by mutation of *ROR2*, *Nat. Genet.*, 25, 419, 2000.

31 van Bokhoven, H. et al., Mutation of the gene encoding the ROR2 tyrosine kinase causes autosomal recessive Robinow syndrome, *Nat. Genet.*, 25, 423, 2000.

32. Patton, M. A. and Afzal, A. R., Robinow syndrome, *J. Med. Genet.*, 39, 305, 2002.

33. Schwabe, G. C. et al., The *Ror2* knockout mouse as a model for the developmental pathology of autosomal recessive Robinow syndrome, *Dev. Dyn.*, 229, 400, 2003.

34. Masiakowski, P. and Yancopoulos, G. D., The Wnt receptor CRD domain is also found in MuSK and related orphan receptor tyrosine kinases, *Curr. Biol.*, 8, R407, 1998.

35. Rehn, M. et al., The frizzled motif: In how many different protein families does it occur? *Trends Biochem. Sci.*, 23, 415, 1998.

36. Oishi, I. et al., The receptor tyrosine kinase Ror2 is involved in non-canonical Wnt5a/JNK signalling pathway, *Genes Cells*, 8, 645, 2003.

37. Kani, S. et al., The receptor tyrosine kinase Ror2 associates with and is activated by casein kinase Iε, *J. Biol. Chem.*, 279, 50102, 2004.

Index